EXPERIMENT OPTIMIZATION
IN CHEMISTRY AND CHEMICAL ENGINEERING

С. Л. Ахназарова,
В. В. Кафаров

ОПТИМИЗАЦИЯ ЭКСПЕРИМЕНТА В ХИМИИ И ХИМИЧЕСКОЙ ТЕХНОЛОГИИ

ИЗДАТЕЛЬСТВО
«ВЫСШАЯ ШКОЛА»
МОСКВА

S. Akhnazarova
V. Kafarov

EXPERIMENT OPTIMIZATION IN CHEMISTRY AND CHEMICAL ENGINEERING

Translated from Russian
by
Vladimir M. Matskovsky
and
Alexander P. Repyev

MIR PUBLISHERS
MOSCOW

First published 1982
Revised from the 1978 Russian edition

The Greek Alphabet

$A\ \alpha$	Alpha	$I\ \iota$	Iota	$P\ \varrho$	Rho
$B\ \beta$	Beta	$K\ \varkappa$	Kappa	$\Sigma\ \sigma$	Sigma
$\Gamma\ \gamma$	Gamma	$\Lambda\ \lambda$	Lambda	$T\ \tau$	Tau
$\Delta\ \delta$	Delta	$M\ \mu$	Mu	$\Upsilon\ \upsilon$	Upsilon
$E\ \varepsilon$	Epsilon	$N\ \nu$	Nu	$\Phi\ \varphi$	Phi
$Z\ \zeta$	Zeta	$\Xi\ \xi$	Xi	$X\ \chi$	Chi
$H\ \eta$	Eta	$O\ o$	Omicron	$\Psi\ \psi$	Psi
$\Theta\ \theta$	Theta	$\Pi\ \pi$	Pi	$\Omega\ \omega$	Omega

На английском языке

Printed in Hungarian People's Republic

CONTENTS

PART TWO

DESIGN OF EXPERIMENTS

INTRODUCTION

Experiment has always been a major tool in tackling practical problems and testing theoretical hypotheses in chemistry and chemical engineering. Traditional experimentation, however, involves a good deal of effort and time, especially where complex processes are involved. A very efficient way to enhance the value of research and to cut down the process development time is through designed experiment, that is, by optimizing experiment at every stage from inception, through research and development, to engineering and production.

At present, process analysis, design, optimization and performance prediction in chemical engineering are above all based on mathematical modelling [1]. If the designer has at his disposal complete information (thermodynamic, kinetic and hydrodynamic data) about the process or plant of interest, he can develop a deterministic mathematical model as a set of ordinary or partial differential equations. In order to determine the coefficients appearing in the equations and also to verify the model, he runs an experiment.

If the available information is not complete, the designer undertakes a functional study of the process or plant; to this end, he runs an experiment and observes the input and output process or plant variables.

In Fig. I.1, the measurable or *controllable* input variables are designated $x_1, \ldots, x_k$; *uncontrollable* or *random* process variables (noise) are designated $w_1, \ldots, w_l$; and the output variables (responses) are designated $y_1, \ldots, y_m$.

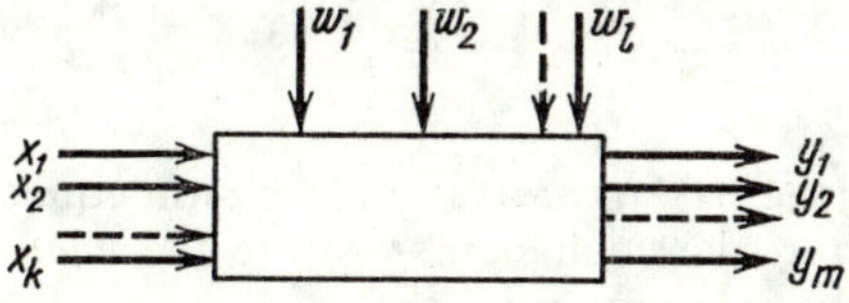

Fig. I. 1. Multivariate process with multiple inputs

The term "random" applies to variables that can be accounted for with difficulty, if at all. This is true, for example, of the fall in the activity of a catalyst, variations in the state of a heat transfer surface, variations in the ambient temperature, etc. The input variables $x_1, \ldots, x_k$ form the basis (or basic) set of variables, because they control the experiment. Of course, this division of process variables into random and basic is arbitrary. The designer may class as random any variable not included in the basis, even though it may be known well. Depending on the objective sought or the experimental capabilities available, some measurable variables may be included in process (or plant) noise. This will of course impair the accuracy of the mathematical model. An output variable may be any technical or economic index of the process or plant. Applying regression or correlation analysis to experimental data, the designer can establish relationships between the various variables and determine the conditions of an optimum.

A typical mathematical model is

$$y = \varphi(x_1, x_2, \ldots, x_k) \tag{I.1}$$

where y is the dependent variable, or the *response*, and x_i's are the independent variables, or *factors*. They occupy what is known as the *factor space*, and the graphical representation of the response function is called the *response surface*.

In process analysis by statistical methods, the mathematical model is most frequently a polynomial—a truncated Taylor series into which the response function, Eq. (I.1), is expanded:

$$y = \beta_0 + \sum_{j=1}^{k} \beta_j x_j + \sum_{\substack{u,j=1 \\ u \neq j}}^{k} \beta_{uj} x_u x_j + \sum_{j=1}^{k} \beta_{jj} x_j^2 + \ldots \tag{I.2}$$

where

$$\beta_j = \partial\varphi/\partial x_j \qquad \beta_{uj} = \partial^2\varphi/\partial x_u \, \partial x_j \qquad \beta_{jj} = \partial^2\varphi/2\,\partial x_j^2$$

In any real process, there always are uncontrollable and unmeasurable variables, and the response y usually varies at random. This is why the evaluation of experimental data yields *sample regression coefficients* b_0, b_j, b_{uj}, and b_{jj}, which are estimates of the theoretical coefficients β_0, β_j, β_{uj}, and β_{jj}. The estimated regression equation developed on the basis of an experiment will then take the form

$$y = b_0 + \sum_{j=1}^{k} b_j x_j + \sum_{\substack{u,j=1 \\ u \neq j}}^{k} b_{uj} x_u x_j + \sum_{j=1}^{k} b_{jj} x_j^2 \tag{I.3}$$

The coefficient b_0 is the *free term* of the regression equation, the coefficients b_j are the *linear terms*, the coefficients b_{jj} are the *quadratic terms*, and the coefficients b_{uj} are the *interaction terms*.

Any physical quantity associated with a process (temperature, pressure, flow rate, etc.) varies with time in a random manner and is therefore a *random process*. Over an observation interval, a random process takes on some specific form unknown in advance, which we shall refer to as a *realization* of the random process. By definition, a random process implies the existence of an infinite number of such realizations. By noting the realizations of the random process at regular intervals, we obtain a set of random variables or sample functions. The time interval must be sufficiently large for the sample functions, or random variables, to be taken from independent experiments.

Random processes can be *stationary* (Fig. I.2) and *nonstationary* (Fig. I.3). Stationary random processes can be described as those processes which are independent of the choice of zero on the time axis. Another idea of stationarity is that a time translation of a sample function results in a similar sample function of the random process. Analysis of a stationary

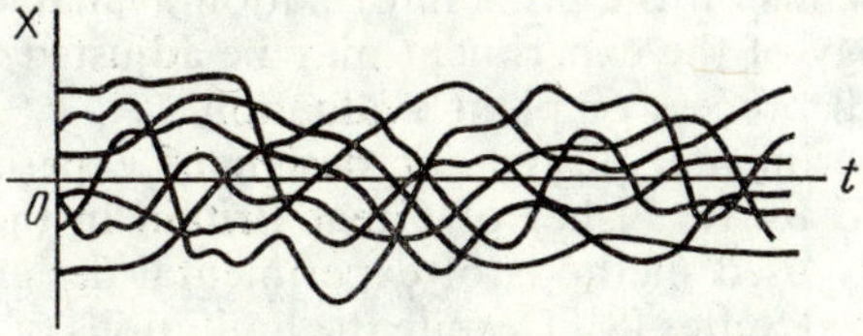

Fig. I. 2. Stationary random process

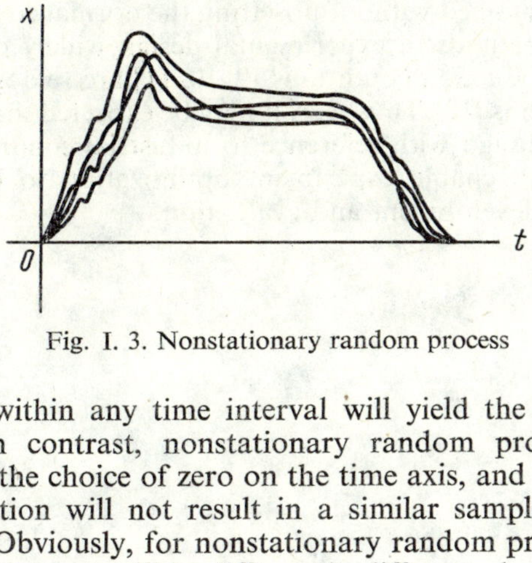

Fig. I. 3. Nonstationary random process

random process within any time interval will yield the same probability characteristics. In contrast, nonstationary random processes are those which depend on the choice of zero on the time axis, and a time translation of a sample function will not result in a similar sample function of the random process. Obviously, for nonstationary random processes the statistical properties of interest will be different in different observation intervals. If the physical quantities representative of some chemical-engineering process vary in the manner of a nonstationary random process, a model in the form of Eq. (I.3), which is an algebraic equation with constant coefficients, cannot be built in principle. This applies, for example, to catalytic reactors if the characteristics of the catalyst change abruptly in the course of service.

The statistical data necessary for analysis (model input specifications) are gathered by conducting an experiment directly on the process or plant of interest. This may be a passive or an active experiment. During a *passive experiment*, the more traditional one of the two, a large series of measurements is carried out, with the values of each of the independent variables altered from measurement to measurement in turn. This procedure can also be used in the course of normal service of the process or plant. The data thus obtained are then processed by the methods of classical regression or correlation analysis [2-7]. An *active experiment* is conducted to a predetermined plan or design (this is known as experimental design). In an active experiment, the independent variables are altered all at the same time, so that their interaction is evaluated at once, and the amount of experimentation can substantially be cut down. The choice of an experimental design depends on the prior information available and the objective sought. The strategy of the experiment may be adjusted to suit a particular stage in the overall process or plant evaluation.

The use of experimental design (or design of experiment) was for the first time proposed by R. Fisher of Great Britain in the 1930s, but credit for the now widely used methods of experimental design goes to Box and Wilson of the United States [8]. Despite the limitations of passive experiment and classical regression analysis [2], they are still widely used under industrial conditions, because information about the behaviour of the process or plant can be obtained without upsetting the normal course of production. At present, the methods of experimental design widely used under laboratory and pilot-production conditions [9, 10, 11] are seldom employed under industrial conditions [12]. However, it may be expected that further advances in experimental design with reference to industrial conditions and progress in technology will enable experiment optimization to be applied at any stage of process development and evaluation.

Part One

ANALYSIS OF EXPERIMENT BY STATISTICAL METHODS

Chapter One

MAIN STATISTICAL CHARACTERISTICS OF RANDOM VARIABLES

1.1. Random Variables. Axioms of Probability Theory. Distribution Laws

A *random variable* is that which, in a trial, takes on a value unpredictable from the conditions of the experiment. In fact, a random variable possesses a set of allowable values, but it takes on only one at each particular trial. In contrast to non-random quantities which vary in value only when the conditions (parameters) of a trial are changed, a random variable can take on different values even though the parameters remain unchanged. Variations in a random variable from measurement to measurement are related to unobservable (random) factors.

It is convenient to characterize a random variable by specifying the set of values that it can take on. In this respect, random variables can be *discrete* and *continuous*. The values that a discrete random variable can take on can be counted in advance; in other words, a discrete variable can take on only distinct values in an interval. The values that a continuous random variable can take on cannot be counted in advance, because they fill the interval continuously, and the variable has the probability of taking any value.

It is not sufficient to specify the set of allowable values in order to characterize a random variable. For complete specification, it is necessary to state which values it can take on and how often. Suppose that a discrete random variable X can take on the values $x_1, x_2, \ldots, x_k$ as a result of an experiment. The ratio between the number m of observations or trials when the random variable X takes on the value x_i and the total number n of trials is called the *relative frequency of the event* $X = x_i$. The relative frequency m/n itself is a random variable and changes according to the number of trials made. With a very long series of trials, the relative frequency tends to stabilize about some value, p_i, called the *probability of the event* $X = x_i$, or statistically,

$$p_i = P(X = x_i) \approx m/n \tag{1.1}$$

According to Bernoulli's theorem, the probability P of the relative frequency of an event, m/n, differing from the probability p_i of the event by more than ε tends to zero as n tends to infinity for any positive value of ε:

$$\lim_{n \to \infty} P\{|m/n - p| > \varepsilon\} = 0 \quad (\varepsilon > 0) \tag{1.1a}$$

Kolmogorov of the Soviet Union has formulated the following axioms of probability theory.

1. The probability of a random event A is a nonnegative number:

$$P(A) \geqslant 0 \tag{1.2}$$

2. The probability of a certain event U is equal to unity:

$$P(U) = 1 \tag{1.3}$$

and the probability of an impossible event V is equal to zero:

$$P(V) = 0 \tag{1.4}$$

Thus,

$$0 \leqslant P \leqslant 1 \tag{1.5}$$

The sum of several events $(A_1 + A_2 + \ldots + A_n)$ is the event consisting in the occurrence of at least one of these events.

3. The probability that at least one of several disjoint events $A_1, A_2, \ldots A_n$ will occur (union is designated by the symbol $\cup$) is equal to the sum of the probabilities of these events:

$$P(A_1 \cup A_2 \cup \ldots \cup A_n) = P(A_1) + P(A_2) + P(A_3) \tag{1.6}$$

This is the addition law of probability.

The product of several events (intersection is designated by the symbol $\cap$), $A_1 \cap A_2 \cap \ldots \cap A_n$, is the event consisting of all points common to $A_1, A_2, \ldots, A_n$.

Random events $A_1, A_2, \ldots, A_n$ are called *statistically independent*, if the probability of any one of them is independent of the probability of any other of these events. The probability of the product of several independent events (joint event) is equal to the product of the probabilities of these events:

$$P(A_1 \cap A_2 \ldots, \cap A_n) = P(A_1)\, P(A_2) \ldots P(A_n) \tag{1.7}$$

Event A is said to be dependent on (related to) event B, if the probability of event A varies according as event B has occurred. The revised probability of A when it is known that B has occurred is called the *conditional probability of A given B* and is denoted by $P(A|B)$.

For related (or mutually dependent) events, the probability of the product of two events is equal to the product of the probability of one by the conditional probability of the other given the first event occurs:

$$P(A \cap B) = P(A)\, P(B|A) \tag{1.8}$$

This is called the multiplication law of compound probabilities. Similarly, if event B precedes event A and affects it in some way, then

$$P(A \cap B) = P(B)\, P(A|B) \tag{1.9}$$

Example 1.1. The probability of trouble-free operation for a computer depends on three series-connected units each of which may fail independently of the other two. The probability of trouble-free operation for the first unit is $P(A_1) = 0.9$, for the second $P(A_2) = 0.8$, and for the third, $P(A_3) = 0.8$. Determine the probability of trouble-free operation for the computer as a whole.

Solution. According to the multiplication law of compound probabilities, Eq. (1.7),

$$P(A) = P(A_1)\,P(A_2)\,P(A_3) = 0.9\times 0.8\times 0.8 = 0.576$$

The sum of the probabilities of all likely values of a random variable is equal to unity

$$\textstyle\sum_{i=1}^{n} p_i = 1 \tag{1.10}$$

because it is true that the random variable will take on one of its values during an experiment. This total probability is distributed among the individual values in a certain definite manner.

A discrete random variable can completely be specified by giving its values arranged in a probability sequence, that is, with the probability p_i stated for each value x_i:

x_i	x_1	x_2	x_3	...	x_n
p_i	p_1	p_2	p_3	...	p_n

Any relation connecting all likely values of a random variable to their respective probabilities is known as the *probability distribution* of that variable. The probability sequence given above is a form of representation of probability distribution.

The distribution of a continuous random variable cannot be specified by giving the probabilities of only a few values. A continuous variable can take so many values that for most of them the probability to assume the given values would be zero; that is, an event can occur, although its probability has been predicted to be zero. For continuous random variables, one is interested in the probability that, as an outcome of an experiment, the value of the variable will occur within a predetermined interval of numbers. It is convenient to use the probability that $X < x$, where x is an arbitrary real number and X is the random variable. This probability is a function of x

$$P(X < x) = F(x) \tag{1.11}$$

and is called the *distribution function* of the random variable.

A distribution function can be used to specify the distribution of a discrete as well as a continuous random variable. By definition, $F(x)$ is a nondiminishing function of x; if $x_1 \leqslant x_2$, then $F(x_1) \leqslant F(x_2)$ (Fig. 1.1*a*). The ordinate of the curve corresponding to point x_1 represents the probability that a trial will show the random variable value, X, less than x_1. The difference of the ordinates corresponding to points x_1 and x_2 gives the probability that the values of the random variable will lie in the interval between x_1 and x_2:

$$P(x_1 \leqslant X \leqslant x_2) = F(x_2) - F(x_1) \tag{1.12}$$

At the extreme values of the argument, the distribution function takes on values 0 and 1:

$$F(-\infty) = 0 \quad F(+\infty) = 1 \tag{1.13}$$

The distribution function of a discrete random variable is always a discontinuous step function which jumps at points corresponding to the likely values of the random variable and is equal to the probabilities of these values (Fig. 1.1*b*). The sum of all jumps is 1.

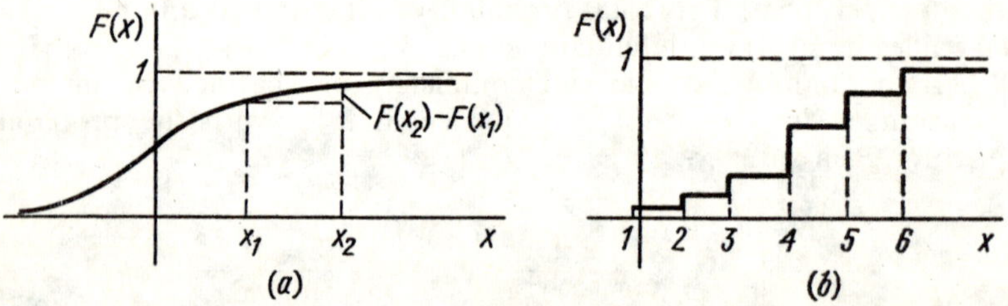

Fig. 1. 1. Distribution function of: (*a*) a continuous random variable, and (*b*) a discrete random variable

For a continuous random variable, use is more frequently made of a derivative of the distribution function, known as the *probability density function* of the random variable X. If $F(x)$ is continuous and differentiable, then

$$f(x) = F'(x) \tag{1.14}$$

The probability density function $f(x)$ also specifies the random variable completely.

The probability density function is a nonnegative function (Fig. 1.2.) The area bounded by the x-axis, lines x_1 and x_2, and the density curve represents the probability that the random veriable will take on values in the interval $[x_1, x_2]$:

$$P(x_1) \leqslant X \leqslant x_2) = \int_{x_1}^{x_2} f(x)\,dx = F(x_2) - F(x_1) \tag{1.15}$$

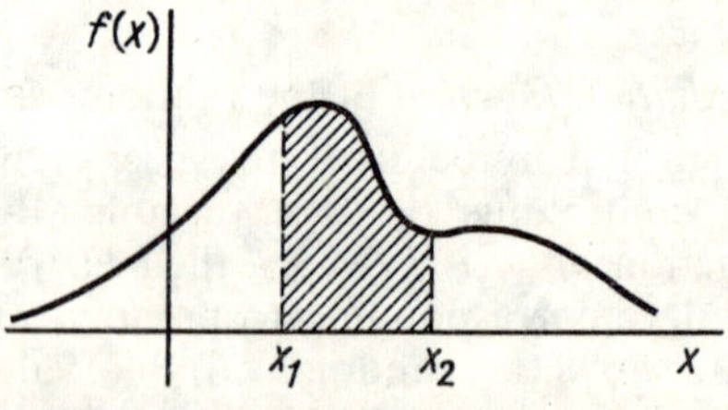

Fig. 1. 2. Probability density function of a continuous random variable

Notably,

$$F(x) = P(-\infty \leqslant X \leqslant x) = \int_{-\infty}^{x} f(x)\,dx \tag{1.16}$$

Hence comes another important property of the probability density function, namely

$$\int_{-\infty}^{\infty} f(x)\,dx = 1 \tag{1.17}$$

because the occurrence of the random variable within the interval $-\infty < X < +\infty$ is a true event.

1.2. Numerical Characteristics

Instead of specifying a random variable by giving its probability distributions, in applied problems this is often done by giving numerical characteristics—real numbers that describe the salient features of the random variable. They are called the *moments* of a random variable. For a random variable X or the associated distribution function, the kth moment μ_k about a is the expected value of $(x-a)^k$ whenever this exists. For a discrete random variable with values $\{x_i\}$ and probabilities $\{p_i\}$

$$\mu_k = \sum (x_i - a)^k p_i$$

and for a continuous random variable with probability density function f,

$$\mu_k = \int_{-\infty}^{\infty} (x-a)^k f(x)\,dx.$$

The moments about zero ($a = 0$) are called raw moments. For a discrete random variable, the kth raw moment is given by

$$\alpha_k = \sum_{i=1}^{n} x_i^k p_i \qquad k = 1, 2, \ldots \tag{1.18}$$

and for a continuous random variable,

$$\alpha_k = \int_{-\infty}^{\infty} x^k f(x)\,dx \tag{1.19}$$

The first raw moment ($k = 1$) of a random variable is known as its arithmetic mean, designated variously as $E\{X\}$, μ_x, or μ. For discrete random variables, it is

$$\alpha_1 = \mu = E\{X\} = \sum_{i=1}^{n} x_i p_i \tag{1.20}$$

and for continuous random variables,

$$\alpha_1 = \mu = E\{X\} = \int_{-\infty}^{\infty} x f(x)\,dx \tag{1.21}$$

If a is the mean, then μ_k is the central moment of order k. The kth central moment of a discrete random variable is given by

$$\mu_k = \sum_{i=1}^{n} (x_i - \mu)^k p_i \tag{1.22}$$

and for a continuous random variable,

$$\mu_k = \int_{-\infty}^{\infty} (x-\mu)^k f(x)\, dx \tag{1.23}$$

The first central moment is zero always, $\mu_1 = 0$. The second central moment is called the *variance* of a random variable; it is a measure or variability or dispersion of the random variable. The variance of a random variable is the expected value of the square of the deviation of the random variable from its expected value or mean, and is designated as Var $\{X\}$, σ_x^2, or σ^2. Thus,

$$\text{Var}\{X\} = E\{(X-\mu)^2\} \tag{1.24}$$

For a discrete random variable, the variance is

$$\text{Var}\{X\} = \mu_2 = \sum_{i=1}^{n} (x_i-\mu)^2 p_i \tag{1.25}$$

and for a continuous random variable,

$$\text{Var}\{X\} = \int_{-\infty}^{\infty} (x-\mu)^2 f(x)\, dx \tag{1.26}$$

The square root of the second central moment is called the *standard deviation* of a random variable:

$$\sigma_x = \sqrt{\text{Var}\{X\}} = \sqrt{\mu_2} \tag{1.27}$$

The third central moment divided by σ^3 is called the *coefficient of skewness:*

$$\gamma_1 = \mu_3/\sigma^3 \tag{1.28}$$

The third central moment is a measure of the symmetry of the distribution of a random variable with respect to the mean. In terms of the raw moments, it is expressed as

$$\mu_3 = \alpha_3 - 3\alpha_1\alpha_2 + 2\alpha_1^3 \tag{1.29}$$

The fourth central moment is given by

$$\mu_4 = \alpha_4 - 4\alpha_1\alpha_3 + 6\alpha_1^2\alpha_2 - 3\alpha_1^4 \tag{1.30}$$

It characterizes the sharpness of the peak about the mode, also called *peakedness* or *kurtosis*. The *coefficient* of *kurtosis* or *excess* is defined as

$$\gamma_2 = \mu_4/\sigma^4 - 3 \tag{1.31}$$

Plots of density functions with nonzero coefficients of skewness and excess are shown in Fig. 1.3. For comparison, it shows a dashed curve having the same mean μ and variance σ^2, but with the coefficients of skewness and excess equal to zero.

The moments exist if the respective integrals (for continuous random variables) exist or if the respective sequences (for discrete random variables) converge absolutely. For random variables with limited values, the moments exist always.

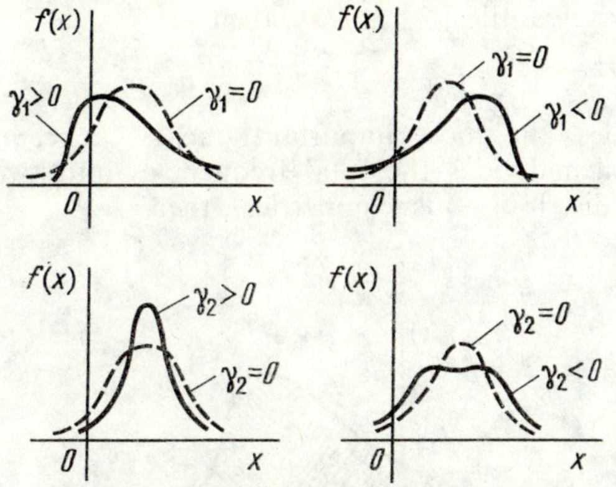

Fig. 1. 3. Probability density function with nonzero coefficients of scewness and excess

If a random variable X has the first and second moments, we can deduce a standardized random variable:

$$X_0 = (X - \mu)/\sigma \tag{1.32}$$

For a standardized random variable,

$$E\{X_0\} = 0 \quad \text{and} \quad \text{Var}\,\{X_0\} = 1 \tag{1.33}$$

Many tables are compiled for standardized random variables. The distribution functions are related to the standardized variable X_0 and the nonstandardized variable X in the following manner:

$$f(x) = \frac{1}{\sigma} f_1(x_0) = \frac{1}{\sigma} f_1\left(\frac{x-\mu}{\sigma}\right) \tag{1.34}$$

$$f_1(x_0) = \sigma f(x) = \sigma f(\mu + \sigma x_0) \tag{1.35}$$

$$F(x) = F_1(x_0) = F_1\left(\frac{x-\mu}{\sigma}\right) \tag{1.36}$$

$$F_1(x_0) = F(x) = F(\mu + \sigma x_0) \tag{1.37}$$

The moments are the total (integral) characteristics of a distribution. The individual values of a distribution function are characterized by a second group of parameters which includes quantiles. *Quantiles,* or fractiles, is a general term for quantities defining points on the x-scale which divide the total frequencies in specified proportions.

The quantile x_p of the distribution (frequency) of a random variable X for which the distribution function is $F(x)$ refers to the solution of the equation

$$F(x_p) = p \tag{1.38}$$

That is, x_p is the value of the random variable such that

$$P(X < x_p) = p \tag{1.39}$$

If we know two quantiles, x_p and x_q, then

$$P(x_p \leqslant X \leqslant x_q) = q-p \tag{1.40}$$

Of all quantiles, the most important one is the *central quantile*, or *median*, $x_{0.5}$, which divides the total frequencies into two equal groups. (Fig. 1.4). If the distribution is symmetrical, then

$$x_{0.5} = \mu \tag{1.41}$$

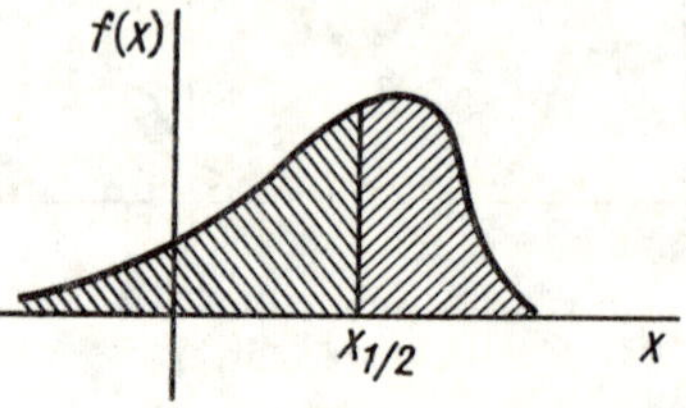

Fig. 1. 4. Median of a distribution

The quantiles x_p and x_{1-p} are called symmetrical. For a distribution symmetrical about the origin (zero), it is always that

$$x_p = -x_{1-p} \tag{1.42}$$

Most frequently, applications of mathematical statistics have recourse to the expected value as it characterizes the position of random variables on the number axis, and the variance (or standard deviation), as it gives a measure of the spread, or dispersion, of the random variable about the mean.

1.3. Properties of Expectation and Variance

We shall give, without proof, the following properties of the mean of random variables.

1. The expectation (expected value or mean) of a nonrandom variable is equal to the value of that variable:

$$E\{c\} = c \tag{1.43}$$

2. A nonrandom variable may be placed outside the sign of expectation:

$$E\{cX\} = cE\{X\} \tag{1.44}$$

3. The expectation of the sum of random variables is equal to the sum of the expectations of these random variables:

$$E\{X_1+X_2+\ldots+X_n\} = E\{X_1\}+E(X_2\}+\ldots+E\{X_n\} \tag{1.45}$$

4. The expectation of the product of independent random variables is equal to the product of the expectations of the variables:

$$E\{X_1 \cdot X_2 \cdot \ldots \cdot X_n\} = E\{X_1\}\,E\{X_2\}\ldots E\{X_n\} \tag{1.46}$$

Random variables are called independent, if each has a distribution of its own, in no way related to those of the other random variables.

Also without proof, the properties of variance may be summed up as follows.

1. The variance of a nonrandom variable is zero:

$$\text{Var}\{c\} = 0 \tag{1.47}$$

2. A nonrandom variable may, on being squared, be placed outside the symbol of variance:

$$\text{Var}\{cX\} = c^2\,\text{Var}\{X\} \tag{1.48}$$

3. The variance of a random variable is equal to the expectation of the random variable squared minus the square of its expectation:

$$\text{Var}\{X\} = E\{X^2\} - \mu^2 \tag{1.49}$$

4. The variance of the sum of independent random variables is equal to the sum of the variances of these variables:

$$\text{Var}\{X_1+X_2+\ldots+X_n\} = \text{Var}\{X_1\}+\text{Var}\{X_2\}+\ldots+\text{Var}\{X_n\} \tag{1.50}$$

Using the properties of the expectation and variance of a random variable, we can show that Eq. (1.33) holds for a standardized random variable; that is, if

$$X_0 = (X-\mu)/\sigma$$

then

$$E\{X_0\} = 0, \qquad \text{Var}\{X_0\} = 1$$

The proof is as follows:

$$E\{X_0\} = E\left\{\frac{X-\mu}{\sigma}\right\} = \frac{1}{\sigma}E\{X-\mu\} = \frac{1}{\sigma}(E\{X\}-\mu) = 0$$

and

$$\text{Var}\{X_0\} = \text{Var}\left\{\frac{X-\mu}{\sigma}\right\} = \frac{1}{\sigma^2}\text{Var}\{X-\mu\} = \frac{1}{\sigma^2}(\text{Var}\{X\}-0)$$
$$= \text{Var}\{X\}/\sigma^2 = 1$$

Example 1.2. Tests of two flow meters have revealed the following probability of error in meter readings (on a two-point scale):

Error level, points	Probability of error observation at a given level	
	flow meter 1	flow meter 2
1	0.20	0.03
2	0.065	0.15

We are to decide, on the basis of the above data, which flow meter has a lower mean error level and gives more consistent readings.

Solution. Let the random error level of the flow meters be denoted as X. The mean error level for each flow meter can then be found by Eq. (1.20):

$$E_1\{X\} = 0.20\times 1 + 0.065\times 2 = 0.33$$
$$E_2\{X\} = 0.03\times 1 + 0.15\times 2 = 0.33$$

As is seen, both flow meters have the same mean error level, and on this basis we cannot say which meter is the better. The next step is, therefore, to determine which meter gives more consistent readings. To this end, we determine the variance of the error level for each flow meter by Eq. (1.25).

$$\mathrm{Var}_1\{X\} = (1-0.33)^2\times 0.2 + (2-0.33)^2\times 0.065 = 0.11$$
$$\mathrm{Var}_2\{X\} = (1-0.33)^2\times 0.03 + (2-0.33)^2\times 0.15 = 0.43$$

Thus, the first flow meter is the better.

1.4. The Uniform Distribution

One of the simplest continuous distributions is the *uniform* (or rectangular) *distribution*. Let us determine its basic numerical measures.

A uniform distribution is that for which the probability density is constant on an interval and is zero outside that interval. In other words, any values within the interval are equally likely. Referring to Fig. 1.5, it is seen that $f(x)$ is constant and equal to c on the interval $[a, b]$ and is zero outside this interval:

$$f(x) = \begin{cases} c & \text{at } a \leqslant x \leqslant b \\ 0 & \text{at } x < a \text{ or } x > b \end{cases}$$

Since the area bounded by the distribution curve is equal to unity,

$$c(b-a) = 1$$

it follows then that

$$c = 1/(b-a)$$

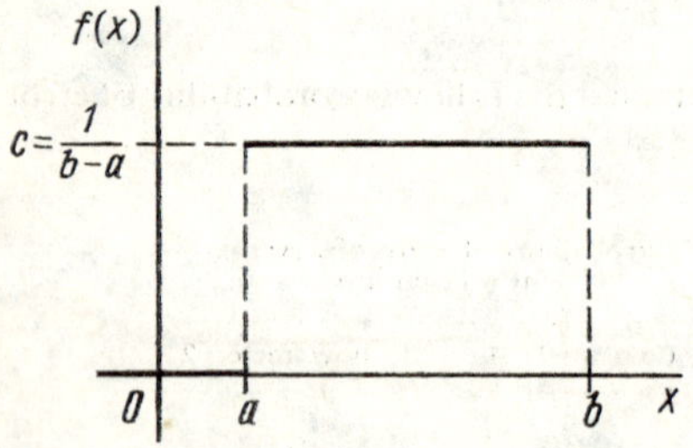

Fig. 1. 5. Uniform probability density function

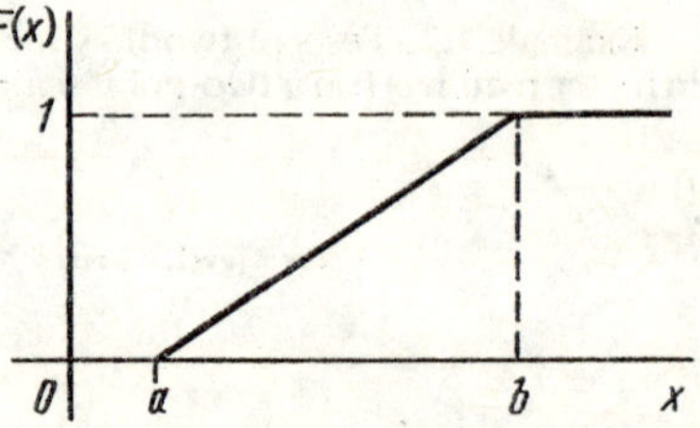

Fig. 1. 6. Uniform distribution function

and the density function $f(x)$ has the form

$$f(x) = 1/(b-a) \quad \text{at} \quad a \leqslant x \leqslant b$$
$$f(x) = 0 \quad \text{at} \quad x < a \text{ or } x > b \tag{1.51}$$

The distribution function is represented by the area under the distribution curve to the left of the point x. Therefore,

$$F(x) = \begin{cases} 0 & \text{at } x < a \\ \dfrac{x-a}{b-a} & \text{at } a \leqslant x \leqslant b \\ 1 & \text{at } x > b \end{cases} \tag{1.52}$$

A plot of $F(x)$ is shown in Fig. 1.6.

The expectation of a random variable having a uniform distribution on the interval $[a, b]$ is

$$\mu = \int_a^b \frac{x}{b-a}\,\mathrm{d}x = (a+b)/2 \tag{1.53}$$

Because the uniform distribution is symmetrical, the median of X is likewise equal to $(a+b)/2$.

The variance of this distribution can be found by Eq. (1.26):

$$\mathrm{Var}\{X\} = \frac{1}{b-a}\int_a^b \left(x-\frac{a+b}{2}\right)^2 \mathrm{d}x = \frac{(b-a)^2}{12} \tag{1.54}$$

The coefficient of skewness (the third moment ratio)

$$\gamma_1 = \mu_3/\sigma^3$$

is equal to zero which indicates that the distribution is symmetrical.

To determine the coefficient of peakedness or excess, we first find the fourth central moment:

$$\mu_4 = \frac{1}{b-a}\int_a^b \left(x-\frac{a+b}{2}\right)^4 \mathrm{d}x = \frac{(b-a)^4}{80}$$

Hence, the coefficient of excess (the fourth moment ratio) is

$$\gamma_2 = \mu_4/\sigma^4 - 3 = -1.2$$

The probability that a uniformly distributed random variable X will fall within the interval $[\alpha, \beta]$, which is a relatively small range of values within the interval $[a, b]$, is defined by the ratio of the length of the interval to the length of the entire interval (Fig. 1.7):

$$P(\alpha \leqslant X \leqslant \beta) = (\beta-\alpha)/(b-a) \tag{1.55}$$

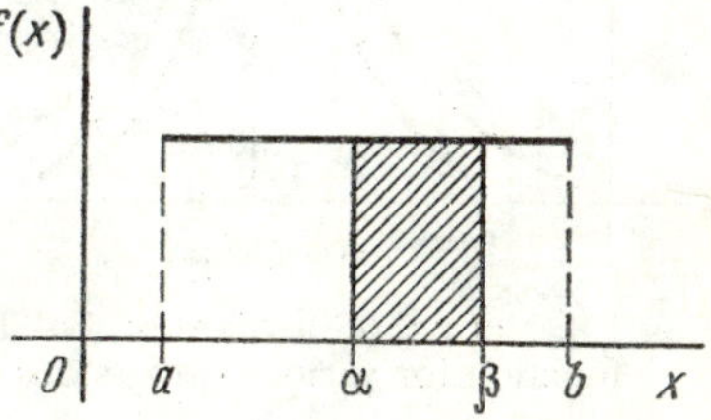

Fig. 1.7. Probability that a uniformly distributed random variable will fall within a given interval

1.5. The Normal (Gaussian) Distribution

A random variable X is said to be normally distributed, if its probability density has the form

$$f(x) = \frac{1}{\sqrt{2\pi}\,\sigma} \exp\left[-(x-\mu)^2/2\sigma^2\right] \quad -\infty < x < \infty \tag{1.56}$$

where μ is the expectation (mean) and σ^2 is the variance of the random variable X, and the normal distribution function has the form

$$F(x) = \frac{1}{\sqrt{2\pi}\,\sigma} \int_{-\infty}^{x} \exp\left[-(x-\mu)^2/2\sigma^2\right] dx \tag{1.57}$$

The normal distribution is the most important distribution in applied statistics and has been most fully investigated theoretically. Under certain conditions, the normal distribution is the limiting one for the sum of a large number n of independent random variables each of which obeys any arbitrary distribution (Lyapunov's theorem). The principal constraint is that all terms play a relatively small part in the total sum.

Many events occur at random due to a large number of independent (or weakly related) factors. Such events can usefully be described by a nearly normal distribution or can be said to be normal as a first approximation. For this reason, the normal distribution is widely used in processing observation data.

By resort to information theory, it can be shown that the normal distribution contains a minimum of information as compared with any other distribution having the same variance. Therefore, replacement of a distribution by the equivalent normal distribution cannot lead to an exaggeration of observation accuracy. Graphically, the normal distribution density is depicted by what is called the *normal*, or *Gaussian*, *curve* (Fig. 1.8).

The normal distribution of a standardized random variable is called the *standard normal distribution*. Its distribution function has the form

$$F_0(x) = \frac{1}{\sqrt{2\pi}} \int_{-\infty}^{x} \exp(-x^2/2)\, dx \tag{1.58}$$

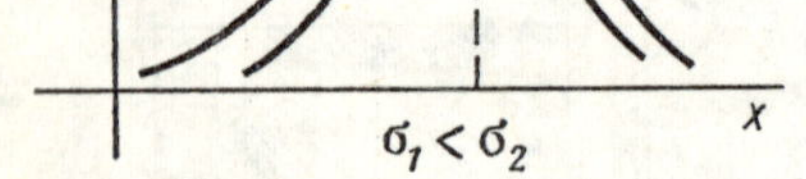

Fig. 1.8. Normal (Gaussian) probability function for various dispersions

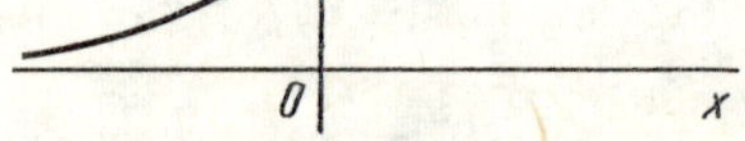

Fig. 1.9. Standardized normal distribution function

A plot of this function appears in Fig. 1.9. In this case,

$$P(x_1 \leqslant X_0 \leqslant x_2) = F_0(x_2) - F_0(x_1) \tag{1.59}$$

The equation

$$\Phi(x) = F_0(x) - 1/2 \tag{1.60}$$

is called the *error function* or *integral*, erf (x):

$$\Phi(x) = F_0(x) - F_0(0) = (\sqrt{2\pi})^{-1} \int_0^x \exp(-x^2/2)\, dx \tag{1.61}$$

The values of this function are given in Table 1 of the Appendix.

The Laplace function is an odd function which implies that

$$\Phi(-x) = -\Phi(x) \tag{1.62}$$

Therefore, the values of $\Phi(x)$ are often tabulated only for $x > 0$.

In view of Eqs. (1.59) and (1.60), for a standard random variable we have

$$\begin{aligned} P(x_1 \leqslant X_0 \leqslant x_2) &= \Phi(x_2) + 1/2 - \Phi(x_1) - 1/2 \\ &= \Phi(x_2) - \Phi(x_1) \end{aligned} \tag{1.63}$$

In the general case,

$$\begin{aligned} P(x_1 \leqslant X \leqslant x_2) &= P\left(\frac{x_1-\mu}{\sigma} \leqslant X_0 \frac{x_2-\mu}{\sigma}\right) \\ &= \Phi\left(\frac{x_2-\mu}{\sigma}\right) - \Phi\left(\frac{x_1-\mu}{\sigma}\right) \end{aligned} \tag{1.64}$$

In many practical problems, x_1 and x_2 are symmetrical about the mean. This is true, for example, of the problem of the absolute deviation. The *absolute deviation* is defined as

$$\Delta X = |X - \mu| \tag{1.65}$$

It is required to determine the probability that the absolute deviation of a random variable will not exceed some predetermined number ε:

$$P(\Delta X \leqslant \varepsilon) = P(\mu - \varepsilon \leqslant X \leqslant \mu + \varepsilon) \tag{1.66}$$

In particular, for a standard random variable,

$$\begin{aligned} P(\Delta X_0 \leqslant \varepsilon) &= P(-\varepsilon \leqslant X_0 \leqslant +\varepsilon) \\ &= \Phi(\varepsilon) - \Phi(-\varepsilon) = 2\Phi(\varepsilon) \end{aligned} \tag{1.67}$$

For a normally distributed random variable with mean μ_x and variance σ,

$$P(\Delta X \leqslant \varepsilon) = P(\Delta X_0 \leqslant \varepsilon/\sigma) = 2\Phi(\varepsilon/\sigma) \tag{1.68}$$

On setting $\varepsilon/\sigma = k$, we develop the following relation from Eq. (1.68):

$$P(\Delta X \leqslant k\sigma) = 2\Phi(k) \tag{1.69}$$

Hence,

$$\begin{aligned} P(\Delta X \leqslant \sigma) &= 2\Phi(1) \doteq 0.6826 \\ P(\Delta X \leqslant 2\sigma) &= 2\Phi(2) = 0.9544 \\ P(\Delta X \leqslant 3\sigma) &= 2\Phi(3) = 0.9973 \end{aligned} \tag{1.70}$$

Thus, deviations greater than three times the standard deviation cannot practically occur.

The normal distribution shows the property of linearity: if independent random variables X_1 and X_2 are normally distributed, then for arbitrary numbers α and β the quantity

$$Y = \alpha X_1 + \beta X_2 \tag{1.71}$$

will likewise be normally distributed. Also, from the properties of expectation and variance, it follows that

$$\mu_y = \alpha\mu_{x1} + \beta\mu_{x2} \tag{1.72}$$

and

$$\sigma_y = (\alpha^2\sigma_{x1}^2 + \beta^2\sigma_{x2}^2)^{1/2} \tag{1.73}$$

Example 1.3. The diameter of bushings fabricated by a shop may be treated as a normal random variable with a standard deviation $\sigma = 0.5$ mm. What is the probability of rejection, if bushings whose diameter deviates from the prescribed value (the mean) by more than 0.8 mm are rejected?

Solution. The problem reduces to determining the absolute deviation for the random variable d, the diameter of the bushings. We are to determine $P(\Delta d \geqslant 0.8)$. It is convenient to proceed by finding the probability of the opposite event, using Eq. (1.68):

$$P(\Delta d \leqslant 0.8) = 2\Phi(0.8/0.5) = 2\Phi(1.6) = 2\times 0.4452 = 0.8904$$

Hence,

$$P(\Delta d \geqslant 0.8) = 1 - 0.8904 = 1.1096$$

Chapter Two
PARAMETERS OF THE DISTRIBUTION FUNCTION

2.1. Population and Sample

In practice, the experimenter has always to be content with a limited number of observations or measurements of the random variable in question. The data set thus obtained is usually referred to as a *sample* from a (statistical) population or universe.

A *statistical population* or *universe* is the complete set of all possible values that the random variable in question might take if the experimenter repeated his observation an infinite number of times. In an analysis of a random process quantity varying with time continuously (say, temperature, pressure, etc.), the observed values of the variables are those taken at discrete instants of time, separated by an interval such that adjacent values can be regarded as obtained from independent trials.

A sample is said to be *representative*, if it gives a sufficiently complete idea about the population involved. If the experimenter has no prior information about the population, the only way to ensure representativity is by random sampling or by impartial selection which is given the statistical name *randomization*. Random sampling can be simulated with the aid of random-number tables [6]. Suppose we are to draw a random sample of 10 objects from a population of 100. Our first step is to number the objects from 00 to 99. Then, referring to a random number table, we arbitrarily select a page, a row, and a column of the table, and read the last two digits of ten numbers appearing in succession. Let these numbers (starting with the first) be as follows:

82 49 18 48 09 50 17 10 37 51

We ignore any repeated number when it appears a second time. The digits thus drawn indicate which objects should be selected. No changes may be made in the selected sequence of numbers as this might lead to a distorted result. Randomization is carried out in a similar manner, except that random numbers are drawn until they cover all the given objects.

From the randomness of samples it immediately follows that any judgement passed on the population on the basis of a sample is random as well. Suppose that an experiment has yielded a sample of values, x_1, x_2, $\ldots$, x_n, of a random variable X. Let x be some point on the x (number) axis, and n_x be the number of selected values lying to the left of the point x on the same axis. The ratio n_x/n is the frequency of the observed values of X less than x. This frequency is a function of x, which we denote $F_n(x)$:

$$F_n(x) = n_x/n \tag{2.1}$$

and call the *empirical* or *sample distribution* function in contrast to the population, or theoretical, distribution.

For each sample, the sample distribution function will be different, but all sample distribution functions of one and the same random variable will have something in common, and this "something" is information about the distribution function of the random variable involved.

It can be proved (Glivenko's theorem) that at n tending to infinity the maximum difference between $F_n(x)$ and $F(x)$ tends to zero with probability 1:

$$P\left(\sup_{n\to\infty}\left|F(x) - \underset{-\infty\leq x\leq+\infty}{F_n(x)}\right| \to 0\right) = 1 \tag{2.2}$$

Practically, this means that, given a sufficiently large sample, the population distribution function can be approximated by a sample distribution function.

Let $x_1 < x_2 < x_3 \ldots < x_n$ be an *ordered sample* drawn from the population of observations of a random variable X. All members of the sample have the same probability of being selected, equal to $1/n$. Therefore, according to the definition of $F_n(x)$, we get

$$\begin{aligned} F_n(x) &= 0 && \text{at } x < x_1 \\ F_n(x) &= k/n && \text{at } x_k \leqslant x < x_{k+1} \\ & && k = 1, 2, \ldots, n-1 \\ F_n(x) &= 1 && \text{at } x > x_n \end{aligned} \tag{2.3}$$

A plot of $F_n(x)$ appears in Fig. 2.1. All members of the sample are discontinuities in this function. At a discontinuity $x = x_k$, $F_n(x)$ abruptly changes

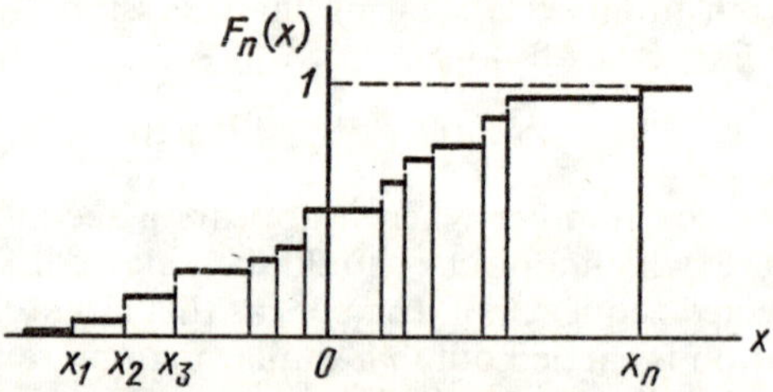

Fig. 2.1. Sampling distribution

in value from $(k-1)/n$ (in the interval $x_{k-1} \leqslant x < x_k$) to k/n and retains the last value over the next interval.

With large-size samples, resort is made to the grouped-data method. By this method, our first step is to find the minimum and the maximum value in the data set, or sample: $x_{\min}$ and $x_{\max}$. Then we choose a number k of what are called class intervals of equal length that will cover the sample range, $x_{\min}$-$x_{\max}$, without overlapping. The number of class intervals can be found by a semiempirical equation of the form

$$k = 1+3.2 \log_{10} n \tag{2.4}$$

and is rounded off to the nearest integer. The width h of a class interval then is

$$h = (x_{\max}-x_{\min})/k \tag{2.5}$$

Now we count the number of observations (measurements, objects, etc.) in the data that belong to each class interval. This count, n_i, is the *class frequency*. Division of the class frequency by the total number of observations, or *sample size*, *n*, yields the relative frequency of a class:

$$p_i^* = n_i/n \tag{2.6}$$

The relative frequency of a class is the fraction of the total observations belonging to that class. All points within a class are referred to its midpoint, x_i^*:

$$x_i^* = (x_{i-1}^*+x_i)/2 \tag{2.7}$$

Table 2.1. Frequency Distribution

Class interval No.	Class interval width	Class midpoint	Number of points in class interval	Relative frequency
1	(x_{min}, x_1)	x_1^*	n_1	p_1^*
2	(x_1, x_2)	x_2^*	n_2	p_2^*
...	...	...	...	...
i	(x_{i-1}, x_i)	x_i^*	n_i	p_i^*
...	...	...	...	...
k	(x_{k-1}, x_{max})	x_k^*	n_k	p_k^*
$\sum$			n	1

The data are then recorded in the form of a frequency table like Table 2.1. A *frequency distribution histogram* for the data given in Table 2.1 is shown in Fig. 2.2. A plot of $F_n(x)$ based on grouped data is shown in Fig. 2.3.

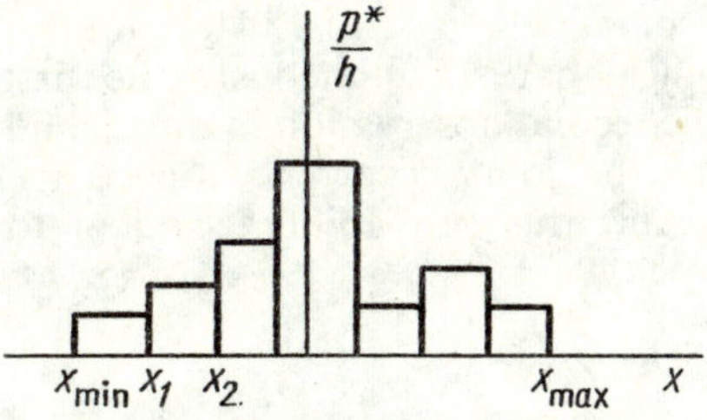

Fig. 2.2. Relative frequency histogram

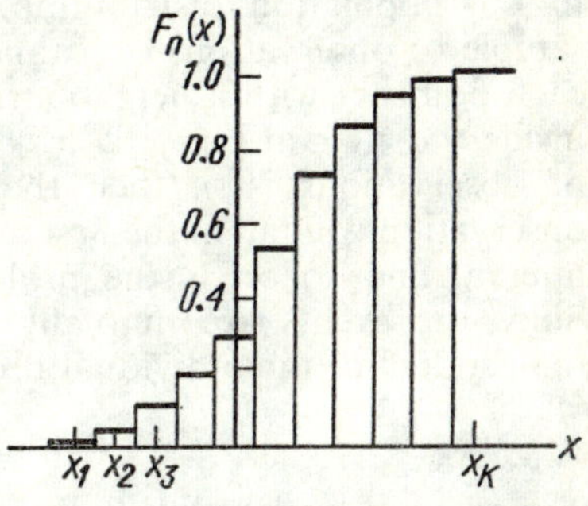

Fig. 2.3. Grouped data distribution

While the analyst would like to obtain the sample distribution function for a process variable, this is difficult. In most cases, this function remains unknown. However, even a very simple analysis can ascertain the form and type of the distribution. In the circumstances, the analyst settles for some numerical measures, or parameters, of the distribution. Basing himself on the data set, or sample, he can find the sample mean, the sample variance, etc. These are all estimates of the respective population parameters. Although sample parameters are themselves random variables, the desired accuracy can be achieved with a smaller sample size, n, than Glivenko's theorem suggests.

To obtain "good" estimates, it is necessary that they, so far as possible, be *consistent* and *unbiased*. An estimate a^* of a parameter a is said to be consistent if it tends (in probability) to approach the population parameter as the sample size n is increased. The sample moments are consistent estimates of the respective population moments. An estimate of a parameter is said to be unbiased if its expected value is equal to the population value at any sample size, $E\{a^*\} = a$. It is also required that estimates be efficient. An estimate is said to be more efficient than any other if the former has a smaller variance. Thus, the efficiency of various unbiased estimates of one and the same parameter at a fixed sample size is inversely proportional to the variances of the estimates.

2.2. The Method of Maximum Likelihood

Estimates of probability parameters can be derived in any one of several methods. A well-known and desirable estimation procedure (when it can be carried out) is that of maximum likelihood. It leads to optimal estimates which meet most of the above requirements.

Basically, the method consists in finding estimates for the unknown parameters such that the *likelihood function* has a maximum value for a random sample of size n. Suppose we know the general form of the probability density function $f(x, a)$ of a population; a is an unknown parameter in the population distribution. By an experiment, we have obtained a sample of observations of a random variable, $x_1, x_2, \ldots, x_n$. We surround each point x_1 with a neighbourhood of length ε. The probability of falling inside the interval $[x_i-\varepsilon/2, x_i+\varepsilon/2]$ approximately is $f(x_i)\varepsilon$. If the number of observations is n, then the probability that simultaneously the first observation will fall in the first interval, the second observation in the second interval and so on is the probability of the joint occurrence of events. Since the events are mutually independent, this probability is equal to the product of the individual probabilities:

$$P(x, a) = f(x_1)\,f(x_2)\,\ldots\,f(x_n)\varepsilon^n \tag{2.8}$$

The event of probability P does occur. It is natural to expect that the event occurring on the very first trial has a maximum probability. Therefore, we must choose as the estimate of a that value of a^* from the range of permissible values of a for which this probability is a maximum. This is the root of the equation

$$\partial P(x, a^*)/\partial a^* = 0 \tag{2.9}$$

which defines the necessary condition for the probability P to be a maximum. The sufficient condition for this maximum to exist is the inequality

$$\partial^2 P(x, a^*)/\partial a^{*2} < 0 \tag{2.10}$$

If there are several maxima, the largest of all must be chosen. The solution is easier to obtain if we use the function

$$L(x, a^*) = \ln \frac{P(x, a^*)}{\varepsilon^n} = \sum_{i=1}^{n} \ln f(x_i, a^*) \tag{2.11}$$

which is called the *likelihood function*. The probability P and the function

L have a maximum value at the same values of the unknown parameter, because

$$\frac{\partial}{\partial a^*} \ln P = (1/P)(\partial P/\partial a^*) \qquad P > 0 \tag{2.12}$$

In the general case, estimation involves several parameters in a one- or multi-dimensional distribution. If a and x be construed as vectors, the principle of maximum likelihood can be stated as before: we seek to find a set of permissible values $a_1^*, a_2^* \ldots, a_k^*$ of the distribution parameters which maximize the likelihood function. The necessary conditions for maximization are defined by

$$\frac{\partial L(x, a_1^*, \ldots, a_k^*)}{\partial a_j^*} = 0 \quad j = 1, 2, \ldots, k \tag{2.13}$$

whereas the nonnegative determinancy of the matrix

$$\frac{\partial^2 L}{\partial a_i^* \, \partial a_j^*} \qquad i, j = 1, 2, \ldots, k \tag{2.14}$$

furnishes a sufficient condition for this local extremum to be a maximum of the likelihood function.

Using the maximum likelihood method and a sample $x_1, x_2, \ldots, x_n$, let us estimate the parameter λ of an exponential distribution for which the density function is

$$f(x) = \lambda \exp(-\lambda x)$$

For this distribution, the likelihood function has the form

$$L = \sum_{i=1}^{n} \ln [\lambda \exp(-\lambda x_i)] = n \ln \lambda - \sum_{i=1}^{n} \lambda x_i, \tag{2.15}$$

Replacing the parameter λ by its estimate λ^* in Eq. (2.15), differentiating the resultant expression with respect to λ^*, and equating to zero, we obtain

$$n/\lambda^* - \sum_{i=1}^{n} x_i = 0$$

Hence,

$$\lambda^* = n/\sum_{i=1}^{n} x_i \tag{2.16}$$

or $\lambda^* = 1/\bar{x}$, where $\bar{x}$ is the sample mean.

Now, let a random variable X be normally distributed:

$$f(x) = (\pi 2\sigma^2)^{-1/2} \exp [-(x-\mu)^2/2\sigma^2]$$

Then the probability of the joint occurrence of n independent events $X = x_j$ $(j = 1, 2, \ldots, n)$ is

$$P(x, \mu, \sigma^2) = (2\pi\sigma^2)^{-n/2} \exp \left[-(1/2\sigma^2) \sum_{i=1}^{n} (x_i - \mu)^2\right] \varepsilon^n \tag{2.17}$$

and the likelihood function is

$$L(x, \mu, \sigma^2) = \ln (P/\varepsilon^n) = -\frac{n}{2} \ln 2\pi - \frac{n}{2} \ln \sigma^2 - \frac{1}{2\sigma^2} \sum_{i=1}^{n} (x_i - \mu)^2 \tag{2.18}$$

Let us replace μ and σ^2 in Eq. (2.18) by their estimates, $\bar{x}$ and s^2 and differentiate with respect to $\bar{x}$ and s^2:

$$\partial L/\partial \bar{x} = (2/2s^2) \textstyle\sum_{i=1}^{n} (x_i - \bar{x}) = 0 \tag{2.19}$$

Since $1/s^2 \neq 0$, we have from Eq. (2.19)

$$\textstyle\sum_{i=1}^{n} (x_i - \bar{x}) = 0$$

Hence,

$$\bar{x} = \frac{1}{n} \textstyle\sum_{i=1}^{n} x_i$$

Differentiating the likelihood function with respect to s^2 gives

$$\frac{\partial L}{\partial s^2} = -\frac{n}{2}\frac{1}{s^2} + \frac{1}{2(s^2)^2} \textstyle\sum_{i=1}^{n} (x_i - \bar{x})^2 = 0,$$

or

$$-\frac{1}{2s^2}\left[n - \frac{1}{s^2} \textstyle\sum_{i=1}^{n} (x_i - \bar{x})^2\right] = 0 \tag{2.20}$$

Since, $1/2s^2 \neq 0$, it follows that

$$n - (1/s^2) \textstyle\sum_{i=1}^{n} (x_i - \bar{x})^2 = 0$$

Hence,

$$s^2 = \frac{1}{n} \textstyle\sum_{i=1}^{n} (x_i - \bar{x})^2 \tag{2.21}$$

2.3. The Sample Mean and Variance

The maximum likelihood method always leads to consistent, although sometimes biased estimates having the smallest variance. For a normally distributed random variable, it is used to find the estimate $\bar{x}$ of the population mean μ

$$\bar{x} = \textstyle\sum_{i=1}^{n} x_i/n$$

called the sample mean, and the estimate s_1^2 of the population variance, Var $\{X\}$ or σ^2

$$s_1^2 = \textstyle\sum_{i=1}^{n} (x_i - \bar{x})^2/n$$

called the sample variance. The sample variance is a somewhat biased estimate:

$$E\{s_1^2\} = \frac{n-1}{n} \text{Var}\,\{X\} \tag{2.22}$$

In order to make it unbiased, s_1^2 must be multiplied by $n/(n-1)$

$$s^2 = s_1^2 n/(n-1) = \frac{1}{n-1} \textstyle\sum_{i=1}^{n} (x_i - \bar{x})^2 \tag{2.23}$$

The argument for using $(n-1)$ instead of n in the denominator is that the sample mean $\bar{x}$ about which the deviation is taken depends on the sample members itself. Each quantity dependent on the sample members and included in the equation for the sample variance is called a *constraint*. It can be proved that the denominator of the sample variance is always equal to the difference between the sample size n and the number l of constraints placed on the sample. This difference

$$v = n-l \tag{2.24}$$

gives the number of *degrees of freedom*. Practically, the sample variance s^2 can best be found by the equation

$$s^2 = \frac{1}{n-1}\left[\sum_{i=1}^{n} x_i^2 - \left(\sum_{i=1}^{n} x_i\right)^2/n\right] \tag{2.25}$$

readily deducible from Eq. (2.23).

$$s^2 = \frac{1}{n-1}\sum_{i=1}^{n}(x_i-\bar{x})^2 = \frac{1}{n-1}\sum_{i=1}^{n}(x_i^2-2x_i\bar{x}+\bar{x}^2)$$

$$= \frac{1}{n-1}[(x_1^2-2x_1\bar{x}+\bar{x}^2)+(x_2^2-2x_2\bar{x}+\bar{x}^2)+\ldots$$

$$+(x_n^2-2x_n\bar{x}+\bar{x}^2)] = \frac{1}{n-1}[(x_1^2+x_2^2+\ldots+x_n^2)$$

$$-2\bar{x}(x_1+x_2+\ldots+x_n)+(\bar{x}^2+\bar{x}^2+\ldots+\bar{x}^2)]$$

$$= \frac{1}{n-1}\left[\sum_{i=1}^{n} x_i^2 - \frac{2}{n}\left(\sum_{i=1}^{n} x_i\right)^2 + n\left(\sum_{i=1}^{n}\frac{x_i}{n}\right)^2\right]$$

$$= \frac{1}{n-1}\left[\sum_{i=1}^{n} x_i^2 - \left(\sum_{i=1}^{n} x_i\right)^2\Big/n\right]$$

An advantage of Eq. (2.25) is that close numbers are subtracted only once, whereas Eq. (2.23) involves this operation more than once, thereby leading to a loss of accuracy.

On the basis of grouped data (see Table 2.1), the sample mean and variance are found by the equations

$$\bar{x} = \sum_{i=1}^{k} n_i x_i^*/n = \sum_{i=1}^{k} p_i^* x_i^* \tag{2.26}$$

$$s^2 = \frac{1}{n-1}\sum_{i=1}^{k} n_i(x_i^*-\bar{x})^2 - h/12$$

$$= \frac{1}{n-1}\sum_{i=1}^{k}(x_i^*-\bar{x})^2 p_i^* - h^2/12 \tag{2.27}$$

The term $h^2/12$ is called the Sheppard correction; it is applied to account for the bias of the variance due to grouping.

2.4. Classification of Measurement Errors

No matter how careful the experimenter may be, values of the measured variables will always differ from their true value. This difference is called an *error of measurement* or *observation.*

If we denote the true value as a, the error as ΔX, and the result as X, then

$$X-a = \Delta X \tag{2.28}$$

In the general case, a measurement may lead to three kinds of error or uncertainty, namely:

(1) *Blunders* which occur because the experimenter has not been careful in carrying out a measurement or observation. A blunder is a large, one of a kind, isolated error. This distinction serves as the basis for the exclusion of blunders.

(2) *Systematic errors* which appear continuously from measurement to measurement and remain constant in magnitude or vary in a predictable manner. Their detection calls for a special effort, but once they are detected, they can readily be compensated for by applying suitable corrections to the results.

(3) *Random errors* which remain after all blunders and systematic errors have been eliminated or allowed for. This definition excludes factors that are continuous or constant in their effect (and lead to systematic errors) and also factors acting only once but strongly (and lead to blunders). Random errors are an outcome of many factors whose individual effects may be so small that they cannot be identified on an individual basis (at least, at the existing state of the art). Random errors are distributed symmetrically about zero: errors of the same magnitude but differing in sign have the same frequency of occurrence. From the symmetry of distribution it follows that the true value of the result is the expectation of the corresponding random variable. Since, according to Eq. (2.28),

$$X = a+\Delta X$$

and assuming the absence of blunders and systematic errors,

$$E\{\Delta X\} = 0 \tag{2.29}$$

it follows that

$$E\{\Delta X\} = a \tag{2.30}$$

In our subsequent discussion we shall solely be concerned with random errors of measurement.

2.5. Law of Error Addition

For independent random variables, the property of additivity holds only for their variances, and not for the standard deviation. If $X_1, X_2, \ldots, X_n$ are independent random variables, $a_1, a_2, \ldots, a_n$ are nonrandom variables, and

$$Z = a_1X_1 + a_2X_2 + \ldots + a_nX_n \tag{2.31}$$

then the sample variance of Z will be defined as

$$s_Z^2 = a_1^2 s_{X_1}^2 + a_2^2 s_{X_2}^2 + \ldots + a_n^2 s_{X_n}^2 \tag{2.32}$$

If we set $a_1 = a_2 = \ldots = a_n = 1/n$, then

$$Z = \frac{X_1 + X_2 + \ldots + X_n}{n} = \bar{X}$$

and

$$s_{\bar{X}}^2 = \frac{s_{X_1}^2 + s_{X_2}^2 + \ldots + s_{X_n}^2}{n^2} = \overline{s_X^2}/n \tag{2.33}$$

where

$$\overline{s_X^2} = \textstyle\sum_{i=1}^{n} s_{X_i}^2/n.$$

If we interpret $X_1, X_2, \ldots, X_n$ as n independent observations of one and the same random variable X, then

$$s_{X_1}^2 = s_{X_2}^2 = \ldots = s_{X_n}^2 = s_X^2$$

and

$$s_{\bar{X}}^2 = s_X^2/n \tag{2.34}$$

From Eq. (2.34) it may be concluded that, in selecting between two analytical methods differing in accuracy, the analyst should give preference to that with which each observation takes a shorter time to complete. For example, using the less accurate proximate analysis, he will be able to carry out a greater number of determinations and achieve a higher accuracy than with the more accurate ultimate analysis.

2.6. Errors of Indirect Measurements

Measurements can be *direct* and *indirect* (or inferential). With a direct measurement, the experimenter determines the unknown by measuring the unknown itself. With an indirect measurement, the experimenter first measures some other quantity related to the unknown, then infers the value of the unknown from the measured value.

Let a random variable z be a function of observations $x_1, \ldots, x_n$ such that

$$z = f(x_1, x_2, \ldots, x_n)$$

The true value of z, instead of coinciding with its expectation, may be a function obeying the same law

$$a_z = f(\mu_{x1}, \mu_{x2}, \ldots, \mu_{xn})$$

The quantity a_z is the mean of an indirect measurement or observation.

The variance of an indirect measurement, σ_z^2, is found in the same manner as the population variance, but the deviations are taken about the mean of the indirect measurement, a_z. It can be evaluated, if we know the variances of the individual observations and the form of the function f. In practice, one usually finds the sample variances s_{xi}^2 and uses them to find the cumulative sample variance of the indirect measurement, s_z^2, which is the estimate of the population variance σ_z^2.

In order to find s_z^2, we expand the function $z = f(x_1, x_2, \ldots, x_n)$ into a Taylor series about the point $(\mu_{x1}, \mu_{x2}, \ldots, \mu_{xn})$ and retain only the first-order terms of the series

$$z \approx f(\mu_{x1}, \mu_{x2}, \ldots, \mu_{xn}) + (\partial f/\partial x_1)(x_1 - \mu_{x1}) + (\partial f/\partial x_2)(x_2 - \mu_{x2}) + \ldots + (\partial f/\partial x_n)(x_n - \mu_{xn}) \quad (2.35)$$

By the law of variance addition,

$$s_z^2 = \sum_{i=1}^{n} (\partial f/\partial x_i)^2 s_{xi}^2 \quad (2.36)$$

Equation (2.36) is sometimes called the *law of error accumulation*.

Example 2.1. Estimate the error in the measured linear velocity of a gas in a pipeline, v, using the following measured data: the volume flow rate of gas is $G = 3\,000$ m^3 h^{-1}; the error of flow rate measurement is $s_G = 10$ m^3 h^{-1}; the cross-sectional area of the pipeline is $A = 0.1$ m^2; the error of cross-section measurement is $s_A = 1$ cm^2.

Solution. Treating the linear velocity as an outcome of an indirect measurement

$$v = G/A = 3\,000 \div 0.1 = 30\,000 \text{ m h}^{-1} = 8.82 \text{ m s}^{-1}$$

we can find s_v by Eq. (2.36):

$$s_v = [(\partial v/\partial G)^2 s_G^2 + (\partial v/\partial A)^2 s_A^2]^{1/2} = (s_G^2/A^2 + G^2 s_A^2/A^4)^{1/2}$$
$$= (1\times10^{-2}\times10^2 + 9\times10^6\times10^{-8})^{1/2} \div (1\times10^{-2}\times3\,600)$$
$$= 0.03 \text{ m s}^{-1}.$$

2.7. Determining the Variance from a Large Number of Small Samples

The sample mean and the sample variance approach the mean and variance of the population ever closer as the sample size is increased. The sample mean characterizes the accuracy of the measurement (by showing how close the average value of the measured data is to the "true" value), and the sample variance defines the precision of the measurement (by showing how widely the individual data points are scattered about their average value). Therefore, we shall call it the *replication variance* or the *error mean square*. Its square root is usually referred to as the *sample standard deviation* or the *standard error*.

If we carry out m replicate experiments (experiments conducted under identical sets of basic conditions) and obtain a sample $(y_1, y_2, \ldots, y_m)$ of readings of the measured quantity, the sample or replication variance will be

$$s_e^2 = \frac{\sum_{r=1}^{m}(y_r-\bar{y})^2}{m-1}$$

where

$$\bar{y} = \frac{1}{m}\sum_{r=1}^{m} y_r$$

is the sample mean, and the standard error (experimental accuracy)

$$s_e = \sqrt{s_e^2}$$

Frequently, in order to estimate the certainty of a particular experimental technique, a long series of observations is conducted in which one and the same specimen is analyzed repeatedly. Of course, this is a time-consuming procedure, but the more serious limitation is that the sample mean may change in an unpredictable manner. It is more reliable to take a large number of small samples with the same total number of determinations. This we shall call the sampling distribution method.

Suppose there are n specimens to be analyzed. For each specimen, a different number of replicate determinations are made: $m_1, m_2, \ldots, m_n$. Let us find the sample variance for each of these data samples, $s_1^2, s_2^2, \ldots, s_n^2$, separately. The respective numbers of degrees of freedom are: $\nu_1 = m_1-1$, $\nu_2 = m_2-1, \ldots, \nu_n = m_n-1$. The pooled replication variance of all experiments is equal to the weighted mean of the individual variances (the weights being the degrees of freedom):

$$s_e^2 = \frac{\nu_1 s_1^2+\nu_2 s_2^2+\ldots+\nu_n s_n^2}{\nu_1+\nu_2+\ldots+\nu_n} = \frac{(m_1-1)s_1^2+(m_2-1)s_2^2+\ldots+(m_n-1)s_n^2}{m_1+m_2+\ldots+m_n-n} \quad (2.37)$$

The number of degrees of freedom for the pooled variance is equal to the total number of observations minus the number of constraints used to determine n means:

$$\nu_e = m_1+m_2+\ldots+m_n-n = \sum_{i=1}^{n} m_i - n \quad (2.38)$$

Since the individual variances are found from replicate readings by the equation

$$s_i^2 = \frac{\sum_{r=1}^{m_i}(y_{ir}-\bar{y}_i)^2}{m_i-1}, \quad i = 1, 2, \ldots, n$$

we may write on the basis of Eq. (2.37):

$$s_e^2 = \frac{(m_1-1)s_1^2+(m_2-1)s_2^2+\ldots+(m_n-1)s_n^2}{m_1+m_2+\ldots+m_n-n}$$

$$= \frac{\sum_{i=1}^{n}(m_i-1)s_i^2}{\sum_{i=1}^{n}m_i-n} = \frac{\sum_{i=1}^{n}(m_i-1)\frac{\sum_{r=1}^{m_i}(y_{ir}-\bar{y}_i)^2}{m_i-1}}{\sum_{i=1}^{n}m_i-n}$$

$$= \frac{\sum_{i=1}^{n}\sum_{r=1}^{m_i}(y_{ir}-\bar{y}_i)^2}{\sum_{i=1}^{n}m_i-n} \qquad (2.39)$$

If the number of replicate determinations is the same for each of the specimens, $m_1 = m_2 = \ldots = m_n = m$, the expression for the replication variance is greatly simplified:

$$s_e^2 = \frac{(m-1)(s_1^2+s_2^2+\ldots+s_n^2)}{mn-n} = \frac{(m-1)\sum_{i=1}^{n}s_i^2}{n(m-1)} = \frac{\sum_{i=1}^{n}s_i^2}{n} \qquad (2.40)$$

Thus, with an equal number of replicate determinations in each experiment, the total variance is equal to the mean of the individual variances. The number of degrees of freedom for the total variance is

$$\nu_e = n(m-1) \qquad (2.41)$$

And finally,

$$s_e^2 = \frac{\sum_{i=1}^{n}\sum_{r=1}^{m}(y_{ir}-\bar{y}_i)^2}{n(m-1)} \qquad (2.42)$$

The number of degrees of freedom for the total reproducibility variance, as found by Eqs. (2.39) and (2.42), is markedly greater than for each variance taken separately. Therefore, the total replication variance gives a far better estimate of the population variance, σ_e^2.

In finding the replication variance from a large number of small samples, it is permissible to combine only those specimens which can be treated as samples drawn from populations having the same variance. Then each of the sample variances, $s_1^2, s_2^2, \ldots, s_m^2$, may be treated as an estimate of one and the same population variance.

Example 2.2. The percent concentration of P_2O_5 in the system $(NH_4)_2HPO_4$—K_2CO_3—H_2O, as found by the colorimetric method, is given in the accompanying table. Determine the standard error of the colorimetric method on the basis of the sampling distribution.

Solution. Referring to the table, we shall find s_i^2 by the equation given below

$$s_i^2 = \frac{1}{m_i-1}\left[\sum_{r=1}^{m_i}y_r^2-\frac{1}{m_i}\left(\sum_{r=1}^{m_i}y_r\right)^2\right]$$

To find the pooled variance by Eq. (2.37), we shall need terms of the form

$$s_i^2(m_i-1) = \sum_{r=1}^{m_i}y_r^2-\frac{1}{m_i}\left(\sum_{r=1}^{m_i}y_r\right)^2$$

Experiment No.	Specimen No.							
	1	2	3	4	5	6	7	8
1	27.9	19.3	4.5	22.3	10.8	16.3	8.8	12.6
2	27.2	19.7	5.2	23.5	8.9	15.8	8.7	13.5
3	26.8	—	4.8	21.7	—	17.2	9.2	13.3
$\sum y_r$	81.9	39.0	14.5	67.5	19.7	49.3	26.7	39.4
$\sum y_r^2$	2236.49	760.58	70.69	1520.43	195.85	811.17	237.77	517.90
m_i	3	2	3	3	2	3	3	3

For the first specimen,

$$(m_1-1)s_1^2 = 2\,236.49 - \frac{6\,707.61}{3} = 0.62$$

Similarly,

$$(m_2-1)s_2^2 = 0.08$$
$$(m_3-1)s_3^2 = 0.61$$
$$(m_4-1)s_4^2 = 1.68$$
$$(m_5-1)s_5^2 = 1.805$$
$$(m_6-1)s_6^2 = 1.01$$
$$(m_7-1)s_7^2 = 0.14$$
$$(m_8-1)s_8^2 = 0.045$$

The number of degrees of freedom for the pooled sample (replication) variance is

$$\nu_e = \sum_{i=1}^{8} \nu_i = \sum_{i=1}^{8} m_i - 8 = 22 - 8 = 14$$

The sample (replication) variance is

$$s_e^2 = \frac{\sum_{i=1}^{8} s_i^2 \nu_i}{\nu_e} = 5.99 \div 14 = 0.427\,9$$

Hence, the standard error of the colorimetric method, as found on the basis of the sampling distribution, is

$$s_e = \sqrt{s_e^2} = \sqrt{0.427\,9} = 0.654$$

2.8. Confidence Intervals and Level of Significance

Sample parameters are random variables, and their deviations from the respective population parameters (errors) are likewise random. These deviations, or errors, can only be estimated in probabilistic sense, that is by giving the probability of a particular deviation or error. For this purpose, mathematical statistics uses two concepts—*confidence intervals* and *level of significance* α.

Let an experiment yield an unbiased estimate a^* of a population parameter a. We seek to determine the probability of an error in the estimate. We specify a sufficiently high probability level $\mathrm{p} = 1-\alpha$ such that an event with probability $1-\alpha$ can be taken as practically true, and find $\varepsilon = f(\mathrm{p}) = \varepsilon_{1-\alpha}$ such that

$$P(|a^*-a| \leqslant \varepsilon_{1-\alpha}) = 1-\alpha \tag{2.43}$$

The range of practical values for the error arising from the replacement of a by a^* will be $\pm\varepsilon_{1-\alpha}$ and errors of large magnitude will only occur with a low probability

$$\alpha = 1-\mathrm{p} \tag{2.44}$$

called the level of significance. Often, the significance level is expressed as a percentage. Alternatively, Eq. (2.43) may be interpreted as the probability that the true value of the parameter a lies between the limits

$$a^*-\varepsilon_{1-\alpha} \leqslant a \leqslant a^*+\varepsilon_{1-\alpha} \tag{2.45}$$

The probability $1-\alpha$ is called the *confidence coefficient*; it characterizes the degree of assurance in the estimate obtained. The interval

$$I = a^* \pm \varepsilon_{1-\alpha}$$

is called the *confidence interval*, and the boundaries of the interval, $a' = a^*-\varepsilon_{1-\alpha}$ and $a'' = a^*+\varepsilon_{1-\alpha}$, are called the *confidence limits*. At a given confidence level, the confidence interval defines the certainty of the estimate. The magnitude of the confidence interval depends on the confidence coefficient with which the parameter a is guaranteed to lie inside the confidence interval: the greater the value of $1-\alpha$, the higher the value of $\varepsilon_{1-\alpha}$. In other words, the greater the assurance with which we wish to guarantee the obtained result, the larger the interval within which it may lie. An increase in the number of observations leads to a decrease in the confidence interval if the confidence coefficient is held constant, or to an increase in the confidence coefficient if the confidence interval is held constant. In practice, it is usual to fix the confidence coefficient at 0.9 (90%), 0.95 (95%), or 0.99 (99%), and the confidence interval $I_{1-\alpha}$ is then found accordingly. In constructing the confidence interval, one solves the problem of the absolute deviation.

$$P(|a^*-a| \leqslant \varepsilon_{1-\alpha}) = P(\Delta a \leqslant \varepsilon_{1-\alpha}) = F(\varepsilon_{1-\alpha})-F(-\varepsilon_{1-\alpha}) = \int_{-\varepsilon}^{+\varepsilon} f(x)\,\mathrm{d}x \tag{2.46}$$

Thus, if we knew the distribution of the estimate a^*, the problem of the confidence interval would be simple to solve. Consider as an example the construction of the confidence interval for the expected value of a normally distributed random variable X with a known population standard deviation σ.

Let there be a sample of size n of values of this random variable. The best estimate of the expected value (mean) is the sample mean, $\bar{x}$:

$$\bar{x} = \frac{1}{n}\sum_{i=1}^{n} x_i$$

In order to construct the confidence interval, we must know the distribution of the estimate. For samples from a normally distributed population, it can be shown (by using, say, the linearity of a normal distribution) that $\bar{x}$ is likewise normally distributed with a mean μ and a standard deviation $\sigma_{\bar{x}} = \sigma/\sqrt{n}$. Then, using the error function according to Eq. (1.68), we get

$$P(|\bar{x}-\mu| < \varepsilon_{1-\alpha}) = 1-\alpha = 2\Phi\,(\varepsilon_{1-\alpha}/\sigma_{\bar{x}}) \tag{2.47}$$

Choosing a particular value for α, we can look up $k_{1-\alpha} = \varepsilon_{1-\alpha}/\sigma_{\bar{x}}$ in tables of the Laplace functions. Then the confidence interval for the mean will have the form

$$\bar{x}-k_{1-\alpha}\sigma_{\bar{x}} \leqslant \mu \leqslant \bar{x}+k_{1-\alpha}\sigma_{\bar{x}}$$

or

$$\bar{x}-k_{1-\alpha}(\sigma/\sqrt{n}) \leqslant \mu \leqslant \bar{x}+k_{1-\alpha}(\sigma/\sqrt{n}) \tag{2.48}$$

It is seen that a decrease in the confidence interval is inversely proportional to the square root of the number of observations. In other words, if we wish to halve the likely error, the number of observations must be increased four-fold.

If we know the population variance σ^2, we can estimate the mean even from a single observation. If for a normally distributed random variable X an experiment yields a value x_1, then the confidence interval with $p = 1-\alpha$ confidence coefficient for the mean will have the form*

$$x_1-\sigma u_{1-\alpha/2} \leqslant \mu \leqslant x_1+\sigma u_{1-\alpha/2} \tag{2.49}$$

where $u_{1-\alpha/2}$ is the quantile of a standard normal distribution. A standard normal distribution is symmetrical about zero, so

$$u_{\alpha/2} = -u_{1-\alpha/2} \tag{2.50}$$

Example 2.3. The mean temperature of a furnace found from four measurements by an optical pyrometer is 2 250 °C. The measurement by this method is 10 °C in error. Find a 95% confidence interval on the true temperature.

Solution. To begin with, we assume that the error of measurement is a known population standard deviation, $\sigma = 10$ °C, and that the random variable X (the furnace temperature) is normally distributed. Then, using Eq. (2.49), we have

$$2\,250-k_{1-\alpha}(10/\sqrt{4}) \leqslant \mu \leqslant 2\,250+k_{1-\alpha}(10/\sqrt{4})$$

At $1-\alpha = 0.95$, $k_{1-\alpha} = 1.96$, so the 95% confidence interval for the true temperature is

$$2\,240.2 \leqslant \mu \leqslant 2\,259.8$$

The manner in which the estimate a^* is distributed depends on the distribution of the random variable X and, notably, the parameter a itself. To circumvent this difficulty, mathematical statistics usually resorts to any one of two methods: (1) at $n \geqslant 30$, the unknown parameters in the expression for $\varepsilon_{1-\alpha}$ are replaced by their estimates or (2) the random variable a^* is replaced by another random variable whose distribution is inde-

---

* Usually stated as "a 100 $(1-\alpha)$ percent confidence interval".—*Translator's note.*

pendent of that of the parameter a being estimated, but is solely decided by the sample size n and the distribution of X. Random variables of this kind are well studied for the normal distribution of X. As the confidence limits a' and a'', it is usual to select symmetrical quantiles

$$a^*_{(1-p)/2} \leqslant a \leqslant a^*_{(1+p)/2} \tag{2.51}$$

or, in view of Eq. (2.44),

$$a^*_{\alpha/2} \leqslant a \leqslant a^*_{1-\alpha/2} \tag{2.52}$$

2.9. Hypothesis Testing

A *statistical hypothesis* is a statement about the population distribution of some random variable. Hypothesis testing consists in comparing some statistical measures called *test* (or *significance)* criteria deduced from a data sample with the values of these criteria taken on the assumption that a given hypothesis is correct. In conducting a hypothesis test, a certain *test hypothesis*, H_0, is set against an *alternative hypothesis*, H_1, which may be stated explicitly or implicitly. There may be several alternative hypotheses.

To reach a decision about the hypothesis, we select the significance level, α, before collecting the sample. Commonly it is set at 0.05, 0.02, 0.01, 0.10, or 0.001. The significance level α is related to the confidence coefficient as $1-\alpha$. Proceeding from this confidence coefficient and the assumed hypothesis about the distribution of the estimate θ^* (the significance criterion), we then find the quantile *confidence limits*, which are ordinarily symmetrical, $\theta_{\alpha/2}$ and $\theta_{1-\alpha/2}$. The numbers $\theta_{\alpha/2}$ and $\theta_{1-\alpha/2}$ are called the *critical values* of the hypothesis. The values of θ^* smaller than $\theta_{\alpha/2}$ and larger than $\theta_{1-\alpha/2}$ form the *critical* (or *rejection) regions* (Fig. 2.4). If the value θ_0 defined by the observations included in the sample falls between $\theta_{\alpha/2}$ and $\theta_{1-\alpha/2}$, that is, within the *acceptance region*, the hypothesis is judged to be tenable; it is accepted. If the value θ_0 falls within the critical region, the hypothesis is rejected.

In applying a hypothesis test, the investigator faces the possibility of two kinds of errors:

(1) An *error of the first kind* (or Type I error). This error is caused by falsely rejecting the hypothesis when it is true. The probability of this error does not exceed the adopted level of significance. For example, at

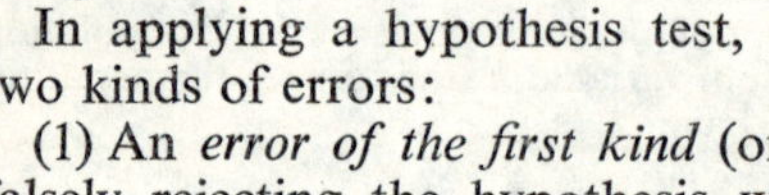

Fig. 2.4. Critical region of a hypothesis

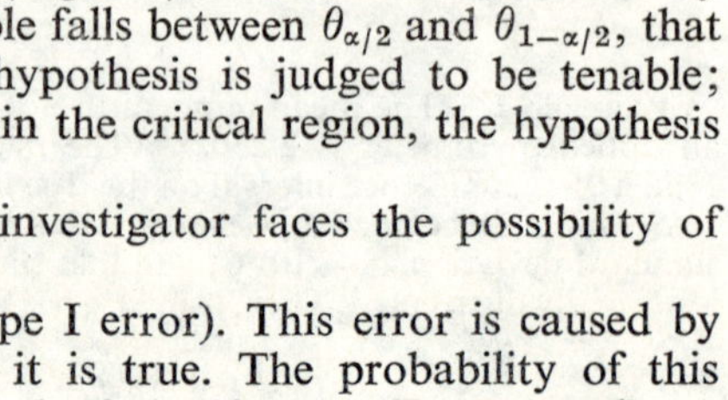

Fig. 2.5. Relationship of Type I and Type II errors to the null hypothesis

$\alpha = 0.05$, the correct hypothesis will be rejected in five cases out of a hundred.

(2) An *error of the second kind* (or Type II error). This error is caused by not rejecting the hypothesis when it is false. The probability of a Type II error depends on the character of the hypothesis being tested, the manner of testing, and many other factors, which fact complicates the procedure. This probability decreases with increasing level of significance, because this leads to an increase in the number of hypotheses being rejected.

If the probability of a Type II error is β, then the probability $(1-\beta)$ is termed the *power of the test*. Figure 2.5. shows two probability density curves for a random variable θ, with curve H_0 applying to the hypothesis under test and the curve H_1 to the alternative hypothesis. If the experiment yields $\theta > \theta_\alpha$, the hypothesis H_0 is rejected and the alternative hypothesis H_1 is accepted. If, however, $\theta < \theta_\alpha$, the converse is true. The area under the probability density curve associated with the hypothesis H_0 to the right of θ_α represents the significance level α, that is, the probability of an error of the first kind. The area under the curve associated with H_1 to the left of θ_α represents the probability of an error of the second kind; and to the right, the power of the test. Thus, as α increases, $(1-\beta)$ also increases.

In hypothesis testing, of all likely criteria, it is sought to choose that which at the specified significance level has the lowest probability of a Type II error. As an illustration, suppose that we have found two values, a_1^* and a_2^*, of some sample parameter. These values may be treated as the estimates of the population parameters a_1 and a_2. We set up the hypothesis that the difference between a_1^* and a_2^* is accidental and that the population parameters are equal, that is, $a_1 = a_2$ or $a_1 - a_2 = 0$. According to Fisher, this is the *null hypothesis*. To test it, we determine if the difference between a_1^* and a_2^* is significant under the null hypothesis. For this purpose, we check the random variable $\Delta a^* = a_1^* - a_2^*$ to see if it is other than zero. Sometimes, it may be more convenient to compare the ratio a_1^*/a_2^* against unity.

By rejecting the null hypothesis, we thereby accept its alternative. The alternative hypothesis may offer two choices: $a_1^* > a_2^*$ and $a_1^* < a_2^*$. If any one of the two inequalities is untenable in advance, the alternative hypothesis is *one-sided*; the values of the parameter a^* under the alternative hypothesis lie on one side of those under the null hypothesis. If, on the other hand, both inequalities are tenable, the hypothesis is *two-sided.*

In hypothesis testing, it is important to take into account any prior information about the likely values of the parameters being estimated, notably if one of the parameters being compared is greater than the other. For example, studying the effects of variations in the purity of a reactant, we know in advance that its purity may decrease with time due to exposure to light. With knowledge of this prior information, we may limit ourselves to a one-sided test which leads to a lower probability of an error of the second kind than a two-sided test.

If we know in advance that any one of the two inequalities, $a_1^* > a_2^*$ or $a_1^* < a_2^*$, is untenable, we only need to analyze one half of the critical region (see Fig. 2.4). For example, in the case of a two-sided test $\alpha = 0.05$ corresponds to two critical values, $\theta_{0.025}$ and $\theta_{0.975}$. This implies that the significant (nonrandom) values θ^* may be those which are less than $\theta_{0.025}$ or greater than $\theta_{0.975}$. With a single-sided test, one of these inequalities, say $\theta^* < \theta_{0.025}$, is known to be untenable in advance, so only $\theta^* > \theta_{0.975}$ is significant. The probability of the last inequality is 0.025, so its significance level is 0.025. Thus, if we use the same critical numbers in a single-sided test as in a two-sided test, the significance level will be halved. Ordinarily, the significance level chosen for a single-sided test is the same as for a two-sided test. In the circumstances, both tests lead to the same error of the first kind, provided the single-sided criterion is derived from a two-sided one corresponding to a significance level which is twice as great as the level adopted. For example, if we want to retain $\alpha = 0.05$ for a single-sided test, we must specify $\alpha = 0.10$ for a two-sided test. The critical values are then $\theta_{0.05}$ and $\theta_{0.95}$. Of the two critical values, only one is retained in a single-sided test, say $\theta_{0.95}$, and the significance level for the single-sided test becomes 0.05. In the case of a two-sided test, the critical value obtained with the same significance level will be $\theta_{0.975}$. However, $\theta_{0.95} < \theta_{0.975}$, which implies that in a single-sided test a greater number of hypotheses will be rejected, leading to a smaller error of the second kind.

2.10. The "Student" *t* Distribution

In the absence of blunders and systematic errors, the expected value of a random variable coincides with the true result of observations. Therefore, the estimate of the mean is important in processing observation data.

The expected value (mean) is the easiest to estimate when we know the variance of the population (see Sec. 2.8). However, no observations can lead to the population variance σ^2; it can only be estimated in terms of the sample variance, s^2. The error resulting from the replacement of the population variance by the sample variance decreases as the sample size n is increased. In practice, the likely error is ignored at $n \geqslant 50$, and the population standard deviation σ in Eq. (2.49) defining the confidence interval is replaced by the sample standard deviation. In our further discussion, we shall assume that the observed random variable is normally distributed.

With small sample sizes, the confidence interval for the expected value is constructed by having recourse to *Student's t distribution*, or simply the *t distribution*. The Student t distribution holds for a random variable (or test statistic)

$$t = \frac{\bar{x}-\mu}{s/\sqrt{n}} \tag{2.53}$$

Its probability density function has the form

$$\varphi(t) = \frac{1}{\sqrt{\pi\nu}} \frac{\Gamma\left(\frac{\nu+1}{2}\right)}{\Gamma\left(\frac{\nu}{2}\right)} \left(1+\frac{t^2}{\nu}\right)^{-\frac{\nu+1}{2}}, \qquad -\infty < t < +\infty \tag{2.54}$$

where $\Gamma(\nu)$ is the gamma function*, and ν is the number of degrees of freedom of the sample. If s^2 and $\bar{x}$ are derived from the same sample, then $\nu = n-1$.

As is seen, Student's distribution depends only on the number of degrees of freedom, ν, with which the sample variance has been determined. A plot of the probability density function for Student's t distribution at $\nu = 1$ and $\nu = 5$ and the standard normal distribution curves are shown in Fig. 2.6. The t-distribution curves have about the same shape as the normal curves, but at low ν they flatten out more slowly towards the axis of abscissae as $|t|$ tends to infinity. At $\nu \to \infty$, the sample variance tends towards the

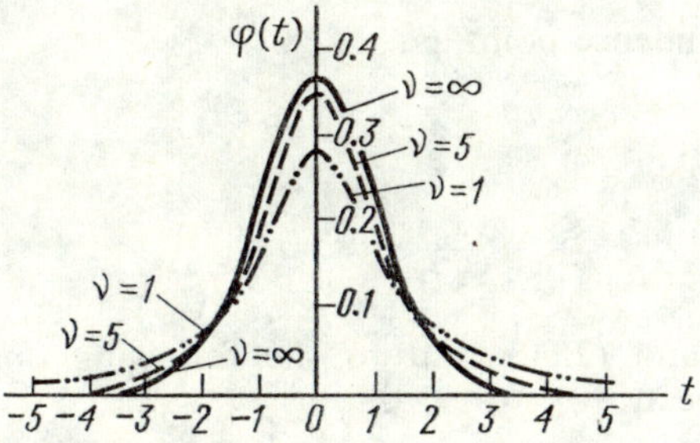

Fig. 2.6. Student's probability density

population variance, and the t distribution tends to the standard normal distribution. At $\nu = \infty$, the t distribution coincides with the standard normal distribution.

The probability that a random variable will fall in some interval $(t_{\alpha/2}, t_{1-\alpha/2})$ is given by

$$P(t_{\alpha/2} \leqslant t \leqslant t_{1-\alpha/2}) = 1-\alpha \tag{2.55}$$

The t distribution is symmetrical about $t = 0$, therefore

$$t_{\alpha/2} = -t_{1-\alpha/2} \tag{2.56}$$

In view of this symmetry, resort is often made to the designation $t_{\alpha,\nu}$, where ν is the number of degrees of freedom and α is the probability that t lies outside the interval $(t_{\alpha/2}, t_{1-\alpha/2})$. On substituting Eq. (2.53) for t in Eq. (2.56), we get

$$-t_{1-\alpha/2} \leqslant \frac{\bar{x}-\mu}{s/\sqrt{n}} \leqslant t_{1-\alpha/2} \tag{2.57}$$

Hence, on re-arranging,

$$\bar{x}-(s/\sqrt{n})t_{1-\alpha/2} \leqslant \mu \leqslant \bar{x}+(s/\sqrt{n})t_{1-\alpha/2} \tag{2.58}$$

$^*\Gamma(\nu) = \int_0^\infty t^{\nu-1}e^{-t}\,dt$ for $\nu > 0$. If ν is an integer, then $\Gamma(\nu) = (\nu-1)$.

The values of the quantiles $t_{1-\alpha/2}$ for various numbers of degrees of freedom v and various significance levels α are given in Appendix Table 3.

In some problems, it is sought to find a one-sided estimate of the mean, that is, either only the upper α (confidence) or only the lower α (confidence) point. At the confidence level α the upper confidence point for the random variable t is defined as

$$t \leqslant t_{1-\alpha}$$

or

$$\frac{\bar{x}-\mu}{s/\sqrt{n}} \leqslant t_{1-\alpha} \tag{2.59}$$

and the lower confidence point as:

$$t \geqslant -t_{1-\alpha}$$

or

$$\frac{\bar{x}-\mu}{s/\sqrt{n}} \geqslant -t_{1-\alpha} \tag{2.60}$$

Equations (2.59) and (2.60) lead to the following one-sided confidence estimates of the mean:

—the upper estimate:

$$\mu \leqslant \bar{x}+(s/\sqrt{n})t_{1-\alpha} \tag{2.61}$$

—the lower estimate:

$$\mu \geqslant \bar{x}-(s/\sqrt{n})t_{1-\alpha} \tag{2.62}$$

Example 2.4. Fluorine is an undesirable impurity in fodder phosphates. Determine the upper safe limit for the fluorine content of phosphates, if repeated analyses have given the following percentages of fluorine in 100 kg of the finished product: 0.18%, 0.12%, 0.13%, and 0.15%. The confidence coefficient is $1-\alpha = 0.95$.

Solution. Let X denote the fluorine percentage in 100 kg of fodder phosphates. On the basis of four replicate determinations, the average fluorine percentage is $\bar{x} = 0.15\%$. The standard error of the mean is $s = 0.03$. The number of degrees of freedom for the standard error is 3. To find the upper safe limit for the fluorine content of the finished product, μ, we shall use Eq. (2.61). At $1-\alpha = 0.95$ and $v = 3$, Appendix Table 3 for $2\alpha = 0.10$ gives $t_{0.95} = 2.35$. Hence,

$$\mu \leqslant 0.15+\frac{0.03}{4}\times 2.35 = 0.17\%$$

Example 2.5 Diammonium phosphate, if graded as analytically pure, must carry at least 99% of the compound. Three determinations of the reactant gave the following percentages for diammonium phosphate: 98.3%, 97.3%, and 97.8%. Test the hypothesis that the difference between the analytical determinations and the certified value is statistically significant.

Solution. Let X denote the analysis result. The mean of the three replicate determinations is $\bar{x} = 97.8\%$. The sample standard deviation is $s = 0.50$. The number of degrees of freedom for the sample standard deviation is $v = 2$. The null hypothesis H_0 is that $\mu = 99\%$ (which implies that the reactant is good). The alternative hypothesis H_1 is that $\mu \neq 99\%$.

To begin with, using the t distribution, we find the critical region for a two-sided test. At $\alpha = 0.05$, the quantile $t_{1-\alpha/2} = 4.30$ for $\nu = 2$ (see Appendix Table 3). By Eq. (2.58), the critical values under the null hypothesis are

$$\bar{x} \leqslant \mu - (s/\sqrt{n})t_{1-\alpha/2}$$
$$\bar{x} \geqslant \mu + (s/\sqrt{n})t_{1-\alpha/2}$$

Only the first of the two inequalities has a physical sense:

$$\bar{x} \leqslant 99 - (0.50/\sqrt{3}) \times 4.30 = 97.76$$

The value $\bar{x} = 97.8\%$ does not fall in the above critical region. Thus, the two-sided test does not give enough ground for rejecting the null hypothesis and declaring the reactant to be of poor quality. According to the physical meaning of the problem, we can apply a one-sided test—diammonium phosphate decomposes when exposed to light—so the sample estimate needs to be compared with only those values which are smaller than 99%

At $\alpha = 0.05$ and $\nu = 2$, from Appendix Table 3 for $2\alpha = 0.10$ we have $t_{0,95} = 2.92$. The critical value under the null hypothesis is

$$\bar{x} \leqslant \mu - (s/\sqrt{n})t_{1-\alpha}$$
$$\bar{x} \leqslant 99 - (0.50/\sqrt{3}) \times 2.92 = 98.16$$

The value $\bar{x} = 97.8\%$ is less than the critical value and, therefore, falls within the critical region. Thus, the one-sided test, as the more accurate of the two, shows with the same specification data that the reactant is of poor quality.

2.11. The Chi-Square Distribution

Inference about the population variance σ^2 of a normally distributed random variable can be made on the basis of its sample analog, that is, the sample variance, s^2. This is done by recourse to what is known as the chi-square (χ^2) distribution.

Let there be a sample of n independent observations $x_1, x_2, \ldots, x_n$ from a normal population having variance σ^2. Then the statistic

$$\chi^2 = \sum_{i=1}^{n} \left(\frac{x_i - \bar{x}}{\sigma}\right)^2 \quad (2.63)$$

is said to have the *chi-square* (χ^2) *distribution with* $\nu = n-1$ *degrees of freedom*.

The chi-square distribution density function depends only on the number of degrees of freedom ν:

$$\varphi(\chi^2) = \frac{1}{2^{\nu/2}\Gamma(\nu/2)} (\chi^2)^{\nu/2-1} \exp(-\chi^2/2), \qquad 0 \leqslant \chi^2 < \infty \quad (2.64)$$

where $\Gamma(\nu)$ is the gamma function.

Figure 2.7 shows probability density curves of the chi-square distribution for several values of ν. The curves are asymmetrical, with the degree of asymmetry decreasing with increasing ν. For a significance level α, the two-sided confidence interval for χ^2 has the form

$$\chi^2_{\alpha/2} \leqslant \chi^2 \leqslant \chi^2_{1-\alpha/2} \quad (2.65)$$

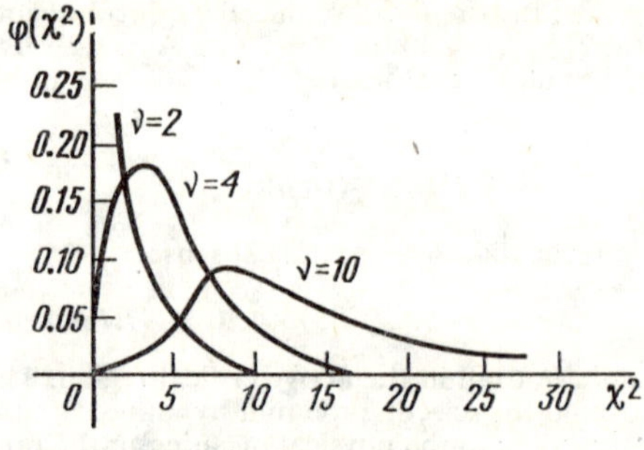

Fig. 2.7. The χ^2 probability density

and the confidence limits are defined as

$$\chi^2 \leqslant \chi^2_{1-\alpha}, \qquad \chi^2 \geqslant \chi^2_{\alpha} \tag{2.66}$$

The quantiles $\chi^2_{1-\alpha}$ for different values of α and ν are given in Appendix Table 4.

Since in terms of sample elements the sample variance is defined as

$$s^2 = \frac{\sum_{i=1}^{n}(x_i-\bar{x})^2}{n-1} = \frac{\sum_{i=1}^{n}(x_i-\bar{x})^2}{\nu}$$

from Eq. (2.63) we get

$$\chi^2 = \nu s^2/\sigma^2 \tag{2.67}$$

On substituting Eq. (2.67) in Eq. (2.65), we obtain

$$\chi^2_{\alpha/2} \leqslant \nu s^2/\sigma^2 \leqslant \chi^2_{1-\alpha/2} \tag{2.68}$$

Solving the above inequality for σ^2 gives the two-sided confidence limits for σ^2:

$$\frac{\nu s^2}{\chi^2_{1-\alpha/2}} \leqslant \sigma^2 \leqslant \frac{\nu s^2}{\chi^2_{\alpha/2}} \tag{2.69}$$

The one-sided confidence limits are obtained in a similar manner:

$$\sigma^2 \leqslant \frac{\nu s^2}{\chi^2_{\alpha}} \qquad \sigma^2 \geqslant \frac{\nu s^2}{\chi^2_{1-\alpha}} \tag{2.70}$$

Example 2.6. Three determinations of P_2O_5 in a complex fertilizer by the sulphuric acid method gave 17.2%, 16.3%, and 15.5%. Determine the standard error of the analysis.

Solution. The sample variance is

$$s^2 = \frac{\sum_{i=1}^{3}(x_i-\bar{x})^2}{3-1} = 0.73$$

The number of degrees of freedom for the sample variance is $\nu = 2$. Assuming a confidence coefficient $1-\alpha = 0.90$, we read from Appendix Table 4 for $\nu = 2$ that $\chi^2_{0.05} = 6.0$ and $\chi^2_{0.95} = 0.103$. Using Eq. (2.69), we find the confidence interval for the population variance

$$\frac{0.73\times 2}{6.0} \leqslant \sigma^2 \leqslant \frac{0.73\times 6}{0.103}$$

$$0.24 \leqslant \sigma^2 \leqslant 14.1$$

Extracting the square root of all terms of the inequality gives the confidence interval for the population standard deviation

$$0.49 \leqslant \sigma \leqslant 3.61$$

Due to the small number of degrees of freedom, the confidence interval turns out to be sharply asymmetrical.

As the number of degrees of freedom is increased, the asymmetry of the distribution curves decreases, and so does the asymmetry of confidence limits. It can be shown that at $n \geqslant 30$ the sample standard deviation s approaches the normal distribution with an expectation $\mu_s = \sigma$ and a sampling standard deviation

$$\sigma_s = \sigma/\sqrt{2\nu} \tag{2.71}$$

The unknown population standard deviation in Eq. (2.71) at $n \geqslant 30$ is replaced by the sample standard deviation

$$\sigma_s \approx s/\sqrt{2\nu} \tag{2.72}$$

Similar to Eq. (2.43), the confidence limits for the population standard deviation are defined by the following inequality:

$$s-(s/\sqrt{2\nu})u_{1-\alpha/2} \leqslant \sigma \leqslant s+(s/\sqrt{2\nu})u_{1-\alpha/2} \tag{2.73}$$

Example 2.7. Determine the standard deviation σ for a sample of 31 observations with a sample standard deviation $s = 0.85$. Take the confidence coefficient as $1-\alpha = 0.9$.

Solution. Let us construct the confidence interval for the standard deviation, using the chi-square distribution. In Appendix Table 4, at $\nu = 30$ and $1-\alpha = 0.90$ we read $\chi^2_{0.05} = 43.8$ and $\chi^2_{0.95} = 18.5$.

Using Eq. (2.69), we find the two-sided confidence estimate of (the confidence interval for) the variance to be

$$\frac{0.85^2\times 30}{43.8} \leqslant \sigma^2 \leqslant \frac{0.85^2\times 30}{18.5}$$

$$0.48 \leqslant \sigma^2 \leqslant 1.13$$

Hence, the confidence limits for the standard deviation are given by

$$0.69 \leqslant \sigma \leqslant 1.05$$

Using the normal distribution, let us determine the confidence limits for σ. Eq. (2.72) gives σ to be:

$$\sigma \approx s/\sqrt{2\nu} = \frac{0.85}{\sqrt{2\times 30}} = 0.11$$

Reading Appendix Table 2 at $\alpha = 0.10$, we find $u_{0.95} = 1.64$. According to Eq (2.73), the confidence limits for the standard deviation are defined by the inequality

$$0.85-0.11\times 1.64 \leqslant \sigma \leqslant 0.85+0.11\times 1.64$$

Finally,

$$0.67 \leqslant \sigma \leqslant 1.03$$

The confidence limits obtained from the normal distribution differ but little from those given above.

2.12. The *F* Distribution

In processing observation data, it is often necessary to compare two or more sample variances. The basic hypothesis tested in the circumstances is whether the sample variances being tested are the estimates of the same population variance. Consider two samples

$$x_1', x_2', \ldots, x_{n_1}'$$
$$x_1'', x_2'', \ldots, x_{n_2}''$$

with means $\bar{x}_1$ and $\bar{x}_2$, respectively. Their sample variances

$$s_1^2 = \frac{\sum_{i=1}^{n_1} (x_i' - \bar{x}_1)^2}{n_1 - 1}$$

$$s_2^2 = \frac{\sum_{i=1}^{n_2} (x_i'' - \bar{x}_2)^2}{n_2 - 1}$$

are found with degrees of freedom $\nu_1 = n_1 - 1$ and $\nu_2 = n_2 - 1$. We are to determine whether the sample variances s_1^2 and s_2^2 are significantly different or the samples obtained may be regarded as drawn from populations having the same variances.

Suppose that the first sample has been drawn from a population with a variance σ_1^2, and the second from a population with a variance σ_2^2. We are to test the null hypothesis under which the population variances are equal, H_0: $\sigma_1^2 = \sigma_2^2$. In order to reject this hypothesis, we must prove beyond doubt that the difference between s_1^2 and s_2^2 at the specified significance level is significant. This is done by applying the *Fisher test*. The sampling distribution involved in this test is called the *Fisher* (*F*) *distribution* defined as

$$F = (s_1^2/\sigma_1^2)/(s_2^2/\sigma_2^2) \tag{2.74}$$

The probability density of the *F* distribution is

$$\varphi(F) = \frac{\Gamma\left(\frac{\nu_1+\nu_2}{2}\right)}{\Gamma\left(\frac{\nu_1}{2}\right)\Gamma\left(\frac{\nu_2}{2}\right)} \nu_1^{\nu_2/2} \cdot \nu_2^{\nu_1/2} \frac{F^{(\nu_1-2)/2}}{(\nu_2+\nu_1 F)^{(\nu_1+\nu_2)/2}} \tag{2.75}$$

where $\Gamma(\nu)$ is the gamma function.

The *F* distribution depends only on the number of degrees of freedom. Figure 2.8 shows the probability density curves of *F* for several values of ν_1 and ν_2. As is seen, the curves are asymmetrical. Appendix Table 5 gives the quantiles $F_{1-\alpha}$ for various significance levels

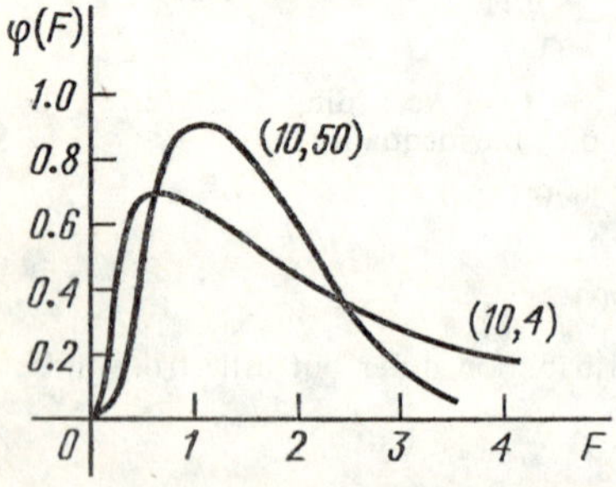

Fig. 2.8. Probability density of *F* for various values of ν_1, ν_2

α, and various degrees of freedom ν_1 and ν_2. The quantiles F_α for α's not included in Appendix Table 5 can be found by the relation

$$F_\alpha(\nu_1, \nu_2) = 1/F_{1-\alpha}(\nu_2, \nu_1) \tag{2.76}$$

Under the null hypothesis, $\sigma_1^2 = \sigma_2^2$ and $\sigma_1^2/\sigma_2^2 = 1$. Therefore, the F distribution can directly be used to make inference about the ratio of the sample variances, s_1^2/s_2^2. A 100 $(1-\alpha)$ per cent confidence interval for F is given by

$$F_{\alpha/2}(\nu_1, \nu_2) \leqslant F \leqslant F_{1-\alpha/2}(\nu_1, \nu_2)$$

In view of Eq. (2.76),

$$1/F_{1-\alpha/2}(\nu_2, \nu_1) \leqslant F \leqslant F_{1-\alpha/2}(\nu_1, \nu_2) \tag{2.77}$$

Under the null hypothesis the F ratio reduces to the statistic

$$F = s_1^2/s_2^2$$

Hence, a 100 $(1-\alpha)$ per cent confidence interval for the variance ratio may be defined as

$$1/F_{1-\alpha/2}(\nu_2, \nu_1) \leqslant s_1^2/s_2^2 \leqslant F_{1-\alpha/2}(\nu_1, \nu_2) \tag{2.78}$$

and any one of the confidence limits as

$$\begin{aligned} s_1^2/s_2^2 &\leqslant F_{1-\alpha}(\nu_1, \nu_2) \quad \text{(the upper limit)} \\ s_1^2/s_2^2 &\geqslant 1/F_{1-\alpha}(\nu_2, \nu_1) \quad \text{(the lower limit)} \end{aligned} \tag{2.79}$$

The probability of inequalities opposite to (2.78) and (2.79) is equal to the significance level α; they form the critical region for the null hypothesis. If the variance ratio obtained falls within the critical region, the difference between the variances is significant. Let, for convenience, s_1^2 denote the larger variance of the two. In testing the null hypothesis, H_0: $\sigma_1^2 = \sigma_2^2$, the one-sided test is applied if the alternative hypothesis is H_1: $\sigma_1^2 > \sigma_2^2$, that is, if we know in advance that the larger sample variance s_1^2 cannot be the estimate of the smaller population variance. In the circumstances, the difference between the variances should, in accord with Eq. (2.79), be regarded as significant if

$$s_1^2/s_2^2 > F_{1-\alpha}(\nu_1, \nu_2) \tag{2.80}$$

The value of $F_{1-\alpha}(\nu_1, \nu_2)$ can be found from Appendix Tables 5 through 7.

The two-sided test, Eq. (2.78), is applied for the alternative hypothesis, H_1: $\sigma_1^2 \neq \sigma_2^2$, that is, when the relative magnitude of the population variances is not known. In the circumstances we need only to test the right-hand side in (2.78), because the left-hand side is satisfied always: according to the statement of the problem, $s_1^2/s_2^2 > 1$, and $1/F_{1-\alpha/2}(\nu_2, \nu_1) < 1$, for small values of α. The difference between the variances should be taken as significant if

$$s_1^2/s_2^2 > F_{1-\alpha/2}(\nu_1, \nu_2) \tag{2.81}$$

Example 2.8. The percentage of available P_2O_5 in a complex fertilizer was determined by the sulphuric acid method; the sample variance of the analysis was $s_1^2 = 0.73$, for two degrees of freedom ($\nu_1 = 2$). Four determinations by the more accurate citrate method gave the following P_2O_5 percentages: 16.5%, 15.9%, 16.6%, and 15.8%. Compare the two methods for accuracy.

Solution. The sample variance (replication error) for the citrate method is

$$s_2^2 = \frac{\sum_{i=1}^{4}(x_i - \bar{x})^2}{4-1} = 0.16$$

for $\nu_2 = 3$. In view of the statement of the problem, we may apply a one-sided test, Eq. (2.80). The variance ratio

$$F = 0.73/0.16 = 4.5$$

must be compared with the tabulated value for $\alpha = 0.05$ and for $\nu_1 = 2$ and $\nu_2 = 3$.

$$F_{1-\alpha}(\nu_1, \nu_2) = 9.6$$

Thus, the computed sample variance ratio is smaller than the tabulated value, so the experimental data do not permit the two methods to be regarded as significantly different in accuracy.

The F test can be used to compare variances one of which is a population variance. The population variance is deemed to have an infinite number of degrees of freedom.

2.13. Comparison of Several Variances

In estimating the variance of a large number of small samples (the sampling distribution method) by the equation

$$s_y^2 = \frac{\nu_1 s_1^2 + \nu_2 s_2^2 + \ldots + \nu_n s_n^2}{\nu_1 + \nu_2 + \ldots + \nu_n} = \frac{\sum_{i=1}^{n} \nu_i s_i^2}{\nu}$$

we adopted the null hypothesis that the respective population variances were equal. For samples varying in size, this hypothesis can be tested by *Bartlett's test*. He has shown that under the null hypothesis the ratio B/C, where

$$B = 2.303\left(\nu \log_{10} s_y^2 - \sum_{i=1}^{n} \nu_i \log_{10} s_i^2\right)$$
$$C = 1 + \frac{1}{3(n-1)}\left(\sum_{i=1}^{n} 1/\nu_i - 1/\nu\right) \tag{2.82}$$

is distributed approximately as χ^2 with $n-1$ degrees of freedom, if all ν_i are greater than two. The hypothesis that all population variances are equal is accepted if

$$B/C \leq \chi_{1-\alpha}^2 \tag{2.83}$$

for the specified significance level α.

The difference between the sample variances may be taken as insignificant and the sample variances themselves as homogeneous. Since $C > 1$

always, the null hypothesis must be accepted if $B \leqslant \chi^2_{1-\alpha}$. If $B > \chi^2_{1-\alpha}$, Bartlett's test is applied in full.

Example 2.9. In the production of phosphorus by sublimation from phosphates, the degree of reduction of the phosphate was measured at four different temperatures. The results of a statistical analysis of variance homogeneity at various temperatures are given in the accompanying table.

Temperature	s_i^2	ν_i	$\nu_i s_i^2$	$\log_{10} s_i^2$	$\nu_i \log_{10} s_i^2$	$\frac{1}{\nu_i}$
T_1	1.72	5	8.60	0.2355	1.177	0.200
T_2	1.60	4	6.40	0.2041	0.816	0.250
T_3	1.97	6	11.82	0.2945	1.767	0.167
T_4	2.37	8	18.96	0.3747	2.995	0.125
$\sum$		23	45.78		6.755	0.742

Determine if the accuracy of analysis varies with temperature.

Solution. From the tabulated data, the sample (reproducibility) variance is

$$s_y^2 = 45.78/23 = 1.99$$

$$\log_{10} s_y^2 = 0.288\,9$$

$$\nu \log_{10} s_y^2 = 23 \times 0.288\,9 = 6.874$$

$$1/\nu = 1/23 = 0.043\,5$$

$$B = 2.303 \times (6.874 - 6.755) = 0.278$$

$$C = \frac{0.742 - 0.043\,5}{3(4-1)} = 1 + 0.077 = 1.077$$

For three degrees of freedom and a significance level $\alpha = 0.05$, Appendix Table 4 gives $\chi^2_{0.95} = 7.8$. Hence, $B < \chi^2_{0.95}$, and at the significance level $\alpha = 0.05$ the hypothesis that the population variances are equal may be accepted. The term C need not have been calculated.

Thus, Bartlett's test shows that the accuracy of analysis is independent of temperature. The sample variances are homogeneous, therefore, we may take as the estimate of the population variance the pooled sample variance s_y^2 with 23 degrees of freedom.

If the sample variances are obtained from samples of the same size, $m_1 = m_2 = \ldots = m_n = m$, resort is made to the more convenient and accurate *Cochran's test*. Cochran has investigated the distribution of the ratio of the maximum sample variance to the sum of all other variances:

$$G = s^2_{\max} / \sum_{i=1}^{n} s_i^2 \qquad (2.84)$$

The G distribution depends only on the number n of variances and the degrees of freedom ν with which each variance is determined: $\nu = m - 1$. The quantiles $G_{1-\alpha}$ for significance levels $\alpha = 0.05$ and $\alpha = 0.01$ are given in Appendix Table 8. If the value of the Cochran statistic as found from the sample variances is smaller than the tabulated one

$$G < G_{1-\alpha}(n, \nu) \qquad (2.85)$$

where n is the number of variances and $v = m-1$, the difference between the variances at the selected significance level is explained by chance. If we also determine the estimated reproducibility variance, the homogeneous variances can be averaged. The number of degrees of freedom, v, for an averaged variance is $v = n(m-1)$.

2.14. Comparison of Two Means

The means of two samples drawn from normally distributed populations can be compared by applying Student's t test.

Let there be two random samples, $x_1, x_2, \ldots, x_n$ and $y_1, y_2, \ldots, y_n$. The first sample has been drawn from a normally distributed population with a mean μ_x and a variance σ_x^2, and the second from a normally distributed population with a mean μ_y and a variance σ_y^2. The respective estimates formed from the sample values are $\bar{x}, s_x^2$, and $\bar{y}, s_y^2$. We are to test the null hypothesis, H_0: $\mu_x = \mu_y$ on the condition that $\sigma_x^2 = \sigma_y^2 = \sigma^2$. Consider the random variable

$$z = \bar{x} - \bar{y} \tag{2.86}$$

Owing to the linearity of a normal distribution (see Sec. 1.5), z is normally distributed with a mean

$$\mu_z = \mu_x - \mu_y$$

and with a variance

$$\sigma_z^2 = \sigma_{\bar{x}}^2 + \sigma_{\bar{y}}^2 = \sigma_x^2/n_1 + \sigma_y^2/n_2 = \sigma^2(1/n_1 + 1/n_2) \tag{2.87}$$

Let us develop the standard random variable

$$\frac{z-\mu_z}{\sigma_z} = \frac{(\bar{x}-\bar{y})-(\mu_x-\mu_y)}{\sigma\sqrt{1/n_1+1/n_2}} \tag{2.88}$$

which has a standard normal distribution.

Replacing the population standard deviation by the sample standard deviation yields a test statistic

$$t = \frac{(\bar{x}-\bar{y})-(\mu_x-\mu_y)}{s\sqrt{1/n_1+1/n_2}} \tag{2.89}$$

which is distributed as Student's t with $v = n_1+n_2-2$ degrees of freedom.

The homogeneity of the sample variances s_x^2 and s_y^2 can be determined by applying Fisher's test. The $100(1-\alpha)$ per cent confidence interval for the difference $\mu_x - \mu_y$ is

$$\bar{x}-\bar{y}-t_{1-\alpha/2}s(1/n_1+1/n_2)^{1/2} \leqslant \mu_x-\mu_y \leqslant \bar{x}-\bar{y}+t_{1-\alpha/2}s(1/n_1+1/n_2)^{1/2} \tag{2.90}$$

and the respective (upper and lower) critical values

$$\mu_x-\mu_y \leqslant \bar{x}-\bar{y}+t_{1-\alpha}s(1/n_1+1/n_2)^{1/2} \tag{2.91}$$

$$\mu_x-\mu_y \geqslant \bar{x}-\bar{y}-t_{1-\alpha}s(1/n_1+1/n_2)^{1/2} \tag{2.92}$$

Under the null hypothesis, the inequalities (2.91) and (2.92) give evidence for accepting or rejecting the hypothesis. The null hypothesis is rejected in the case of a two-sided test if

$$|\bar{x}-\bar{y}| \geqslant t_{1-\alpha/2}s(1/n_1+1/n_2)^{1/2} \tag{2.93}$$

and rejected in the case of a one-sided test if

$$|\bar{x}-\bar{y}| \geqslant t_{1-\alpha}s(1/n_1+1/n_2)^{1/2} \tag{2.94}$$

The above tests cannot be applied if the population variances σ_x^2 and σ_y^2 are not equal. In such a case, recourse can be had to any one of several approximate tests for comparing two means. At $n_1 = n_2 = n$, resort can be made to the approximate t-test; the test statistic has the form

$$t \approx \frac{(\bar{x}-\bar{y})\sqrt{n}}{(s_x^2+s_y^2)^{1/2}} \tag{2.95}$$

with degrees of freedom given by

$$\nu = \frac{n-1}{c^2-(1-c)^2} \tag{2.96}$$

where

$$c = s_x^2/(s_x^2+s_y^2) \tag{2.97}$$

If s_x^2 has $\nu_1 = n_1-1$ degrees of freedom, and s_y^2 has $\nu_2 = n_2-1$ degrees of freedom, use may be made of another approximate test. Find the ratios

$$v_1 = s_x^2/n_1, \qquad v_2 = s_y^2/n_2$$

Take the quantiles $t_{1-\alpha/2}(\nu_1)$ and $t_{1-\alpha/2}(\nu_2)$ from Appendix Table 3. Calculate the test statistic

$$T = \frac{v_1 t_{1-\alpha/2}(\nu_2)+v_2 t_{1-\alpha/2}(\nu_2)}{(v_1+v_2)^{1/2}} \tag{2.98}$$

The null hypothesis H_0: $\mu_x = \mu_y$ is rejected, if

$$|\bar{x}-\bar{y}| > T \tag{2.99}$$

This is a two-sided test, but it reduces to a one-sided test, if we replace $\alpha/2$ by α.

Example 2.10. Radical polymerization has been investigated with reference to 4-vinyl-piridine and two haloid alkyls, butyl iodide (C_4H_9I) and ethyl bromide (C_2H_5Br). For ether alkyl, the sample size was $n = 8$. The mean yield with the former, $\bar{x}$, was 67.72%, and with the latter, $\bar{y}$, was 91.61%. The sample standard deviation (replication error) of the polymerization process was $s = 16.6\%$. Compare the reactivity of the alkyls.

Solution. The null hypothesis is that the haloid alkyls used have an equal reactivity. The sample variance has a distribution with degrees of freedom equal to

$$\nu = 8+8-2 = 14$$

Since the averages give ground for assuming that ethyl bromide has a higher reactivity, we may apply a one-sided test, Eq. (2.95). For 14 degrees of freedom, the one-sided 5% critical value is $t_{0.95} = 1.76$. Therefore,

$$t_{1-\alpha}s(1/n_1+1/n_2)^{1/2} = 1.76\times16.6\times(1/8+1/8)^{1/2} = 14.15$$

$$\bar{y}-\bar{x} = 23.89 > 14.15$$

Thus, at the 5% significance level, the null hypothesis is rejected, and the difference in reactivity between the two alkyls should be taken as significant. If we do not presume that ethyl bromide has a higher reactivity, a two-sided test must be applied to the null hypothesis, Eq. (2.94). For 14 degrees of freedom, the two-sided 5-% critical value is $t_{0.975} = 2.15$. Therefore,

$$t_{1-\alpha/2}s(1/n_1+1/n_2)^{1/2} = 2.15\times16.6\times(1/8+1/8)^{1/2} = 17.2$$

$$\bar{y}-\bar{x} = 23.89 > 17.2$$

Thus, the two-sided test rejects the null hypothesis, too.

As often as not, a sample in practice is built from several subgroups drawn in some particular order (say, from different parts of a population). Before such subgroups may be combined in a single sample, it is important to prove that the means of the subgroups are homogeneous. This is done by applying the t-test in order to see if the difference between the subgroup means and the common sample mean is significant.

Let there be k subgroups of values of a random variable X, the size of the subgroups being $m_1, m_2, \ldots, m_k$. The subgroup means found from these values are $\bar{x}_1, \bar{x}_2, \ldots, x_j, \ldots, \bar{x}_k$, the sample mean is $\bar{x}$, and the sample standard deviation is s, defined as

$$\bar{x}_j = \frac{\sum_{i=1}^{m_j} x_{ji}}{m_j}, \qquad \bar{x} = \frac{\sum_{i=1}^{n} x_i}{n}$$

where $n = m_1+m_2+\ldots+m_k$, and

$$s = \sqrt{\frac{\sum_{i=1}^{n}(x_i-\bar{x})^2}{n-1}}$$

Let

$$y_j = (\bar{x}_j-\bar{x})/s \tag{2.100}$$

It can be proved [12] that the test statistic

$$t_j = \frac{y_j\sqrt{m_j(n-2)}}{\sqrt{n-m_j-m_jy_j^2}} \tag{2.101}$$

is distributed as Student's t with $\nu = n-2$ degrees of freedom. This test statistic is used in testing the null hypothesis that the mean of the jth subgroup, μ_j, is equal to the mean of the entire sample, μ:

$$H_0: \mu_j = \mu$$

The null hypothesis is rejected if a two-sided test fails to satisfy the inequality (2.55) and a one-sided test fails to satisfy the inequality (2.60). As a rule, the mean $\bar{x}_j$ is chosen to be the largest (smallest) mean among the means of the subgroups.

2.15. Comparison of Several Means

Where several means are involved, they might be compared by applying the t-test to a pair of the means at a time. A better way to utilize information about all means is to apply *Duncan's multiple range test* [9]. Suppose we have obtained k means from k samples differing in size:

$$\bar{x}_1, \bar{x}_2, \ldots, \bar{x}_j, \ldots, \bar{x}_k$$

$$\bar{x}_j = \frac{\sum_{i=1}^{n_j} x_{ji}}{n_j}$$

The population variances are equal, that is

$$\sigma_1^2 = \sigma_2^2 = \ldots = \sigma_k^2 = \ldots = \sigma_x^2$$

The steps in applying Duncan's test are as follows: (1) arrange the k means in order from low to high; (2) determine the error mean square s^2 with its degrees of freedom ν; (3) obtain the standard error of the mean for each treatment

$$s_{\bar{x}_j} = \sqrt{\text{Error mean square/No. of observations in } \bar{x}_j}$$

$j = 1, 2, \ldots, k$; (4) enter the Duncan table of significant ranges (Appendix Table 10) at the significance level disired, using n_D degrees of freedom for error mean square and $p = 2, 3, \ldots, k$; and list these $k-1$ ranges; (5) multiply these ranges by $s_{\bar{x}_j}$ to form a group of $(k-1)$ least significant ranges; (6) test the $\bar{x}_j$ observed ranges between means, starting with largest versus smallest, which is compared with the least significant range for $p = k$; then test largest versus second smallest with the least significant range for $p = k-1$; etc. Continue this for second largest versus smallest, etc. until all $k(k-1)/2$ possible pairs have been tested. The sole exception to the rule above is that no difference between two means can be declared significant if the two means concerned are both contained in a subset with a nonsignificant range.

Example 2.11. Radical polymerization was studied with reference to 4-vinyl piridine and several haloid alkyls, namely CH_3I, C_2H_5I, C_3H_7I, C_4H_9I, C_2H_5Br, C_3H_7Br, C_4H_9Cl, and C_3H_7Cl. Eight replicate determinations were made for each of the listed haloid alkyls, and the mean yield was determined from these eight replicates in each case. The results are given in the table that follows.

Nos.	0	1	2	3	4	5	6	7
Alkyl	CH_3I	C_2H_5I	C_3H_6I	C_4H_7I	C_2H_5Br	C_3H_7Br	C_4H_9Cl	C_3H_7Cl
Mean yield, %	95.12	85.46	87.77	67.72	91.61	76.04	13.28	11.72

The sample standard deviation in measuring the polymer yield was $s_x = 16.3$. The number of degrees of freedom was $\nu = 56$.

According to the rules of Duncan's test, we arrange the means in ascending order

11.72	13.28	67.72	76.04	85.46	87.77	91.61	95.12
(7)	(6)	(3)	(5)	(1)	(2)	(4)	(0)

The standard error of the mean is

$$s_{\bar{x}} = 16.3/\sqrt{8} = 5.9$$

Referring to the Duncan table (Appendix Table 10), select significant ranges for $n_2 = 56$ and a significance level of 0.05:

p	2	3	4	5	6	7	8
range	2.85	3.0	3.09	3.12	3.21	3.25	3.29

The least significant ranges (LSRs) multiplied by the standard error of the mean give:

p	2	3	4	5	6	7	8
LRS$\times s_{\bar{x}}$	16.96	17.85	18.39	18.56	19.1	19.34	19.58

$(0)-(7) = 95.12-11.72 = 83.4 > 19.58$	a significant difference
$(0)-(6) = 95.12-13.28 = 81.84 > 19.34$	a significant difference
$(0)-(3) = 95.12-67.72 = 27.4 > 91.1$	a significant difference
$(0)-(5) = 95.12-76.04 = 19.08 > 18.56$	a significant difference
$(0)-(1) = 95.12-85.46 = 9.66 < 18.39$	a nonsignificant difference
$(0)-(2) = 95.12-87.77 = 7.35 < 17.85$	a nonsignificant difference
$(0)-(4) = 95.12-91.61 = 3.51 < 16.96$	a nonsignificant difference
$(4)-(7) = 79.89 > 19.34$	a significant difference
$(4)-(6) = 78.33 > 19.10$	a significant difference
$(4)-(3) = 23.89 > 18.56$	a significant difference
$(4)-(5) = 13.57 < 18.39$	a nonsignificant difference
$(4)-(1) = 6.16 < 17.85$	a nonsignificant difference
$(4)-(2) = 3.84 < 16.96$	a nonsignificant difference
$(2)-(7) = 76.05 > 19.10$	a significant difference
$(2)-(6) = 74.49 > 18.56$	a significant difference
$(2)-(3) = 20.05 > 18.39$	a significant difference
$(2)-(5) = 11.76 < 17.85$	a nonsignificant difference
$(2)-(1) = 2.31 < 16.96$	a nonsignificant difference
$(1)-(7) = 73.74 > 18.56$	a significant difference
$(1)-(6) = 72.18 > 18.39$	a significant difference
$(1)-(3) = 17.74 < 17.85$	a nonsignificant difference
$(1)-(5) = 9.42 < 16.96$	a nonsignificant difference
$(5)-(7) = 64.32 > 18.39$	a significant difference
$(5)-(6) = 62.76 > 17.85$	a significant difference
$(5)-(3) = 8.32 < 16.96$	a nonsignificant difference
$(3)-(7) = 56.00 > 17.85$	a significant difference
$(3)-(6) = 54.44 > 16.96$	a significant difference
$(6)-(7) = 1.56 < 16.96$	a nonsignificant difference

2.16. Rejection of Outliers

A trouble in an instrument during an experiment or negligence on the part of the experimenter may remain undetected so that anomalous measurements often termed "blunders", "mavericks", "wilds" or outliers are made. On the other hand, the outlier may be simply one of the extreme values in a probability distribution. When the analyst knows that an abnormal error or blunder has been made, he does not hesitate to discard such an observation. When he does not have enough ground for either accepting or rejecting an extreme observation, he must resort to some kind of statistical judgement. Most conveniently the detection and elimination of blunders may be treated as a test for the homogeneity of observations, that is, testing the hypothesis that all sample elements $x_1, x_2, \ldots, x_n$ have been drawn from the same population. The random variable of interest is presumed, as before, to be normally distributed. Several methods have been proposed for solving this problem.

There is a sample of values $x_1, x_2, \ldots, x_n$ of the random variable X. Let x_{max} (x_{min}) be the largest (smallest) observation. The respective test statistic sensitive to the kind of wildness envisioned is

$$v = \frac{x_{max} - \bar{x}}{s\sqrt{(n-1)/n}} \tag{2.102}$$

or

$$v' = \frac{\bar{x} - x_{min}}{s\sqrt{(n-1)/n}} \tag{2.103}$$

with a special form of distribution which only depends on the number of degrees of freedom, $\nu = n-2$. The values of the test statistic $v(v')$ for significance levels $\alpha = 0.10, 0.05, 0.025$ and 0.01 and for degrees of freedom from 1 to 23 are given in Table 2.2.

Table 2.2. Values of v (v') for Various Significance Levels

Degrees of freedom, ν	Significance levels, α				Degrees of freedom, ν	Significance levels, α			
	0.10	0.05	0.025	0.01		0.10	0.05	0.025	0.01
1	1.406	1.412	1.414	1.414	13	2.326	2.493	2.638	2.800
2	1.645	1.689	1.710	1.723	14	2.354	2.523	2.670	2.837
3	1.791	1.869	1.917	1.955	15	2.380	2.551	2.701	2.871
4	1.894	1.996	2.067	2.130	16	2.404	2.577	2.728	2.903
5	1.974	2.093	2.182	2.265	17	2.426	2.600	2.754	2.932
6	2.041	2.172	2.273	2.374	18	2.447	2.623	2.778	2.959
7	2.097	2.237	2.349	2.464	19	2.467	2.644	2.801	2.984
8	2.146	2.294	2.414	2.540	20	2.486	2.664	2.823	3.008
9	2.190	2.343	2.470	2.606	21	2.504	2.683	2.843	3.030
10	2.229	2.387	2.519	2.663	22	2.520	2.701	2.862	3.051
11	2.264	2.426	2.562	2.714	23	2.537	2.717	2.880	3.071
12	2.297	2.461	2.602	2.759					

The observation x_{max} (x_{min}) is discarded as an outlier (at the significance level α), if the value of v or v', as found by Eq. (2.102) or (2.103), respectively, exceeds the tabulated value.

If the analyst is in doubt about two or three observations, he must proceed as follows. Compute the test statistic v (or v') for all dubious observations, retain the dubious observation having the lowest value of v (or v'), discard the remaining outliers, and carry out the test as explained above. To this end, compute the sample mean $\bar{x}$ and variance s^2 for the reduced sample and the new value or v (or v') for the observation in question. If the retained observation turns out to be wild, this is still more true of the observations discarded prior to the test. If the observation under test does not turn out to be an outlier, it must be included in the sample, and the test must be run again for the observation having the next larger value of v (or v'). To this end, compute again $\bar{x}$, s, etc.

Example 2.12. Five repeated measurements gave a sample mean of $\bar{x} = 85\%$, with $s = 2\%$, for the degree of extraction of alkaloids from a vegetable raw material. The maximum observation, 92%, obtained in one of the replicates seems dubious. Check to see if this observation is an outlier.

Solution. Using Eq. (2.102), compute the test statistic v for the dubious observation:

$$v = \frac{92-85}{2\times\sqrt{4/5}} = 7 \div (2\times0.895) = 3.9$$

Reading Table 2.2 for $\nu = n-2 = 3$ and $\alpha = 0.05$, we find $v_{0.95} = 1.869$. As is seen,

$$v = 3.9 > v_{0.95}$$

Thus, at a significance level of $\alpha = 0.05$, an extraction of 92% is an outlier, and it should be removed from the sample. Now we can proceed to test if there are any other outliers in the remaining observations; for this, we should compute anew $\bar{x}$ and s, and apply the test.

2.17. Comparison of Sampling and Population Distributions

So far we have tested hypotheses about the parameters of a population distribution on the assumption that the observed random variable is normally distributed. In mathematical statistics, the hypothesis that the distribution of interest is a normal one is referred to as the *basic hypothesis*. It can be tested on the basis of sample data by recourse to *goodness of fit tests*. Goodness of fit tests lead to the probability of a deviation of the empirical distribution from the hypothetical distribution and show if this deviation is due to chance or to an error in the hypothesis. If the probability is found to be high, this gives sufficient evidence to claim that the deviation of the empirical distribution from the hypothetical distribution is due to chance and that the hypothesis about the presumed distribution is not rejected.

Because goodness of fit tests are probabilistic in character, they do not enable the analyst flatly either to accept or to reject the test hypothesis. They only permit him to assert that the hypothesis does not run counter to experimental data if the probability of the observed deviation from the hypothetical distribution is high, or that the hypothesis does not fit experimental data, if this probability is low. Most often, two goodness of fit tests are used. They are the *chi-square test* and the *Kolmogorov-Smirnov test*.

For a chi-square test, the entire range of values of the random variable covered by a sample of size n is divided into k class intervals. The number of intervals is usually chosen between 8 and 20, according to the sample size. The number of class intervals can be found by the semiempirical equation (2.4). Let the size of the ith class interval be n_i. Next we construct a frequency distribution histogram (see Sec. 2.1) and on the basis of this histogram or from general considerations, we choose the form of the hypothetical distribution. The parameters of this distribution can be determined likewise either from theoretical considerations or by determining their sample estimates. Once the form of the hypothetical distribution has been chosen, we calculate the probability p_i that the random variable X will fall within the ith class interval. The statistic describing the deviation of the empirical distribution from the hypothesized one is

$$\chi^2 = \sum_{i=1}^{k} \frac{(n_i - np_i)^2}{np_i} \tag{2.104}$$

where k is the number of class intervals and n is the sample size. The statistic defined by Eq. (2.104) is distributed approximately as χ^2 with $\nu = k - c - 1$ degrees of freedom, where c is the number of unknown population parameters of the hypothetical distribution estimated from the given sample.

For a normal distribution $c = 2$, if both $\bar{x}$ and s are found from a given sample. The hypothesis about the assumed form of distribution is accepted at a given significance level α, if

$$\chi^2 < \chi^2_{1-\alpha} \tag{2.105}$$

where $\chi^2_{1-\alpha}$ is found from Appendix Table 4 for the specified significance level and degrees of freedom. If $\chi^2 > \chi^2_{1-\alpha}$, this is evidence that the hypothesis does not fit the empirical distribution.

In applying the chi-square test, it is desirable that n be sufficiently large (from 50 to 150), and the size of each interval be $n_i = 5$ to 8. If any of the class intervals is less than 5 in size, two or more adjacent class intervals must be combined into one. This will of course reduce the number of degrees of freedom.

For a normal distribution, the probability p_i that the values of the random variable will fall within the ith class interval can be found from Eq. (1.64):

$$P(a \leqslant X \leqslant b) = \Phi\left(\frac{b - \bar{x}}{s}\right) - \Phi\left(\frac{a - \bar{x}}{s}\right) \tag{2.106}$$

In calculating the population probabilities p_i, it should be assumed that the leftmost class interval extends up to $-\infty$, and the rightmost class interval to $+\infty$.

In applying Kolmogorov's test, we compute the maximum absolute deviation of the empirical distribution function $F_n(x)$ from the theoretical (population) distribution function $F(x)$:

$$D = \max |F_n(x) - F(x)| \tag{2.107}$$

Then we find the test statistic

$$\lambda = \sqrt{n}\, D \tag{2.108}$$

The quantiles $\lambda_{1-\alpha}$ of Kolmogorov's distribution are given in Table 2.3. If the calculated value of λ is smaller than the tabulated critical value, $\lambda_{1-\alpha}$, the hypothesis that the population distribution $F(x)$ agrees with the sam-

Table 2.3. Quantiles of Kolmogorov's Distribution

α	$\lambda_{1-\alpha}$	α	$\lambda_{1-\alpha}$	α	$\lambda_{1-\alpha}$
0.99	0.44	0.50	0.83	0.15	1.14
0.90	0.57	0.40	0.89	0.10	1.22
0.80	0.64	0.30	0.97	0.05	1.36
0.70	0.71	0.25	1.02	0.02	1.52
0.60	0.77	0.20	1.07	0.01	1.63

pling distribution $F_n(x)$ is not rejected. At $\lambda \geqslant \lambda_{1-\alpha}$, the hypothesis is either rejected or treated as dubious. In applying Kolmogorov's test, the significance level α is usually chosen to be 0.2 or 0.3.

Kolmogorov's test can also be applied to see if two samples of sizes n_1 and n_2 belong to the same population distribution. Then the maximum deviation D_{n_1, n_2} is found from the expression

$$D_{n_1, n_2} = \max |F_{n_1}(x) - F_{n_2}(x)| \tag{2.109}$$

where $F_{n_1}(x)$ and $F_{n_2}(x)$ are the empirical distribution functions constructed for the first and second sample, respectively; and the test statistic is found from

$$\lambda = \sqrt{\frac{n_1 n_2}{n_1 + n_2}} D_{n_1, n_2} \tag{2.110}$$

For a normal distribution, the function $F(x)$ is defined by

$$F(x) = \frac{1}{2} + \Phi\left(\frac{x - \bar{x}}{s}\right) \tag{2.111}$$

For samples less than 20 in size ($n < 20$), the hypothesis about the form of distribution may be tested by simple tests based on a comparison of population distribution parameters with their estimates formed from sample values. It is most convenient to form estimates of moments.

A normal distribution is fully specified by two parameters, namely the population mean μ and the population standard deviation σ. All the other moments of a normal distribution can be expressed in terms of the population mean and the population standard deviation. For example, the coefficient of skewness, defined by Eq. (1.28), for a normal population is given equal to zero

$$\gamma_1 = \mu_3/\sigma^2 = 0 \tag{2.112}$$

because $\mu_3 = 0$. The coefficient of excess, defined by Eq. (1.24), is likewise equal to zero

$$\gamma_2 = (\mu_4/\sigma^4) - 3 = 0 \tag{2.113}$$

because for a normal distribution

$$\mu_4 = 3\sigma^4 \tag{2.114}$$

The estimates of the coefficients of excess and skewness are

$$\gamma_1^* = \mu_3^*/s^3 = \frac{1}{ns^3}\sum_{i=1}^{n}(x_i-\bar{x})^3 \qquad (2.115)$$

$$\gamma_2^* = \mu_4^*/s^4 = \frac{1}{ns^4}\sum_{i=1}^{n}(x_i-\bar{x})^4-3 \qquad (2.116)$$

The distributions of the above estimates are very complicated and little investigated. The respective variances, however, are well known.

$$\text{Var}\,\{\gamma_1^*\} = \frac{6(n-1)}{(n+1)(n+3)} \qquad (2.117)$$

$$\text{Var}\,\{\gamma_2^*\} = \frac{24n(n-2)(n-3)}{(n+1)^2(n+3)(n+5)} \qquad (2.118)$$

If we know Var (γ_1^*) and Var (γ_2^*), we can readily check if the estimated coefficients of skewness and excess differ significantly from zero. If

$$|\gamma_1^*| \leqslant 3\sqrt{\text{Var}\,\{\gamma_1^*\}} \qquad (2.119)$$

and

$$|\gamma_2^*| \leqslant 5\sqrt{\text{Var}\,\{\gamma_2^*\}} \qquad (2.120)$$

the empirical distribution may be taken as being normal.

Example 2.13. Nickel catalyst grains were measured accurate to 1 μm. Using a sample of size $n = 200$, test if the grain size distribution is normal. The deviations of grain size from nominal are given in the accompanying table. The measurements are grouped into 10 class intervals of length $h = 5$ μm.

Interval length, h	Interval boundaries x_{i+1}, x_i	Interval midpoint, x_i	Points per interval, n_i	Relative frequency, p_i^*
1	−20, −15	−17.5	7	0.035
2	−15, −10	−12.5	11	0.055
3	−10, −5	−7.5	15	0.075
4	−5, 0	−2.5	24	0.120
5	0, 5	2.5	49	0.245
6	5, 10	7.5	41	0.205
7	10, 15	12.5	26	0.130
8	15, 20	17.5	17	0.085
9	20, 25	22.5	7	0.035
10	25, 30	27.5	3	0.015

Solution. The hypothesis that the catalyst grain distribution is normal can be tested by finding the coefficients of skewness and excees. The tabulated data can be used to find the sample mean, the sample variance, the third and fourth central moments of the random variable X (the catalyst grain size) by the grouped data method, using the equations

$$\bar{x} = \sum_{i=1}^{10} p_i^* x_i, \qquad s^2 = \sum_{i=1}^{10} p_i^*(x_i-\bar{x})^2$$

$$\mu_3^* = \sum_{i=1}^{10} p_i^*(x_i-\bar{x})^3, \quad \mu_4^* = \sum_{i=1}^{10} p_i^*(x_i-\bar{x})^4$$

where p_i^* is the relative frequency defined by Eq (2.6). The data needed for the calculations are presented in the table that follows.

i	x_i	$p_i^* x_i$	$x_i-\bar{x}$	$(x_i-\bar{x})^2$	$(x_i-\bar{x})^3$	$(x_i-\bar{x})^4$	$p_i^*(x_i-\bar{x})^2$	$p_i^*(x_i-\bar{x})^3$	$p_i^*(x_i-\bar{x})^4$
1	−17.5	−0.6125	−21.8	475.2	−10 360	225 853	17.43	−362.6	7 904.9
2	−12.5	−0.6875	−16.8	282.2	−4 742	79 659	15.52	−260.8	4 381.2
3	−7.5	−0.5675	−11.8	139.2	−1 643	19 388	10.44	−123.2	1 454.1
4	−2.5	−0.3000	−6.8	46.2	−314	2 138	5.54	−37.7	256.6
5	2.5	−0.6125	−1.8	3.2	−6	10	0.78	−1.5	2.4
6	7.5	1.5375	3.2	10.2	30	105	2.09	6.2	21.5
7	12.5	1.6250	8.2	67.2	551	4 125	8.74	71.6	587.7
8	17.5	1.4875	13.2	174.2	2 300	30 300	14.81	195.5	2 580.6
9	22.5	0.7875	18.2	331.2	6 029	109 720	11.59	211.0	3 840.2
10	27.5	0.4125	23.2	532.2	12 487	289 702	7.98	187.3	4 345.5
Totals		4.295					94.92	−114.2	25 375

On calculating, we obtain:

$$\bar{x} = 4.30\ \mu\text{m}, \quad s = 9.71\ \mu\text{m}, \quad \mu_3^* = -114.2\ \mu\text{m}^3, \quad \mu_4^* = 25\,375\ \mu\text{m}^4$$

Using Eqs. (2.115) and (2.116), the estimated coefficients of skewness and excess are found to be

$$\gamma_1^* = \mu_3^*/s^3 = -0.124\,7$$

$$\gamma_2^* = \mu_4^*/s^4 - 3 = -0.145\,5$$

and their variances are given by Eqs. (2.117) and (2.118) to be

$$\text{Var}\,\{\gamma_1^*\} = \frac{6(200-1)}{(200+1)(200+3)} = 0.029, \qquad \sqrt{\text{Var}\,\{\gamma_1^*\}} \approx 0.17$$

$$\text{Var}\,\{\gamma_2^*\} = \frac{24\times 200(200-2)(200-3)}{(200+1)^2(200+3)(200+5)} = 0.113, \qquad \sqrt{\text{Var}\,\{\gamma_2^*\}} \approx 0.34$$

Thus,

$$3\sqrt{\text{Var}\,\{\gamma_1^*\}} = 0.51 \quad \text{and} \quad |\gamma_1^*| = 0.124\,7 < 3\sqrt{\text{Var}\,\{\gamma_1^*\}}$$

$$5\sqrt{\text{Var}\,\{\gamma_2^*\}} = 1.70 \quad \text{and} \quad |\gamma_2^*| = 0.145\,5 < 5\sqrt{\text{Var}\,\{\gamma_2^*\}}$$

To sum up, the empirical (observed) distribution may be taken as being normal.

The above conclusion can be checked by applying the chi-square and Kolmogorov-Smirnov tests. For this purpose, we compile the table that follows.

The probabilities p_i have been found from Eq. (1.64). Instead of the parameters, the respective estimates have been taken: $\bar{x} = 4.3$ μm and $s = 2.71$ μm. As an example, for the second interval,

$$p_2 = \Phi\left(\frac{-10-4.3}{9.71}\right) - \Phi\left(\frac{-15-4.3}{9.71}\right) = \Phi(-14.3/9.71) - \Phi(-19.3/9.71)$$

$$= \Phi(-1.472) - \Phi(-1.987) = -0.429\,2 + 0.476\,7 = 0.047\,5$$

and, upon multiplication,

$$np_2 = 200\times 0.047\,5 = 9.5$$

i	Interval boundaries	n_i	np_i	$nF_n(x)$	$nF(x)$	$n_i\,\vert F_n(x)-F(x)\vert$	$\frac{(n_i-np_i)^2}{np_i}$
1	$-\infty,\ -15$	7	4.7	7	4.7	2.3	1.12
2	$-15,\ -10$	11	9.50	18	14.2	3.8	
3	$-10,\ -5$	15	19.5	33	33.7	0.7	1.18
4	$-5,\ 0$	24	31.6	57	65.3	8.3	1.81
5	$0,\ 5$	49	40.3	106	105.6	0.4	1.87
6	$5,\ 10$	41	38.9	147	144.5	2.5	0.11
7	$10,\ 15$	26	29.2	173	173.7	0.7	0.20
8	$15,\ 20$	17	15.6	190	189.3	0.7	0.002
9	$20,\ 25$	7	7.7	197	197.0	0.0	0.001
10	$25,\ +\infty$	3	3.0	200	200	0.0	

Since np_i is less than 5 for the first and the last intervals, the first two and the last two intervals are combined into one. The test statistic χ^2 is given by

$$\chi^2 = \sum_{i=1}^{8} \frac{(n_i - np_i)^2}{np_i} = 6.296$$

The number of degrees of freedom is $\nu = 8-3 = 5$. Appendix Table 4 gives $\chi^2_{0.95} = 11.1$. Because the test statistic deduced from the sample data is $\chi^2 = 6.296 < \chi^2_{0.95}$, the chi-square test gives evidence to claim that the distribution is normal.

In order to apply the Kolmogorov-Smirnov test, we need to know the difference $n\,|F_n(x)-F(x)|$. In our case,

$$nD = \max n\,|F_n(x)-F(x)| = 8.3$$

Using Eq. (2.108), we get

$$\lambda = nD/\sqrt{n} = 0.59$$

Reading Table 2.3 for the significance level $\alpha = 0.2$, we obtain

$$\lambda_{0.8} = 1.07$$

Thus,

$$\lambda = 0.59 < \lambda_{0.8}.$$

So, the Kolmogorov-Smirnov test, too, gives evidence to claim that the distribution is normal.

2.18. The Omega-Square (ω^2) Test

In contrast to the chi-square test, the omega-square (ω^2) test is based directly on the observed (ungrouped) values of the random variable.

Let there be a sample of n observations of a random variable X. We test the hypothesis that the distribution function of the random variable is $F(x)$. Let us construct the empirical distribution function $F_n(x)$. For a comparison of $F_n(x)$ with $F(x)$, we define a test statistic

$$\omega^2 = \int_{-\infty}^{\infty} [F_n(x) - F(x)]^2 \, dF(x) \qquad (2.121)$$

assuming that $F(x)$ has a derivative, that is, the probability density function

$$dF(x) = F'(x)\,dx = f(x)\,dx \tag{2.122}$$

Let us arrange the sample in the ascending order of magnitude

$$x_1 < x_2 < x_3 < \ldots < x_n$$

and break up the range of integration into intervals

$$(-\infty, x_1),\ (x_1, x_2),\ \ldots,\ (x_{n-1}, x_n),\ (x_n, +\infty)$$

Then, in view of Eq. (2.121), we get

$$\omega^2 = \int_{-\infty}^{x_1} [0-F(x)]^2\,dF + \sum_{k=1}^{n-1} \int_{x_k}^{x_k+1} \left[\frac{k}{n}-F(x)\right]^2 dF + \int_{x_n}^{+\infty} [1-F(x)]^2\,dF \tag{2.123}$$

$$\int_{-\infty}^{x_1} F^2(x)\,dF = \left.\frac{F^3(x)}{3}\right|_{-\infty}^{x_1} = \frac{F^3(x_1)}{3}$$

$$\int_{x_k}^{x_{k+1}} \left[\frac{k}{n}-F(x)\right]^2 dF = \frac{1}{3}\left[F(x)-\frac{k}{n}\right]^3 \Bigg|_{x=x_k}^{x=x_{k+1}}$$

$$= \frac{1}{3}\left[F\left(x_{k+1}\right)-\frac{k}{n}\right]^3 - \frac{1}{3}\left[F(x_k)-\frac{k}{n}\right]^3 \tag{2.124}$$

$$\int_{x_n}^{+\infty} [1-F(x)]^2\,dF = \frac{[1-F(x_n)]^3}{3}$$

Thus, we obtain

$$\omega^2 = \frac{F^3(x_1)}{3} + \frac{1}{3}\sum_{k=1}^{n-1}\left[F(x_{k+1})-\frac{k}{n}\right]^3$$

$$-\frac{1}{3}\sum_{k=1}^{n-1}\left[F(x_k)-\frac{k}{n}\right]^3 + \frac{[1-F(x_n)]^3}{3} \tag{2.125}$$

By combining the terms dependent on $F(x_k)$ (with a given $k = 1, 2, \ldots, n$) in the two sums, Eq. (2.125), we get

$$\omega^2 = \frac{1}{12n^2} + \frac{1}{n}\sum_{k=1}^{n}\left[F(x_k)-\frac{2k-1}{2n}\right]^2 \tag{2.126}$$

Equation (2.126) demonstrates how the omega-square test criterion depends on the individual members of the ordered sequence. The exact distribution of ω^2 is very complex, but investigations have shown that already at n exceeding 40 the distribution of the product $n\omega_n^2$ is close to some limiting distribution for which tabulated data are available. On the basis of these tables, the critical values for $n\omega^2$ have been found. The quantiles $(n\omega^2)_{1-\alpha}$ are presented in Table 2.4.

If the calculated value of $n\omega^2$ is less than the tabulated one, $(n\omega^2)_{1-\alpha}$, the hypothesis that the theoretical distribution $F(x)$ fits the empirical distribution $F_n(x)$ is not rejected. At $n\omega^2 \geqslant (n\omega^2)_{1-\alpha}$, the hypothesis is rejected.

Table 2.4. Quantiles of $n\omega^2$ Distribution

α	$(n\omega^2)_{1-\alpha}$	α	$(n\omega^2)_{1-\alpha}$
0.5	0.1184	0.05	0.4614
0.4	0.1467	0.03	0.5489
0.3	0.1843	0.02	0.6198
0.2	0.2412	0.01	0.7435
0.1	0.3473	0.001	1.1679

The omega-square test utilizes the information contained in the sample data more completely than the chi-square test. The grouping of data carried out in the chi-square test introduces a certain degree of arbitrariness. The grouping itself leads to the loss of some information contained in the sample. Also, the $n\omega^2$ distribution converges to the limit distribution faster than the chi-square test, especially in the region of high values of ω^2 which alone are essential for a probabilistic estimation.

2.19. The Wilcoxon Two-Sample Test

The *Wilcoxon test* is applied to test the hypothesis that two samples belong to the same population. Let there be two samples of sizes m and n of random variables X and Y, respectively. We arrange the samples in ascending order of magnitude

$$x_1 < x_2 < \dots < x_m$$

$$y_1 < y_2 < \dots < y_n$$

The null hypothesis H_0 is that $F(x) = F(y)$. The alternative hypothesis is set up as an inequality of the form $F(x) < F(y)$.

The Wilcoxon test is based on the distribution of the total number of inversions. More specifically, we arrange all $(m+n)$ values in a single sequence in ascending order of magnitude

$$y_1 x_1 y_2 y_3 x_2 y_4 x_3 x_4 y_5 \dots x_m y_n \tag{2.127}$$

If some value x is preceded by any y, the pair gives an inversion. In the sequence (2.127), x_1 gives one inversion with y_1, x_2 gives three inversions (with y_1, y_2 and y_3), etc.

At $m > 10$ and $n > 0$, the total number of inversions u is distributed nearly normally with a mean

$$\mu_u = mn/2 \tag{2.128}$$

and a variance

$$\sigma_u^2 = (mn/12)(m+n-1) \tag{2.129}$$

According to Eq. (2.50), the critical values for the null hypothesis at a significance level $\alpha = 0.05$ will be

$$u \leqslant \mu_u - 1.96\sigma_u$$
$$u \geqslant \mu_u + 1.96\sigma_u \quad (2.130)$$

Example 2.14. The concentration of available P_2O_5 in a complex fertilizer was determined by (x) the citrate method and (y) the sulphuric acid method. The results are presented in the accompanying table. Test the null hypothesis H_0 that either method has the same error distribution.

Method	Specimen No.													
	1	2	3	4	5	6	7	8	9	10	11	12	13	14
x	16.3	15.5	16.7	16.0	13.7	11.0	12.5	13.4	14.4	14.7	16.9	15.7	13.5	14.0
y	16.5	15.9	16.6	15.8	13.3	11.2	12.4	13.6	14.9	14.6	16.8	16.2	13.8	14.3

Solution. Arrange the data in a single ascending sequence:

x	y	y	x	y	x	x	y	x	y	x	y	x	y
11.0	11.2	12.4	12.5	13.3	13.4	13.5	13.6	13.7	13.8	14.0	14.3	14.4	14.6
x	y	x	x	y	y	x	y	x	y	y	x	y	x
14.7	14.9	15.5	15.7	15.8	15.9	16.0	16.2	16.3	16.5	16.6	16.7	16.8	16.9

The number of inversions for x is

$$u = 2+3+3+4+5+6+7+8+8+10+11+13+14 = 94.$$

From Eqs. (2.128) and (2.129), we find

$$\mu_u = (14 \times 14)/2 = 98$$
$$\sigma_u^2 = (14 \times 14 \div 12)(14+14+1) = 474$$
$$\sigma_u = 21.8$$

At $\alpha = 0.05$, the critical values for the null hypothesis, according to Eq. (2.130), will be

$$u \leqslant 98 - 1.96 \times 21.8, \qquad u \leqslant 55.3$$
$$u \geqslant 98 + 1.96 \times 21.8, \qquad u \geqslant 140.7$$

The number of inversions equal to 94 does not fall within the critical region, so we have no evidence to claim that the two methods differ in accuracy significantly.

2.20. Testing the Hypothesis of Normalcy from a Large Number of Small Samples

Let there be a sufficiently large number n of small independent samples of the same size m. We set out to test the hypothesis that the populations from which the samples have been drawn are normal, subject to the condition that the parameters of these populations may be different. The test statistic is

$$\tau = \frac{x_{ik} - \bar{x}_k}{s_k} \tag{2.131}$$

where x_{ik} is the ith member of the kth sample, $\bar{x}_k$ is the sample mean of the kth sample, and s_k is the standard deviation of the kth sample.

It can be shown that the distribution of the test statistic is independent of the population mean and variance, but solely depends on the sample size m. The probability density of the test statistic τ is

$$f(\tau) = \begin{cases} \dfrac{1}{\sqrt{\pi}}\dfrac{1}{\sqrt{v+1}}\dfrac{\Gamma\left(\dfrac{v+1}{2}\right)}{\Gamma\left(\dfrac{v}{2}\right)}\left(1-\dfrac{\tau^2}{v+1}\right)^{\frac{v-2}{2}} & \text{for } |\tau| < \sqrt{v+1} \\ 0 & \text{for } |\tau| \geqslant \sqrt{v+1} \end{cases} \tag{2.132}$$

where the number of degrees of freedom is $v = m-2$. For various values of m, we obtain from Eq. (2.132):

$$\begin{aligned} m &= 3 \quad f(\tau) = \frac{1}{\pi\sqrt{2\left(1-\dfrac{\tau^2}{2}\right)}} && |\tau| < \sqrt{2} \\ m &= 4 \quad f(\tau) = \frac{1}{2\sqrt{3}} && |\tau| < \sqrt{3} \\ m &= 5 \quad f(\tau) = \frac{\sqrt{1-\dfrac{\tau^2}{4}}}{\pi} && |\tau| < 2 \\ m &= 6 \quad f(\tau) = \frac{3\left(1-\dfrac{\tau^2}{5}\right)}{4\sqrt{5}} && |\tau| < \sqrt{5} \end{aligned} \tag{2.133}$$

As follows from Eq. (2.133) at $m = 4$, the fractional deviations in the individual samples are uniformly distributed, if the parent populations are normal. This can be utilized for testing the hypothesis of normalcy, if the number of samples is sufficiently large. At $m \neq 4$, because the appropriate tables are lacking, the test statistic τ must be replaced by another test statistic, η, defined as

$$\eta = \frac{\tau\sqrt{v}}{\sqrt{v+1-\tau^2}} \tag{2.134}$$

It can be shown that if the parent populations are normal, the test statistic η has a Student's t distribution with $\nu = m-2$ degrees of freedom.

In testing the hypothesis of normalcy from a large number of small samples, one value is taken at random from each sample. This can be done in a simplified way, that is, by taking all the first observations, or all the second observations, etc. This selection can be regarded as chance. If the samples are large, say, with a size $m > 10$, we may apply several independent tests, say, for the first and last members of each sample. Then, for each selected value we compute τ by Eq. (2.131) if $m = 4$, or η by Eq. (2.134), if $m \neq 4$. On passing to the test statistics τ and η for testing the hypothesis of the uniform distribution for τ or the Student t distribution for η (and, as a consequence, the normalcy hypothesis for the parent population), we may resort to any of the goodness-of-fit tests discussed earlier.

Example 2.15. Test the hypothesis about the normal distribution of ammonium saltpeter concentration (in grams per litre) in the secondary vapour past the reactor in the production of ammonium saltpeter from four replicate determinations in each of forty specimens (see the accompanying table).

Speciemen No.	x_1	x_2	x_3	x_4	Specimen No.	x_1	x_2	x_3	x_4
1	5.97	6.39	6.05	5.64	21	4.09	4.19	3.96	4.18
2	5.56	6.02	5.14	5.46	22	4.87	5.10	4.38	4.20
3	4.51	5.32	5.06	4.30	23	2.93	4.60	2.93	4.03
4	5.28	4.40	4.88	4.83	24	3.86	4.40	4.92	4.17
5	5.36	5.52	4.60	5.49	25	5.74	5.06	4.81	5.52
6	4.82	4.99	5.42	5.34	26	5.26	6.01	6.09	6.07
7	5.61	4.83	5.37	5.27	27	6.45	5.99	5.77	6.05
8	4.79	4.51	5.54	5.75	28	5.13	5.19	5.08	5.35
9	4.69	5.62	6.77	6.19	29	5.18	4.59	4.90	5.26
10	5.30	5.60	6.16	6.11	30	4.94	4.44	4.66	5.01
11	4.28	4.47	4.10	4.53	31	4.56	4.12	4.63	4.24
12	4.32	4.03	4.49	4.04	32	3.84	3.22	3.28	3.16
13	3.17	4.85	4.43	4.39	33	2.86	3.78	3.75	3.03
14	4.41	4.04	4.01	3.82	34	4.00	3.48	4.00	4.35
15	5.66	6.24	5.95	5.49	35	3.95	4.05	4.20	4.40
16	5.19	5.45	5.14	5.08	36	3.98	5.06	3.94	4.52
17	5.02	4.45	5.22	4.82	37	3.87	3.09	3.86	3.49
18	4.35	4.43	4.32	4.26	38	4.64	4.33	3.94	4.40
19	5.10	4.67	4.57	4.79	39	4.63	4.16	4.24	4.50
20	5.17	4.83	4.68	4.79	40	4.14	5.24	4.11	4.44

Solution. Let us take the first value (x_1) from the four replicate determinations of each specimen and calculate for each of the 40 values the test statistic τ by Eq. (2.131). The results are presented in the table that follows.

No.	$\bar{x}$	s	$x_1-\bar{x}$	$\tau=\frac{x_1-\bar{x}}{s}$	No.	$\bar{x}$	s	$x_1-\bar{x}$	$\tau=\frac{x_1-\bar{x}}{s}$
1	6.01	0.31	−0.04	−0.137	21	4.11	0.11	−0.02	−0.141
2	5.55	0.36	0.01	0.0413	22	4.64	0.42	0.23	0.557
3	4.96	0.34	0.45	−1.316	23	3.62	0.83	−0.69	−0.831
4	4.93	0.24	0.35	1.467	24	4.34	0.45	−0.48	−1.067
5	5.24	0.43	0.12	0.272	25	5.28	0.42	0.46	1.083
6	5.14	0.28	−0.32	−1.131	26	5.86	0.40	−0.60	−1.493
7	5.27	0.33	0.34	1.044	27	6.07	0.28	0.39	1.358
8	5.15	0.59	−0.36	−0.603	28	5.19	0.12	−0.06	−0.486
9	5.82	0.89	−1.13	−1.272	29	4.98	0.30	0.2	0.652
10	5.79	0.41	−0.49	−1.189	30	4.76	0.26	0.18	0.677
11	4.35	0.19	−0.07	−0.334	31	4.39	0.25	0.17	0.702
12	4.22	0.22	0.10	0.445	32	3.38	0.31	0.47	1.481
13	4.21	0.72	−1.04	−1.437	33	3.36	0.48	−0.50	−1.035
14	4.07	0.25	0.34	1.378	34	3.96	0.36	−0.04	0.120
15	5.84	0.33	−0.18	−0.531	35	4.15	0.20	−0.2	−1.022
16	5.22	0.16	−0.03	−0.153	36	4.38	0.53	−0.4	−0.749
17	4.88	0.33	0.14	0.436	37	3.58	0.37	0.29	0.792
18	4.34	0.71	0.01	0.141	38	4.33	0.29	0.31	1.078
19	4.78	0.23	0.32	1.391	39	4.38	0.22	0.25	1.129
20	4.87	0.21	0.30	1.444	40	4.48	0.53	−0.34	−0.651

From Eq. (2.133), it follows that at $m = 4$ the distribution density of the test statistic τ is

$$f(\tau) = \frac{1}{2\sqrt{3}} \quad \text{at } |\tau| < \sqrt{3}$$

if the parent populations are normally distributed. Let us test this hypothesis by applying the omega-square (ω^2) criterion. We arrange the calculated values of τ in ascending order and evaluate the empirical distribution function $F_n(\tau)$ for each member of the ordered sequence from the equation

$$F_n(\tau) = (2k-1)/2n \tag{2.135}$$

where $n = 40$, $k = 1, 2, \ldots, n$, and also the cumulative theoretical function $F(\tau)$. The results are presented in the table that follows. In the table, the values of τ are arranged in ascending order.

Since

$$f(\tau) = 0 \quad \text{at } \tau < -\sqrt{3}$$

we have

$$F(\tau) = \int_{-\sqrt{3}}^{\tau_k} \frac{1}{2\sqrt{3}}\, du = \frac{1}{2\sqrt{3}}(\tau_k + \sqrt{3}) \tag{2.136}$$

$$k = 1, 2, \ldots, n$$

Thus, at $\tau_1 = -1.494$

$$F(-1.494) = \frac{-1.494 + \sqrt{3}}{2\sqrt{3}} = 0.068\,7$$

Determination No.	τ	$F(\tau)$	$F_n(\tau)$	$F(\tau)-F_n(\tau)$	$[F(\tau)-F_n(\tau)]^2$
1	−1.494	0.0687	0.0125	0.0562	0.003161
2	−1.437	0.0852	0.0375	0.0477	0.002273
3	−1.316	0.1201	0.0625	0.0576	0.003318
4	−1.272	0.1328	0.0875	0.0453	0.002053
5	−1.189	0.1568	0.1125	0.0443	0.001959
6	−1.131	0.1735	0.1375	0.0360	0.001297
7	−1.067	0.1920	0.1625	0.0295	0.000869
8	−1.035	0.2012	0.1875	0.0137	0.000188
9	−1.022	0.2050	0.2125	−0.0075	0.000057
10	−0.831	0.2601	0.2375	0.0226	0.000511
11	−0.749	0.2838	0.2625	0.0213	0.000455
12	−0.651	0.3122	0.2875	0.0247	0.000609
13	−0.603	0.3259	0.3125	0.0134	0.000179
14	−0.531	0.3467	0.3375	0.0092	0.000084
15	−0.480	0.3613	0.3625	−0.0012	0.000001
16	−0.334	0.4037	0.3875	0.0162	0.000263
17	−0.153	0.4557	0.4125	0.0432	0.001868
18	−0.141	0.4594	0.4375	0.0219	0.000478
19	−0.137	0.4605	0.4625	−0.002	0.000004
20	−0.041	0.5119	0.4875	0.0244	0.000596
21	0.12	0.5346	0.5125	0.0221	0.000489
22	0.141	0.5408	0.5375	0.0033	0.000011
23	0.272	0.5785	0.5625	0.0160	0.000257
24	0.404	0.6165	0.5875	0.0290	0.000842
25	0.445	0.6285	0.6125	0.0160	0.000257
26	0.557	0.6608	0.6375	0.0233	0.000543
27	0.652	0.6881	0.6625	0.0256	0.000657
28	0.677	0.6954	0.6875	0.0079	0.000063
29	0.702	0.7027	0.7125	−0.0099	0.000097
30	0.792	0.7287	0.7375	−0.0088	0.000077
31	1.044	0.8014	0.7625	0.0389	0.001511
32	1.078	0.8112	0.7875	0.0237	0.000561
33	1.083	0.8126	0.8125	0.0001	0.000000
34	1.129	0.8259	0.8375	−0.0016	0.000134
35	1.358	0.8920	0.8525	0.0295	0.000871
36	1.378	0.8978	0.8875	0.0103	0.000106
37	1.391	0.9015	0.9125	−0.0110	0.000120
38	1.444	0.9168	0.9375	−0.0207	0.000427
39	1.467	0.9235	0.9625	−0.0390	0.001522
40	1.481	0.9275	0.9875	−0.0600	0.003597
$\sum$					0.032057

The empirical distribution function is evaluated by Eq. (2.135). Thus, at $\tau = \tau_1$, we have

$$F_n(\tau_1) = \frac{2\times1-1}{2\times40} = 0.012\,5$$

To evaluate the test statistic $n\omega^2$, we substitute $n = 40$ and the sum from the table in the equation

$$n\omega^2 = (1/12n) + \sum_{k=1}^{n}\left[F(\tau_k) - \frac{2k-1}{2n}\right]^2 \tag{2.137}$$

and obtain

$$n\omega^2 = 1 \div (20\times40) + 0.032\,057 = 0.034\,14$$

For $\alpha = 0.05$, the tabulated value (see Table 2.4) is

$$(n\omega^2)_{1-\alpha} = 0.461\,4$$

Thus, the calculated value is smaller than the tabulated one. As a consequence, the hypothesis that the concentration of ammonium saltpeter in the liquor vapour is normally distributed is not rejected.

2.21. Fitting a Distribution to Observed Data

The normal distribution is well studied, and numerous tables have been compiled for it. Therefore, if a sample distribution is other than normal, an attempt is made to transform the observations x_i in such a way that the transformed function, $y_i = f(x_i)$ has a normal distribution. For example, logarithmic transformation replaces a strongly asymmetrical distribution by a distribution close to normal. If we denote $\ln X = Y$, then

$$X = \exp(Y) \tag{2.138}$$

The estimate for the mean of the normally distributed variable Y is the sample mean $\bar{y}$:

$$\bar{y} = \frac{1}{n}\sum_{i=1}^{n} y_i$$

The estimate for the mean for the original random variable X then takes the form

$$\exp(\bar{y}) = \sqrt[n]{\prod_{i=1}^{n} \exp(y_i)} = \sqrt[n]{\prod_{i=1}^{n} x_i} = \sqrt[n]{x_1 x_2 \ldots x_n} \tag{2.139}$$

Here, the mean of X is the median, $\exp(\mu_y)$. The mean of X is

$$\mu_x = \exp(\mu_y + \sigma^2/2)$$

The confidence interval for the mean of Y takes the form

$$\bar{y} - ts/\sqrt{n} \leqslant \mu_y \leqslant \bar{y} + ts/\sqrt{n}$$

This leads to the following confidence interval for the mean of X:

$$\sqrt[n]{\prod_{i=1}^{n} x_i}\, \exp(-ts/\sqrt{n}) \leqslant \exp(\mu_y)$$

$$\leqslant \sqrt[n]{\prod_{i=1}^{n} x_i}\, \exp(ts/\sqrt{n}) \tag{2.140}$$

At high values of n, Eq. (2.140) may be re-written

$$\sqrt[n]{\prod_{i=1}^{n} x_i (1-ts/\sqrt{n})} \leqslant \exp(\mu_y) \leqslant \sqrt[n]{\prod_{i=1}^{n} x_i (1+ts/\sqrt{n})} \quad (2.141)$$

A suitable transformation can as readily be chosen if the departure of the sample distribution from normal has occurred because the population variance σ^2 changed during observations. Experience shows that observations will be normally distributed, if a given data set has been collected, using specimens in which the concentration of the component being determined differs by a factor or three or four as a maximum. Otherwise, the sample standard deviation will be related to the concentration in a complex manner, $s = f(x)$, and the resultant distribution will be asymmetrical.

Let us replace X by $Y = \varphi(X)$. Then, according to Eq. (2.36), we obtain

$$s_y = \varphi'(x) s_x = \varphi'(x) f(x) \quad (2.142)$$

Choose the transforming function $\varphi(x)$ in such a manner that the variance of Y is constant, that is,

$$\varphi'(x) f(x) = c \quad (c > 0) \quad (2.143)$$

Then the sought transforming function will be defined by

$$\varphi(x) = c \int \mathrm{d}x/f(x) \quad (2.144)$$

Using the transforming function thus found, we may change from direct to indirect observations with a constant variance. Now all observations can be regarded as a sample from a single population.

Example 2.16. In measuring the concentration of potash, it was noted that the sample standard deviation (replication error) decreased with increasing concentration:

Potash concentration, x, %	0.2	10	16	20	30	40
Sample standard deviation, s	0.32	0.25	0.20	0.17	0.11	0.05

Fit a transforming function to the measured quantity (concentration), so as to make the sample variance constant.

Solution. To establish the functional relationship between the sample standard deviation and the concentration, we apply the least squares method to the observation data, assuming that $s = f(x)$ is linear. This leads us to the equation

$$s = 0.31 - 0.007x$$

To make the variance constant, we define the transforming function from Eq. (2.144)

$$y = \varphi(x) = c \int \frac{\mathrm{d}x}{0.31 - 0.007x} = -(c/0.007) \ln(0.007x - 0.031)$$

On setting $c = 0.007 \log_{10} e$, we obtain

$$y = -\log_{10}(0.007x - 0.031)$$

With the concentration found by the above equation, the variance is stabilized.

If the normal distribution does not fit a given sample distribution, a good analytical approximation can sometimes be obtained by using Charlier's A-series. Using the first three terms of Charlier's series, we obtain equations for the probability density and the distribution function:

$$f(u) = (1/6)\,[\varphi(u)-(\gamma_1/6)\varphi'''(u)+(\gamma_2/24)\varphi''''(u)] \qquad (2.145)$$

$$F(u) = 0.5+\Phi(u)-(\gamma_1/6)\varphi''(u)+(\gamma_2/24)\varphi'''(u) \qquad (2.146)$$

where γ_1 and γ_2 are coefficients of skewness and excess, respectively, $\Phi(u)$ is the Laplace function,

$$u = (x-\mu)/\sigma \qquad (2.147)$$

$\varphi''(u)$, $\varphi'''(u)$ and $\varphi''''(u)$ are the second, third and fourth derivatives of the probability density for a normal distribution.

Example 2.17. Using Charlier's A-series, fit a distribution to the observed data of Example 2.13. Also, using the chi-square test, check if this distribution fits the empirical distribution better than the normal distribution.

Solution. In Example 2.13. using a sample of size $n = 200$, we obtained the following sample parameters: $\bar{x} = 4.30$ μm, $s = 9.71$ μm, $\gamma_1^* = -0.124\,7$, and $\gamma_2^* = -0.145\,5$. Replacing the unknown population parameters by their sample estimates in Eq. (2.147)

$$u = (x-4.30)/9.71$$

we find the limits for the standard random variable U. The results are presented in the table that follows.

i	u_{i-1}, u_i	$\Phi(u_i)$	$\varphi'' u_i$	$\varphi'''(u_i)$	$F(u_i)$
1	$-\infty$, −1.99	−0.5, −0.4767	0, 0.1630	0, 0.1062	0, 0.0267
2	−1.99, −1.47	−0.4767, −0.4292	0.1630, 0.1572	0.1062, −1.1670	0.0267, 0.0715
3	−1.47, −0.96	−0.4292, −0.3315	0.1572, −0.0197	−0.1670, −0.5021	0.0715, 0.1712
4	−0.96, −0.44	−0.3315, −0.1700	−0.0197, 0.2920	−0.5021, 0.4472	0.1712, 0.3266
5	−0.44, 0.07	−0.1700, 0.0279	−0.2920, −0.3960	−0.4472, 0.0834	0.3266, 0.5192
6	0.07, 0.59	0.0279, 0.2224	−0.3960, 0.2185	0.0834, 0.5245	0.5192, 0.7147
7	0.59, 1.10	0.2224, 0.3673	−0.2185, 0.0458	0.5245, 0.4290	0.7147, 0.8627
8	1.10, 1.62	0.3643, 0.4474	0.0458, 0.1745	0.4290, 0.0654	0.8627, 0.9506
9	1.62, 2.13	0.4474, 0.4834	0.1745, 0.1460	0.0654, −0.1351	0.9506, 0.9872
10	2.13, ∞	0.4834, 0.5000	0.1460, 0	−0.1351, 0	0.9872, 1.0000

The values of Charlier's distribution function $F(u_i)$ are found by Eq. (2.146). As an example, we shall show how this is done for the second interval. The interval boundaries are $(-1.99, -1.47)$. For these interval boundaries, Appendix Table 1 gives the following values of the Laplace function:

$$\Phi(-1.99) = -\Phi(1.99) = -0.470\,7$$

$$\Phi(-1.47) = -\Phi(1.47) = -0.429\,2$$

Since $\varphi(u)$ is an even function, the second derivative $\varphi''(u)$ must likewise be even, and the third derivative $\varphi'''(u)$ must be odd:

$$\begin{aligned} \varphi''(-u) &= \varphi''(u) \\ \varphi'''(-u) &= -\varphi'''(u) \end{aligned} \qquad (2.148)$$

Therefore,

$$\varphi''(-1.99) = \varphi''(1.99) = 0.163\,0$$
$$\varphi''(-1.47) = \varphi''(1.47) = 0.157\,2$$
$$\varphi'''(-1.99) = -\varphi'''(1.99) = 0.106\,2$$
$$\varphi'''(-1.47) = -\varphi'''(1.47) = -0.167\,0$$

Using the chi-square test, let us evaluate the goodness of fit of Charlier's series to the empirical distribution. The theoretical probabilities p_i that data will fall in the ith interval on the basis of Charlier's distribution can be found by Eq. (1.12):

$$p_i = F(u_i) - F(u_{i-1})$$

where u_i is the right-hand boundary and u_{i-1} is the left-hand boundary of the ith interval. The results are presented in the table that follows. In the table, the last two intervals are

i	p_i	np_i	n_i	$p_i - np_i$	$\frac{(n_i - np_i)^2}{np_i}$
1	0.0267	5.34	7	1.66	0.516
2	0.0484	9.68	11	1.32	0.180
3	0.0901	19.22	15	−4.22	0.926
4	0.1454	29.08	24	−5.08	0.887
5	0.1926	38.52	49	10.48	2.852
6	0.1955	39.10	41	1.90	0.092
7	0.1480	29.60	26	−3.60	0.438
8	0.0879	17.58	17	−0.58	0.019
9	0.0336	7.32 }	7		
10	0.0128	2.56 }	3	0.12	0.001

combined into one, because $np_{10} = 2.56 < 5$. The chi-square statistic is evaluated by the equation

$$\chi^2 = \sum_{i=1}^{9} \frac{(n_i - np_i)^2}{np_i} = 5.911$$

The number of degrees of freedom is $v = 9-4-1 = 4$, because the number of parameters deduced from the sample data is $c = 4(\bar{x}, s, \gamma_1^*,$ and $\gamma_2^*)$. Reading Appendix Table 4 for $v = 4$, we get

$$\chi^2_{0.95} = 9.5$$

Thus

$$\chi^2 < \chi^2_{0.95}$$

and the hypothesis that Charlier's distribution fits the empirical data is not rejected. However, this does not give us evidence to claim that it fits better than the normal distribution examined in Example 2.13.

2.22. Nonparametric Methods

If we know nothing about the distribution of a random variable, we can derive some of the estimates by nonparametric methods. One such method is the construction of a confidence interval for the population mean, using Chebyshev's inequality. The population mean μ can always be estimated in terms of the sample mean $\bar{x}$. The accuracy of the estimate can be stated if we know the population variance σ^2, which may well be replaced by its statistic, the sample variance s^2. This replacement is legitimate on the basis of Glivenko's theorem (see Sec. 2.1).

According to Chebyshev's inequality, the probability that a random variable differs from its expected value by more than any positive number ε must be less than σ^2/ε^2:

$$P(|X-\mu| \geqslant \varepsilon) \leqslant \sigma^2/\varepsilon^2 \quad (2.149)$$

On setting $\varepsilon = k\sigma$, we get

$$P(|X-\mu| \geqslant k\sigma) \leqslant 1/k^2 \quad (2.150)$$

For a significance level α, Chebyshev's inequality defines the following confidence interval for the population mean μ:

$$x-\sigma/\sqrt{\alpha} \leqslant \mu \leqslant x+\sigma/\sqrt{\alpha} \quad (2.151)$$

which enables us to estimate the mean from a single observation. If

$$\bar{x} = \frac{1}{n} \sum_{i=1}^{n} x_i$$

then

$$\sigma_{\bar{x}} = \sigma/\sqrt{n}$$

Hence,

$$\bar{x}-\sigma/\sqrt{np} \leqslant \mu \leqslant \bar{x}+\sigma/\sqrt{np} \quad (2.152)$$

For a significance level $\alpha = 0.05$, we have

$$\bar{x}-4.46\sigma/\sqrt{n} \leqslant \mu \leqslant \bar{x}+4.46\sigma/\sqrt{n} \quad (2.153)$$

In practice, it is usual to round off the factor 4.46 to 5 (which corresponds to $1-\alpha = 0.96$). Let us agree to deem that deviations with the probability $\alpha < 0.04$ cannot practically happen. So we can say that, whatever the population distribution of a random variable X with a variance σ^2, the deviation from the population mean by more than 5σ cannot practically happen (see Eq. (2.120) for the estimate of excess).

Owing to simplicity and universality, Chebyshev's inequality is widely employed in theoretical applications, but for practical uses it proves too approximate. The estimates formed on the basis of Chebyshev's inequality are markedly inferior to those obtained for the normal distribution. For example, at $1-\alpha = 0.95$, when Eq. (2.153) is applied to a normal distribution it contains the factor 1.96 instead of 4.46. The point is that in dealing with a normal distribution we know the distribution density function of the variable in question. In using Chebyshev's inequality, we know nothing about the distribution density. The estimates can substantially be improved, if it is possible to learn anything about the density of the distribution of interest. For example, if we know that the density of the distribution in question falls off on either side of the mean (a symmetrical, single-peak distribution), Chebyshev's inequality holds in an augmented form:

$$P(|x-\mu| \geqslant k\sigma) \leqslant (2/3k)^2 \tag{2.154}$$

At a confidence coefficient $1-\alpha = 0.95$, Eq. (2.154) leads to the following confidence interval

$$\bar{x}-2.96\sigma/\sqrt{n} \leqslant \mu \leqslant \bar{x}+2.96\sigma/\sqrt{n} \tag{2.155}$$

So we can say that the deviation from the expected value in excess of 3σ cannot practically happen (see Eq. (2.119)). The sample coefficient of skewness has a symmetrical single-peak distribution. If we have a sample of n values $(x_1, x_2, \ldots, x_n)$ from a population whose distribution is unknown, it is useful to test the hypothesis that the observed distribution is symmetrical. If the hypothesis is confirmed, we can apply Chebyshev's inequality in its augmented form, Eq. (2.154). The hypothesis that the distribution is symmetrical is true, if the probability of values $x < \bar{x}$ in the sample is one-half.

The estimate for the probability of event $A(x < \bar{x})$ is the frequency of event A (Bernoulli's theorem), which can be determined from the sample data. If in a given sample k members are smaller than the sample mean, the frequency of event A is $\omega = k/n$. The number that a given event occurs is a random variable having a binomial distribution:

$$P(\omega = k/n) = C_n^k p^k(1-p)^{n-k} = \frac{n!}{k!(n-k)!}p^k q^{n-k}$$

where p is the probability of event A in a single trial. The frequency of an event is itself a random variable.

In order to test the hypothesis about the symmetry of the distribution, is necessary to construct a confidence interval for the unknown event it $< \bar{x}$, using the computed frequency. The hypothesis is not rejected if $X = 1/2$ falls within the confidence interval.

In constructing the confidence interval, the point of departure is the fact that at large values of n or at p not very close to zero or unity the binomial distribution differs but little from normal with the same mean $\mu = np$ and variance $\sigma^2 = np(1-p) = npq$. From the linearity of the normal distribution it follows that the distribution of the frequency ω will likewise be close to normal with a mean

$$\mu = m/n = p$$

and a variance

$$\sigma_\omega = \sigma/n = \sqrt{p(1-p)/n} \tag{2.156}$$

Therefore, at a confidence coefficient $1-\alpha = 0.95$, the confidence interval is

$$\omega-1.96\sqrt{\frac{p(1-p)}{n}} \leqslant p \leqslant \omega+1.96\sqrt{\frac{p(1-p)}{n}} \tag{2.157}$$

Both the left- and right-hand sides of the above expression contain the unknown quantity, p, and it is extremely tedious to solve the inequalities for p. Therefore, in the equation

$$\sigma_\omega = \sqrt{p(1-p)/n}$$

the quantity p is replaced by the frequency ω deduced from the sample data. As a result, the confidence interval defined by Eq. (2.157) at $\alpha = 0.05$ takes the form

$$\omega - 1.96\sqrt{\frac{\omega(1-\omega)}{n}} \leqslant p \leqslant \omega + 1.96\sqrt{\frac{\omega(1-\omega)}{n}} \tag{2.158}$$

Example 2.18. An experiment yielded a sample of 20 values of a random variable X with an unknown distribution:

11.2	11.8	12.4	13.3	14.2	15.5	12.1
9.8	14.5	12.5	12.3	16.4	10.1	11.3
15.4	13.7	10.8	14.4	9.6	12.6	

Test the hypothesis about the symmetry of the distribution.

Solution. To begin with, we find the sample mean

$$\bar{x} = \frac{1}{n}\sum_{i=1}^{n} x_i = \frac{1}{20}\sum_{i=1}^{20} x_i = 12.7$$

The number k of sample members less than the mean is 12. Hence, the frequency is

$$\omega = 12 \div 20 = 3/5$$

and at a confidence coefficient $1-\alpha = 0.95$, Eq. (2.158) yields the following confidence interval for the values $x < \bar{x}$:

$$\frac{3}{5} - 1.96\sqrt{\frac{3}{5}\left(1-\frac{3}{5}\right)\frac{1}{20}} \leqslant p \leqslant \frac{3}{5} + 1.96\sqrt{\frac{3}{5}\left(1-\frac{3}{5}\right)\frac{1}{20}}$$

and, finally

$$0.38 \leqslant p \leqslant 0.82$$

Thus, $p = 1/2$ falls within the confidence interval. This gives us evidence to claim that the observed sample does not contradict the hypothesis that the population distribution of the random variable X is symmetrical.

Chapter Three
THE ANALYSIS OF VARIANCE

3.1. The Objective of the Analysis of Variance

In any experiment, the means of the observables vary because of changes in the basic factors (both qualitative and quantitative) that control the conditions of the experiment, and also in accidental factors. It is the objective of the *analysis of variance* to investigate the effect of the various factors on variation about the means [11, 13, 14, 15].

The analysis of variance utilizes the additivity of variances (see Sec. 1.3) of a random variable due to the effect of independent factors. In 1938, R.A. Fisher defined the analysis of variance as the separation of the variation ascribed to one group of sources from the variation ascribed to other groups. According to the number of sources of variation involved, there may be one-way and many-way analyses of variance.

The analysis of variance (ANOVA) is especially powerful in cases involving a multiplicity of factors. A limitation of the classical method of analysis is that only one factor is varied at a time, whereas the remaining ones are held constant. This requires that for each factor a separate series of observations be carried out, not used in the study of other factors. Also, the classical method does not throw light on how the various factors would interact if they were varied all at the same time. In the analysis of variance, each observation serves to evaluate all the factors and their interaction at the same time.

Basically, the analysis of variance starts with resolving the total sample variance of a given random variable into separate components of variance, that is, into a sum of several distinct components, each corresponding to a source of variation. Each of these components is an estimate of the population variance. Before we can say whether the effect of a given factor is significant, we must evaluate the significance of the corresponding sample variance in comparison with the replication variance due to chance factors. The significance of the sample variances (estimates of the population variance) is tested by applying the Fisher (F) test (see 2.11). If the computed value of F test statistic proves smaller than that given in the table, there is no evidence to claim that the effect of the factor in question is significant. If, on the other hand, the calculated value of F test statistic exceeds that given in the table, the factor in question does affect the variation about the means.

In our subsequent discussion we shall assume the following: (1) random errors of observation are normally distributed; (2) factors only affect the variation of the means, and the observation variance remains constant; (3) the experiments are of equal precision.

The requirement of normal distribution for observation errors determines the choice of basic factors in a study of a process by the analysis of variance method. If we are to obtain a normally distributed output (response) variable, it is desirable to class as random only those factors which will affect the response only slightly. We may only make an exclusion to the factors that themselves lead (for some reason) to a normal distribution of the results.

Factors dealt with in the analysis of variance may be of any one of two kinds: (1) those with random levels and (2) those with fixed levels. In the former case, it is presumed that factor levels are chosen from an infinite set of likely levels and that the selection is accompanied by randomization. In the circumstances, the experimental results are more significant, because the conclusion drawn from the experiment can be generalized to the entire population. If all factor levels are chosen at random, we have a *completely randomized model.* If all factors have fixed levels, we have a *fixed level model.* If some of the factors are chosen to have fixed levels and others, randomly selected levels, we have a *mixed model.* Sometimes, there is no difference between the criteria applied to the various models, and the only difference lies in the conclusions drawn. In other cases, there may be difference between the criteria used.

The analysis of variance can be applied in several distinct forms, according to the structure of the process being investigated. The selection of a particular form usually constitutes a major difficulty in the practical application of the analysis of variance.

3.2. One-Way ANOVA

Suppose some factor A which we postulate as having some effect on a response variable of interest y has k levels. We set up an experiment in which n_1 observations are made of the response y at level a_1, n_2 observations at level a_2, and so on, with n_i observations at a given level a_i. The levels a_i are called treatments, there being k treatments in the experimental design. The observations $(y_{1i}, y_{2i}, \ldots, y_{ni})$ constitute a random sample of size n_i at the treatment level a_i, or in tabular form

$$\begin{array}{c} y_{11}\, y_{21} \ldots y_{k1} \\ y_{12}\, y_{22} \ldots y_{k2} \\ \ldots\ldots\ldots\ldots \\ \ldots\ldots\ldots\ldots \\ y_{1n_1}\, y_{2n_2} \ldots y_{kn_k} \end{array}$$

The model of the response to any single treatment can be written

$$y_{ij} = \mu + d_i + \varepsilon_{ij} \tag{3.1}$$

where y_{ij} represents the jth observation ($j = 1, 2, \ldots, n_i$) on the ith treatment ($i = 1, 2, \ldots, k$ levels), μ is a common effect for the whole experiment, d_i represents the effect of the ith treatment, and ε_{ij} represents the random error present in the jth observation on the ith treatment (level).

Suppose that the ith treatment produces its effect and the observations at a fixed factor level are normally distributed about the mean $\mu + d_i$ with a total variance σ^2. The total number of observations is

$$N = n_1 + n_2 + n_3 + \ldots + n_k \tag{3.2}$$

We are now to test the null hypothesis that there is no treatment effect

$$H_0\colon\ \mu_1 = \mu_2 = \ldots = \mu_k = \mu$$

The analysis is easiest to carry out with the same number of observations (replications) on each level or treatment, that is

$$n_1 = n_2 = \ldots = n_k = n$$

The layout of a one-way ANOVA experiment with the same number of replications on all treatments or levels is given in Table 3.1.

Table 3.1. Data Layout for a One-Way ANOVA with an Equal Number of Replications on Each Treatment

Observation No.	Factor A levels (treatments)			
	a_1	a_2	...	a_k
1	y_{11}	y_{21}		y_{k1}
2	y_{12}	y_{22}		y_{k2}
.	.			
.	.			
.	.			...
n	y_{1n}	y_{2n}		y_{kn}
Totals	$A_1 = \sum_{j=1}^{n} y_{1j}$	$A_2 = \sum_{j=1}^{n} y_{2j}$		$A_k = \sum_{j=1}^{n} y_{kj}$

In the circumstances, the total number of observations will be $N = kn$, and those on each treatment or level make up a data set or a (partial) sample.

Let $\bar{y}_i$ denote the mean of the ith (partial) sample, that is, the mean of observations on the ith treatment or level

$$\bar{y}_i = \frac{1}{n} \sum_{j=1}^{n} y_{ij} = A_i/n \tag{3.3}$$

and $\bar{y}$ denote the overall (or grand mean) of the total sample of size N, that is, all k_k (partial) samples:

$$\bar{y} = \frac{1}{N} \sum_{i=1}^{k} \sum_{j=1}^{n} y_{ij} = \frac{1}{k} \sum_{i=1}^{k} \bar{y}_i \tag{3.4}$$

We begin the one-way ANOVA by partitioning the total sample variance

$$s^2 = \frac{\sum_{i=1}^{k} \sum_{j=1}^{n} (y_{ij} - \bar{y})^2}{N-1} = \frac{1}{N-1} \left[\sum_{i=1}^{k} \sum_{j=1}^{n} y_{ij}^2 - \frac{\left(\sum_{i=1}^{k} \sum_{j=1}^{n} y_{ij} \right)^2}{N} \right] \tag{3.5}$$

into two components, one attributed to the identifiable source, factor A, and the other representing the variation due to uncontrolled factors and random errors associated with the response measurements.

The effect of the uncontrolled (error) factor can readily be evaluated because several replicate measurements are made at each level. The sample variance at each level can be written

$$s_i^2 = \frac{1}{n-1} \sum_{j=1}^{n} (y_{ij} - \bar{y}_i)^2 = \frac{1}{n-1} \left[\sum_{j=1}^{n} y_{ij}^2 \frac{1}{n} - \left(\sum_{j=1}^{n} y_{ij} \right)^2 \right] \tag{3.6}$$

$$i = 1, 2, \ldots, k$$

If we are not sure that all the observations are equally precise, the homogeneity of the sample variances $s_1^2, s_2^2, \ldots, s_k^2$ can be checked by applying Cochran's test (see Sec. 2.13). If the sample variances do not differ significantly, the population variance characterizing the uncontrollable (error) factor can be estimated in terms of the error variance:

$$s_e^2 = \frac{1}{k}\sum_{i=1}^{k} s_i^2 = \frac{1}{k(n-1)}\left[\sum_{i=1}^{k}\sum_{j=1}^{n} y_{ij}^2 - \frac{1}{n}\sum_{i=1}^{k}\left(\sum_{j=1}^{n} y_{ij}\right)^2\right] \quad (3.7)$$

The error variance s_e^2 has $k(n-1) = N-k$ degrees of freedom.

An approximate estimate for the variance of factor A may be written

$$\sigma_A^2 \approx s^2 - s_e^2 \quad (3.8)$$

A more accurate estimate for σ_A^2 can be formed by taking the deviations of the means $\bar{y}_i$ at the various levels about the grand mean $\bar{y}$:

$$\frac{1}{k-1}\sum_{i=1}^{k}(\bar{y}_i - \bar{y})^2 \approx \sigma_A^2 + \sigma^2/n \approx \sigma_A^2 + s_e^2/n \quad (3.9)$$

Hence,

$$\sigma_A^2 \approx \frac{1}{k-1}\sum_{i=1}^{k}(\bar{y}_i - \bar{y})^2 - s_e^2/n \quad (3.10)$$

The variance of factor A for a fixed-level model, σ_A^2, is not related to any random variable; this is an arbitrary name for the expected value of the mean square of deviations due to factor A. This notation is convenient because it defines the dispersion caused by factor A in similar terms with that caused by the uncontrolled factor, so that factor A can directly be compared with the uncontrolled factor. Let also introduce the following designation

$$s_A^2 = \frac{n}{k-1}\sum_{i=1}^{k}(y_i - \bar{y})^2 \approx n\sigma_A^2 + s_e^2 \quad (3.11)$$

This variance has $(k-1)$ degrees of freedom. If s_A^2 differs significantly from s_e^2, the null hypothesis, $H_0\colon \mu_1 = \mu_2 = \ldots = \mu_k = \mu$, is rejected, and the effect of factor A is taken as significant. The null hypothesis is tested by applying the F test. Since the alternative hypothesis is set up as an inequality, we may apply a one-sided F test. The effect of factor A is significant, if

$$s_A^2/s_e^2 > F_{1-\alpha}(\nu_1, \nu_2) \quad (3.12)$$

$$\nu_1 = k-1$$

$$\nu_2 = k(n-1) = N-k$$

The analysis of variance can be carried out, using the following algorithm:

(1) take the totals for the columns (treatments)

$$A_i = \sum_{j=1}^{n} y_{ji} \quad (3.13)$$

(2) take the sum of the squares of all observations

$$SS_1 = \sum_{i=1}^{k} \sum_{j=1}^{n} y_{ij}^2 \quad (3.14)$$

(3) take the sum of the squares of the column totals divided by the number of observations (replications) in a column

$$SS_2 = \frac{1}{n} \sum_{i=1}^{k} A_i^2 \quad (3.15)$$

(4) find the square of the sum total divided by the number of all observations (the correcting term)

$$SS_3 = \frac{1}{N} \left(\sum_{i=1}^{k} A_i\right)^2 \quad (3.16)$$

(5) take the sum of the squares for a column

$$SS_A = SS_2 - SS_3 \quad (3.17)$$

(6) take the sum total of the squares equal to the difference between the sum of the squares of all observations and the correcting term

$$SS_{tot} = SS_1 - SS_3 \quad (3.18)$$

(7) find the residual (or error) sum of squares

$$SS_{res} = SS_1 - SS_2 \quad (3.19)$$

(8) find the sample variance for the factor A

$$s_A^2 = SS_A/k(-1) \quad (3.20)$$

(9) find the error variance

$$s_e^2 = SS_{res}/(kn-1) \quad (3.21)$$

The results are usually presented in what is known as an ANOVA (analysis of variance) table (see Table 3.2).

If $s_A^2/s_e^2 \leqslant F_{1-\alpha}$, the effect of factor A should be deemed insignificant. Then the total variance s^2 is only related to the uncontrolled (random) factor and may serve as the estimate for the replication variance. This estimate is better than that defined by Eq. (3.7), because it has a greater number of degrees of freedom, equal to $kn-1$.

In interpreting the results of an ANOVA, it is important to bear in mind that a very low value of the variance ratio may be related to the fact that some important uncontrolled factor was not randomized in the course of the experiment. This may lead to an increased variance within the treatments while leaving the variance between the treatments unchanged and leading thereby to a reduced variance ratio. In the circumstances, the experimental results will not obey the model defined by Eq. (3.1).

If, on the other hand, the inequality (3.12) holds, the difference between s_A^2 and s_e^2 is significant, and so is the effect of factor A. Let us form the estimate for the effect of factor A from Eq. (3.11):

$$\sigma_A^2 \approx (s_A^2 - s_e^2)/n \quad (3.22)$$

Table 3.2. ANOVA Table for One-Way Analysis of Variance with an Equal Number of Replications on All Treatments (Levels)

Source of variation	Degrees of freedom	Sum of squares (SS)	Mean square	Expected mean square
Between treatments (d_i)	$k-1$	$SS_A = SS_2 - SS_3$	$s_A^2 = SS_A/(k-1)$	$n\sigma_A^2 + \sigma_e^2$
Within treatments or error (ε_{ij})	$k(n-1)$	$SS_{res} = SS_1 - SS_2$	$s_e^2 = SS_{res}/k(n-1)$	σ_e^2
Totals	$kn-1$	$SS_{tot} = SS_1 - SS_3$	$SS_{tot}/(kn-1)$	

This gives evidence to reject the null hypothesis that $\mu_1 = \mu_2 = \ldots = \mu_k = \mu$ and to deem the difference between the means $\mu_1, \mu_2, \ldots, \mu_k$ significant. Which means are different can be determined by applying the Student (t) test, the Fisher test, or Duncan's multiple ranges test (see Sec. 2.14).

In interpreting the results of an ANOVA for a completely randomized model, we are usually interested in the components of the total variance rather than in testing hypotheses about the means. In contrast to a fixed-level model, the conclusions drawn from a completely randomized model can be generalized to the entire population of levels.

Consider the analysis sequence for a case in which a different number of replicate measurements is made on each of k treatments. Let n_i be the number of replications carried out on the level (treatment) a_i. The total number of all observations will be

$$N = \sum_{i=1}^{k} n_i$$

We then proceed as follows:

(1) determine the totals for the columns (treatments)

$$A_i = \sum_{j=1}^{n_i} y_{ji} \qquad i = 1, 2, \ldots, k \tag{3.23}$$

(2) take the sum of squares of all observations

$$SS_1 = \sum_{i=1}^{k} \sum_{j=1}^{n_i} y_{ij}^2 \tag{3.24}$$

(3) take the sum of squares of the column (treatment) totals divided by the number of observations in each column (treatment)

$$SS_2 = \sum_{i=1}^{k} A_i^2/n_i \tag{3.25}$$

(4) take the square of the grand total divided by the number of all observations

$$SS_3 = \frac{1}{N}\left(\sum_{i=1}^{k} A_i\right)^2 \tag{3.26}$$

The remaining calculations are carried out, using Eqs. (3.17) through (3.21). If s_A^2 and s_e^2 differ significantly from each other, the variance for factor A should be found by the equation

$$\sigma_A^2 \approx \frac{(k-1)N}{N^2-\sum_{i=1}^{k} n_i}(s_A^2-s_e^2) \tag{3.27}$$

Example 3.1. Use a one-way ANOVA to determine the effect of a haloid alkyl (factor A) on radical polymerization. A total of five haloid alkyls were tried, namely: (a_1) CH_3I, (a_2) C_3H_7I, (a_3) C_4H_9I, (a_4) C_2H_5Br, and (a_5) C_3H_7Br. The response variable y is the percent yield of a polymer. The data layout for a one-way ANOVA is given in the table that follows.

Observation No.	Factor A levels (treatments)				
	a_1	a_2	a_3	a_4	a_5
1	79.80	87.30	42.45	76.0	70.70
2	86.30	69.60	64.3	83.5	64.65
3	86.50	81.75	78.9	72.80	38.50
4	92.30	77.95	61.00	89.00	77.00
5	76.50	83.65	31.30	76.50	91.50
6	87.05	64.80	72.85	87.45	68.00
7	82.50	67.30	58.65	74.50	38.05
8	90.00	75.45	52.50	93.15	79.95
Total	$A_1 = 680.95$	$A_2 = 607.8$	$A_3 = 461.95$	$A_4 = 652.9$	$A_5 = 528.35$

Determine the mean yield for each of the haloid alkyls:

$$\bar{y}_1 = 680.95/8 = 85.1,\quad \bar{y}_2 = 607.8/8 = 75.97,\quad \bar{y}_3 = 461.95/8 = 57.74,$$
$$\bar{y}_4 = 652.9/8 = 81.61,\quad \bar{y}_5 = 528.35/8 = 66.04$$

and the total (or grand mean) for all observations

$$\bar{\bar{y}} = \frac{1}{5}\sum_{i=1}^{5} \bar{y}_i = 73.2$$

To facilitate the calculation, we shall replace y's by their deviations y' from the grand mean rounded off to 73 (see the table that follows). Referring to the table and using Eqs. (3.14) through (3.21), we go through the sequence outlined above. The results are presented in the ANOVA table that follows.

Next we apply the Fisher (F) test to the variances obtained by calculation:

$$F = s_A^2/s_e^2 = 1\,021/150.9 = 6.9$$

Appendix Table 5 gives

$$F_{0.95}(4,\ 35) = 2.65$$

Since $F > 2.65$, the difference between the haloid alkyds is significant.

Observation No.	Factor A levels (treatments)				
	a_1	a_2	a_3	a_4	a_5
1	6.8	14.3	−30.55	3.0	−2.3
2	13.3	−3.4	−8.7	10.5	−8.35
3	13.5	8.75	5.9	−0.2	−34.5
4	19.3	4.95	−12.0	16.0	4.0
5	3.5	10.65	−41.7	3.5	18.5
6	14.05	−8.2	−0.15	14.45	−5.0
7	9.5	−5.7	−14.35	1.5	−34.95
8	17.0	2.45	−20.5	20.15	6.95
Total	96.95	23.8	−122.05	68.9	−55.65

Source of variation	Degrees of freedom	Sum of squares	Mean square
Between treatments (d_i), A	4	4 084.5	1 021
Within treatments (residual)	35	5 281.46	150.9
Total	39	9 365.96	

Now that, by use of the ANOVA, we have found out that the mean yields of the polymer substantially differ from alkyl to alkyl, we proceed to compare the effects of the various alkyls. This can be done by applying the Duncan multiple range test (see Sec. 2.14) for a confidence coefficient $1-\alpha = 0.95$. The standard error of the mean is

$$s_{\bar{y}} = \sqrt{150.9/8} = 4.35$$

Arrange the means from low to high

$\bar{y}_3(C_4H_9I)$	$\bar{y}_5(C_3H_7Br)$	$\bar{y}_2(C_3H_7I)$	$\bar{y}_4(C_2H_5Br)$	$\bar{y}_1(CH_3I)$
57.74	66.04	75.97	81.61	85.1

then enter Appendix Table 10 of significant ranges at a significance level of 0.05, using $\nu = 35$ degrees of freedom for the error mean square and

p	2	3	4	5

and list these

$k-1$ ranges	2.875	3.025	3.11	3.185

Now multiply these ranges by the standard error of the mean, $s_{\bar{y}}$ to form a group of $k-1$ least significant ranges

$rs_{\bar{y}}$	12.5	13.2	13.52	13.9

and test the observed ranges between the means, beginning with the largest versus the smallest, which is compared with the least significant range for $p = k$; then test the largest versus second smallest with the least significant range for $p = k-1$, etc. Continue this for the second largest versus the smallest, etc. until all $k(k-1)/2$ possible pairs have been tested:

$\bar{y}_1-\bar{y}_3 =$	$27.3 > 13.9$	a significant difference
$\bar{y}_1-\bar{y}_5 =$	$19.05 > 13.52$	a significant difference
$\bar{y}_1-\bar{y}_2 =$	$9.13 < 13.2$	an insignificant difference
$\bar{y}_1-\bar{y}_4 =$	$3.49 < 12.5$	an insignificant difference
$\bar{y}_4-\bar{y}_3 =$	$23.81 > 13.52$	a significant difference
$\bar{y}_4-\bar{y}_5 =$	$15.56 > 13.2$	a significant difference
$\bar{y}_4-\bar{y}_2 =$	$5.64 < 12.5$	an insignificant difference
$\bar{y}_2-\bar{y}_3 =$	$18.17 > 13.2$	a significant difference
$\bar{y}_2-\bar{y}_5 =$	$9.92 < 12.5$	an insignificant difference
$\bar{y}_5-\bar{y}_3 =$	$8.25 < 12.5$	an insignificant difference

3.3. Two-Way ANOVA

Now we shall consider the case in which we must set up an experiment to study the effects of two factors A and B on a response variable y. Factor A has k levels $a_1, a_2, \ldots, a_k$, whereas factor B has m levels $b_1, b_2, \ldots, b_m$. For each combination of levels (a_i, b_j), we measure the response y_{ij}, by carrying out n observations. The data layout for a two-way ANOVA experiment is given in Table 3.3. The total number of observations is $N = knm$. The model of the response to each treatment may be written

$$y_{ijq} = \mu+\alpha_i+\beta_j+\alpha_i\beta_j+\varepsilon_{ijq} \tag{3.28}$$

where μ represents an overall mean or a common effect, α_i represents the effect of factor A on the ith level ($i = 1, 2, \ldots, k$), β_j represents the effect of factor B on the jth level ($j = 1, 2, \ldots, m$), $\alpha_i\beta_j$ represents the effect due to the interaction of the two factors.

The interaction term is the deviation of the mean of observations in the (ij)th set from the sum of the first three terms in the model defined by Eq. (3.28), and ε_{ijq} $(q = 1, 2, \ldots, n)$ represents the variation within the set of observations. As before, let ε_{ijq} be normally distributed with zero mean and a variance σ_e^2. If we assume that there is no interaction between the two factors, we may adopt a linear model

$$y_{ij} = \mu+\alpha_i+\beta_j+\varepsilon_{ij} \tag{3.29}$$

This model is ordinarily used in cases with no replications, and the respective data layout is given in Table 3.4. To begin with, we shall try the linear model defined by Eq. (3.29). Let $\bar{y}_i$ denote the column means

$$\bar{y}_i = A_i/m \tag{3.30}$$

$\bar{y}'_j$ the row means

$$\bar{y}'_j = B_j/k \tag{3.31}$$

Table 3.3. Data Layout for a Two-Way ANOVA with Replications

B	*A*: a_1	a_2	$\cdots$	a_i	$\cdots$	a_k	Row totals
b_1	$y_{111}, y_{112}, \ldots, y_{11n}$	$y_{211}, y_{212}, \ldots, y_{21n}$		$y_{i11}, y_{i12}, \ldots, y_{i1n}$		$y_{k11}, y_{k12}, \ldots, y_{k1n}$	B_1
b_2	$y_{121}, y_{122}, \ldots, y_{12n}$	$y_{221}, y_{222}, \ldots, y_{22n}$	$\cdots$	$y_{i21}, y_{i22}, \ldots, y_{i2n}$		$y_{k21}, y_{k22}, \ldots, y_{k2n}$	B_2
$\vdots$	$\vdots$	$\vdots$		$\vdots$		$\vdots$	$\vdots$
b_j	$y_{1j1}, y_{1j2}, \ldots, y_{1jn}$	$y_{2j1}, y_{2j2}, \ldots, y_{2jn}$	$\cdots$	$y_{ij1}, y_{ij2}, \ldots, y_{ijn}$		$y_{kj1}, y_{kj2}, \ldots, y_{kjn}$	B_j
$\vdots$	$\vdots$	$\vdots$		$\vdots$		$\vdots$	
b_m	$y_{1m1}, y_{1m2}, \ldots, y_{1mn}$	$y_{2m1}, y_{2m2}, \ldots, y_{2mn}$		$y_{im1}, y_{im2}, \ldots, y_{imn}$		$y_{km1}, y_{km2}, \ldots, y_{kmn}$	B_m
Column totals	A_1	A_2	$\cdots$	A_i	$\cdots$	A_k	

Table 3.4. Data Layout for a Two-Way ANOVA with No Replications

B	*A*: a_1	a_2	$\cdots$	a_k	Row totals
b_1	y_{11}	y_{21}	$\cdots$	y_{k1}	B_1
b_2	y_{12}	y_{22}	$\cdots$	y_{k2}	B_2
$\vdots$					$\vdots$
b_m	y_{1m}	y_{2m}	$\cdots$	y_{km}	B_m
Column totals	A_1	A_2	$\cdots$	A_k	

and $\bar{\bar{y}}$ the overall (grand) mean

$$\bar{\bar{y}} = \frac{1}{km} \sum_{i=1}^{k} \sum_{j=1}^{m} y_{ij} \tag{3.32}$$

The dispersion of the row means $\bar{y}_1, \bar{y}_2, \ldots, \bar{y}_k$ about the grand mean $\bar{\bar{y}}$ is independent of factor B, because all levels of factor B are averaged out. This dispersion is related to factor A and the random error term. Because the variation of the mean (see Sec. 2.5) is $1/m$ times the variation of a single measurement, we have

$$\frac{1}{k-1} \sum_{i=1}^{k} (\bar{y}_i - \bar{\bar{y}})^2 \approx \sigma_A^2 + \sigma^2/m \tag{3.33}$$

In turn, the variation of the row means is independent of factor A and is related to factor B:

$$\frac{1}{m-1} \sum_{j=1}^{m} (\bar{y}'_j - \bar{\bar{y}})^2 \approx \sigma_B^2 + \sigma^2/k \tag{3.34}$$

From Eqs. (3.33) and (3.34), we can estimate the effects due to factors A and B, if we know the estimate of the population variance σ^2.

In order to evaluate the random error factor in the absence of replicate observations, we can proceed as follows. To begin with, we find the variance of observations in the ith column:

$$s_i^2 = \frac{1}{m-1} \sum_{j=1}^{m} (y_{ij} - \bar{y}_i)^2 \tag{3.35}$$

This variance is due to the effect of factor B and the random error factor

$$s_i^2 \approx \sigma_B^2 + \sigma^2$$

The accuracy of the equation can be enhanced still more, if we replace s_i^2 with the weighted variance averaged over all columns:

$$\sigma_B^2 + \sigma^2 \approx \frac{1}{k} \sum_{i=1}^{k} s_i^2 = \frac{1}{k(m-1)} \sum_{i=1}^{k} \sum_{j=1}^{m} (y_{ij} - \bar{y}_i)^2 \tag{3.36}$$

Subtracting Eq. (3.35) from Eq. (3.34) gives

$$\sigma^2 - \sigma^2/k \approx \frac{1}{k(m-1)} \sum_{i=1}^{k} \sum_{j=1}^{m} (y_{ij} - \bar{y}_i)^2 - \frac{1}{m-1} \sum_{j=1}^{m} (\bar{y}'_j - \bar{\bar{y}})^2 \tag{3.37}$$

Hence,

$$\sigma^2 \approx \frac{1}{(k-1)(m-1)} \left[\sum_{i=1}^{k} \sum_{j=1}^{m} (y_{ij} - \bar{y}_i)^2 - k \sum_{j=1}^{m} (\bar{y}'_j - \bar{\bar{y}})^2 \right] \tag{3.38}$$

Let the above estimate, Eq. (3.38), of the variance be denoted as s_e^2. It has $(k-1)(m-1)$ degrees of freedom. Introduce also the following notation:

$$s_A^2 = \frac{m}{k-1} \sum_{i=1}^{k} (\bar{y}_i - \bar{\bar{y}})^2 \approx m\sigma_A^2 + s_e^2 \tag{3.39}$$

and

$$s_B^2 = \frac{k}{m-1} \sum_{j=1}^{m} (\bar{y}_j' - \bar{\bar{y}})^2 \approx k\sigma_B^2 + s_e^2 \tag{3.40}$$

We may treat s_A^2 and s_B^2 as sample variances with $(k-1)$ and $(m-1)$ degrees of freedom, respectively. The hypothesis about the significance of factors A and B is tested by applying the Fisher (F) test. If the variance ratio

$$F = s_A^2/s_e^2 < F_{1-\alpha}(\nu_1, \nu_2) \tag{3.41}$$
$$\nu_1 = k-1$$
$$\nu_2 = (k-1)(m-1)$$

we accept the hypothesis H_0: $\alpha_i = 0$. If, on the other hand,

$$F = s_A^2/s_e^2 > F_{1-\alpha}(\nu_1, \nu_2) \tag{3.42}$$

we reject the null hypothesis, and the effect due to factor A is taken as significant. Similarly, if

$$F = s_B^2/s_e^2 < F_{1-\alpha}(\nu_1, \nu_2) \tag{3.43}$$
$$\nu_1 = m-1$$
$$\nu_2 = (k-1)(m-1)$$

we accept the null hypothesis H_0: $\beta_j = 0$. If, however,

$$F = s_B^2/s_e^2 > F_{1-\alpha} \tag{3.44}$$

the effect of factor B is taken as significant.

In testing the null hypotheses, we have applied a one-sided F test, because the alternative hypothesis to $\sigma_A^2 = \sigma_e^2$ is set up as an inequality $\sigma_A^2 > \sigma_e^2$.

For a linear model defined by Eq. (3.29), the analysis of variance can conveniently be carried out in the following sequence:

(1) take the column sums

$$A_i = \sum_{j=1}^{m} y_{ij}, \qquad i = 1, 2, \ldots, k \tag{3.45}$$

(2) take the row sums

$$B_j = \sum_{i=1}^{k} y_{ij}, \qquad j = 1, 2, \ldots, m \tag{3.46}$$

(3) take the sum of the squares of all observations

$$SS_1 = \sum_{i=1}^{k} \sum_{j=1}^{m} y_{ij}^2 \tag{3.47}$$

(4) take the sum of the mean squares for the columns

$$SS_2 = \frac{1}{m} \sum_{i=1}^{k} A_i^2 \tag{3.48}$$

(5) take the sum of the row mean squares

$$SS_3 = \frac{1}{k} \sum_{j=1}^{m} B_j^2 \tag{3.49}$$

(6) find the correction term defined as the grand total squared and divided by the total number of observations

$$SS_4 = \frac{1}{mk}\left(\sum_{i=1}^{k} A_i\right)^2 = \frac{1}{mk}\left(\sum_{j=1}^{m} B_j\right)^2 \tag{3.50}$$

(7) take the sum of the column squares

$$SS_A = SS_2 - SS_4 \tag{3.51}$$

(8) take the sum of the row squares

$$SS_B = SS_3 - SS_4 \tag{3.52}$$

(9) find the total sum of squares defined as the difference between the sum of squares of all observations and the correction term

$$SS_{\text{tot}} = SS_1 - SS_4 \tag{3.53}$$

(10) find the residual (error) sum of squares

$$SS_{\text{res}} = SS_{\text{tot}} - SS_A - SS_B = SS_1 - SS_2 - SS_3 + SS_4$$
$$= SS_1 - SS_2 - SS_3 + SS_4 \tag{3.54}$$

(11) find the variance for factor A

$$s_A^2 = SS_A/(k-1) \tag{3.55}$$

(12) find the variance for factor B

$$s_B^2 = SS_B/(m-1) \tag{3.56}$$

(13) find the variance s_e^2

$$s_e^2 = SS_{\text{res}}/(k-1)(m-1) \tag{3.57}$$

The results can conveniently be presented in an ANOVA table (see Table 3.5).

Table 3.5. ANOVA Table for a Two-Way ANOVA without Replications

Source of variation	Degrees of freedom	Sum of squares	Mean square	Expected mean square
A	$k-1$	$SS_A = SS_2 - SS_4$	$s_A^2 = \frac{SS_A}{k-1}$	$m\sigma_A^2 + \sigma_e^2$
B	$m-1$	$SS_B = SS_3 - SS_4$	$s_B^2 = \frac{SS_B}{m-1}$	$k\sigma_B^2 + \sigma_e^2$
Residue	$(k-1)(m-1)$	$SS_{res} = SS_1 - SS_2 - SS_3 + SS_4$	$s_e^2 = \frac{SS_{res}}{(k-1)(m-1)}$	σ_e^2
Totals	$km-1$	$SS_{tot} = SS_1 - SS_4$		

After the significance of a given factor has been evaluated by the analysis of variance, the next step is to establish which means of the response y are different. This is done by applying Student's t test or the Duncan multiple range test.

The linear model defined by Eq. (3.29) holds, if there is no interaction between factors A and B. Otherwise, this interaction will have a variance of its own, σ^2_{AB}, like any other factor. This variance is a measure of how the effect of factor A depends on the level of factor B and, conversely, how the effect due to factor B depends on the level of factor A. In the presence of interaction between the factors, the above sequence is extended to include the interaction variance σ^2_{AB} with the error variance s^2_e. We can separate σ^2_{AB} from the rest of s^2_e only if we run replicate observations.

Let n replicate observations be carried out in each combination of levels of factors A and B. For example, in Table 3.3 the cell at the intersection of the ith column and the jth row contains a series of observations: y_{ij1}, y_{ij2}, ..., y_{ijn}. Let $\bar{y}_{ij}$ denote the mean of observations in a cell. The sample variance for each cell

$$s^2_{ij} = \frac{1}{n-1} \sum_{u=1}^{n} (y_{iju} - \bar{y}_{ij})^2 \qquad (3.58)$$

will have $n-1$ degrees of freedom. If the sample variances for all cells are homogeneous, we may average them and use the weighted average variance

$$s^2_e = \frac{1}{mk} \sum_{i=1}^{k} \sum_{j=1}^{m} s^2_{ij} \qquad (3.59)$$

as the estimate for the population error (replication) variance σ^2. The sample error variance s^2_e has $mk(n-1)$ degrees of freedom. More conveniently, the sample error variance can be found by the equation

$$s^2_e = \frac{\sum\limits_{j=1}^{k} \sum\limits_{i=1}^{m} \sum\limits_{u=1}^{n} y^2_{iju} - \frac{1}{n} \sum\limits_{i=1}^{k} \sum\limits_{j=n}^{m} y^2_{ij}}{mk(n-1)} \qquad (3.60)$$

where y_{ij} is the sum of observations in the ijth cell.

In applying the analysis of variance to a nonlinear model, it is more convenient to use the following sequence.

Referring to Table 3.3:

(1) find the sum of observations for each cell

$$y_{ij} = \sum_{u=1}^{n} y_{iju} \qquad j = 1, 2, \ldots, m \quad i = 1, 2, \ldots, k \qquad (3.61)$$

(2) find the square of the sums of observations for each cell

$$y^2_{ij} = \left(\sum_{u=1}^{n} y_{iju}\right)^2 \qquad (3.62)$$

(3) find the totals for the columns

$$A_i = \sum_{j=1}^{m} \sum_{u=1}^{n} y_{iju} \qquad (3.63)$$

(4) find the totals for the rows

$$B_j = \sum_{i=1}^{k} \sum_{u=1}^{n} y_{iju} \tag{3.64}$$

(5) find the grand total (the total of all observations)

$$\sum_{i=1}^{k} \sum_{j=1}^{m} \sum_{u=1}^{n} y_{iju} = \sum_{i=1}^{k} A_i = \sum_{j=1}^{m} B_j \tag{3.65}$$

(6) find the sum of squares of all observations

$$SS_1 = \sum_{i=1}^{k} \sum_{j=1}^{m} \sum_{u=1}^{n} y_{iju}^2 \tag{3.66}$$

(7) find the sum of column mean squares

$$SS_2 = \frac{1}{mn} \sum_{i=1}^{k} A_i^2 \tag{3.67}$$

(8) find the sum of row mean squares

$$SS_3 = \frac{1}{kn} \sum_{j=1}^{m} B_j^2 \tag{3.68}$$

(9) find the correction term (defined as the grand total squared and divided by the total number of observations)

$$SS_4 = \frac{1}{N}\left(\sum_{i=1}^{k} \sum_{j=1}^{m} \sum_{u=1}^{n} y_{iju}\right)^2 = \frac{1}{mkn}\left(\sum_{i=1}^{k} A_i\right)^2 = \frac{1}{mkn}\left(\sum_{j=1}^{m} B_j\right)^2 \tag{3.69}$$

(10) find the sum of squares for a column

$$SS_A = SS_2 - SS_4 \tag{3.70}$$

(11) find the sum of squares for a row

$$SS_B = SS_3 - SS_4 \tag{3.71}$$

(12) find the sum of squares for error variance

$$SS_e = SS_1 - \frac{1}{n} \sum_{i=1}^{k} \sum_{j=1}^{m} y_{ij}^2 \tag{3.72}$$

(13) find the total sum of squares defined as the difference between the sum of squares of all observations and the correction term:

$$SS_{tot} = SS_1 - SS_4 \tag{3.73}$$

(14) find the residual sum of squares for the interaction term AB

$$SS_{AB} = SS_{tot} - SS_A - SS_B - SS_e \tag{3.74}$$

(15) find the sample variance for factor A

$$s_A^2 = SS_A/(k-1) \tag{3.75}$$

(16) find the sample variance for factor B

$$s_B^2 = SS_B/(m-1) \tag{3.76}$$

(17) find the interaction variance

$$s_{AB}^2 = SS_{AB}/(k-1)(m-1) \tag{3.77}$$

(18) find the error variance

$$s_e^2 = SS_e/mk(n-1) \tag{3.78}$$

The hypothesis about the significance of interaction between factors A and B is tested identically for a random-level and a fixed-level model by applying the Fisher (F) test. However, the hypothesis about the significance of factor A and B is tested differently for different models. An ANOVA table for a two-way ANOVA with replications for a random-level model is given in Table 3.6.

Table 3.6. ANOVA Table for a Two-Way ANOVA with Replications for a Random-Level Model

Source of variation	Degrees of freedom	Sum of squares	Mean square	Expected mean square
A	$k-1$	$SS_A = SS_2 - SS_4$	$s_A^2 = \dfrac{SS_A}{k-1}$	$nm\sigma_A^2 + n\sigma_{AB}^2 + \sigma_e^2$
B	$m-1$	$SS_B = SS_3 - SS_4$	$s_B^2 = \dfrac{SS_B}{m-1}$	$nk\sigma_B^2 + n\sigma_{AB}^2 + \sigma_e^2$
AB	$(k-1)(m-1)$	$SS_{AB} = SS_1 - SS_2 - SS_3 + SS_4$	$s_{AB}^2 = \dfrac{SS_{AB}}{(k-1)(m-1)}$	$n\sigma_{AB}^2 + \sigma_e^2$
Residual (error)	$mk(n-1)$	$SS_e = SS_1 - \dfrac{1}{n}\sum_{i=1}^{k}\sum_{j=1}^{m} y_{ij}^2$	$s_e^2 = \dfrac{SS_e}{mk(n-1)}$	σ_e^2
Totals	$mkn-1$	$SS_{tot} = SS_1 - SS_4$		

As is seen from Table 3.6, in order to evaluate the significance of factor A, we must set up a variance ratio of the form

$$F = s_A^2/s_{AB}^2 \tag{3.79}$$

The effect of factor A is significant if

$$s_A^2/s_{AB}^2 > F_{1-\alpha}(\nu_1, \nu_2) \tag{3.80}$$

where α is the significance level, $\nu_1 = k-1$, and $\nu_2 = (k-1)(m-1)$. Similarly, the effect of factor B is taken as significant if

$$s_B^2/s_{AB}^2 > F_{1-\alpha}(\nu_1, \nu_2) \tag{3.80a}$$

where $\nu_1 = m-1$, and $\nu_2 = (k-1)(m-1)$.

If inequalities (3.80) and (3.80*a*) are not satisfied, the effect of factors A and B should be taken as insignificant.

For a fixed-level model, the interaction terms disappear from the sums of squares for the factors, SS_A and SS_B [14]. Because of this, the variance ratio set up in order to evaluate the significance of factor A takes the form

$$F = s_A^2/s_e^2 \tag{3.81}$$

where the denominator is the estimate of the error variance. This ratio is compared with tabulated values of $F_{1-\alpha}(\nu_1, \nu_2)$, for degrees of freedom $\nu_1 = k-1$ and $\nu_2 = mk(n-1)$. Similarly, for factor B, the variance ratio is

$$F = s_B^2/s_e^2 \tag{3.81a}$$

which is then compared with tabulated values of $F_{1-\alpha}(\nu_1, \nu_2)$ for degrees of freedom $\nu_1 = m-1$ and $\nu_2 = mk(n-1)$.

If

$$s_A^2/s_e^2 > F_{1-\alpha}(\nu_1, \nu_2) \tag{3.81b}$$

and

$$s_B^2/s_e^2 > F_{1-\alpha}(\nu_1, \nu_2) \tag{3.81c}$$

the effect of factors A and B should be taken as significant. If inequalities (3.81*b*) and (3.81*c*) are not satisfied, the effect of factors A and B is insignificant.

In order to evaluate the significance of the effect due to the interaction between the two factors, we set up a variance ratio of the form

$$F = s_{AB}^2/s_e^2$$

which is then compared with tabulated values of $F_{1-\alpha}(\nu_1, \nu_2)$ at a significance level α and degrees of freedom $\nu_1 = (k-1)(m-1)$ and $\nu_2 = mk(n-1)$. If this variance ratio exceeds the tabulated value of the F test statistic

$$s_{AB}^2/s_e^2 > F_{1-\alpha}(\nu_1, \nu_2)$$

the effect due to the interaction of the two factors should be taken as significant. Otherwise, that is, if

$$s_{AB}^2/s_e^2 < F_{1-\alpha}(\nu_1, \nu_2)$$

the effect due to the interaction of the two factors should be taken as insignificant.

Example 3.2. An experiment was conducted to evaluate the effect of two factors on organic synthesis, namely (A) the type of solvent at levels a_1, a_2, a_3, and a_4, and (B) the type of haloid alkyl at levels b_1, b_2, b_3, and b_4. The per cent yields of the polymer are given in the table that follows.

B	A			
	a_1	a_2	a_3	a_4
b_1	13.2	4.7	53.4	13.6
	13.9	5.8	48.3	13.2
b_2	18.9	19.8	14.0	9.5
	21.0	17.9	13.2	8.6
b_3	7.3	38.2	5.1	54.4
	8.5	37.7	5.9	55.2
b_4	20.0	60.1	19.6	58.2
	20.8	60.9	18.5	59.7

Two replicate measurements were made for each combination of solvent and haloid alkyl. It is required to evaluate the significance of the effect due to type of solvent and type of alkyl on the synthesis process.

Solution. A fixed-level model applies to this example, because the levels for factors A and B were not taken at random—the goal of the experiment was to evaluate the effect on the process of only four specific types of solvent and alkyl. Using Eqs. (3.61) through (3.78), we proceed as explained earlier, namely:

1. Referring to the previous table, find the sums of observations in each of the cells:

B	A				Row totals
	a_1	a_2	a_3	a_4	
b_1	27.1	10.5	101.7	26.8	166.1
b_2	39.1	37.7	27.2	18.1	122.1
b_3	15.8	75.9	11.0	109.6	212.3
b_4	40.8	121	38.1	117.9	317.8
Column totals	122.8	245.1	178.0	272.4	818.3

2. Square the results thus obtained and arrange them in tabular form:

B	A			
	a_1	a_2	a_3	a_4
b_1	734.41	110.25	10 342.89	718.24
b_2	1 528.81	1 421.29	739.84	327.61
b_3	249.64	5 760.81	121.0	12 012.16
b_4	1 064.64	14 641	1 451.61	13 900.41

3. Find the totals for the columns. For example,

$$A_1 = 27.1 + 39.1 + 15.8 + 40.8 = 122.8$$

4. Find the totals for the rows. For example,

$$B_2 = 39.1 + 37.7 + 27.2 + 18.1 = 122.1$$

5. Find the grand total—the sum of all observations

$$\sum_{i=1}^{4} \sum_{j=1}^{4} \sum_{u=1}^{2} y_{iju} = \sum_{i=1}^{4} A_i = \sum_{j=1}^{4} B_j = 818.3$$

6. Find the sum of the squares of all observations

$$SS_1 = \sum_{i=1}^{4} \sum_{j=1}^{4} \sum_{u=1}^{2} y_{iju}^2 = 32\,916.43$$

7. Find the sum of the squared column totals divided by the number of observations in a column

$$SS_2 = \frac{1}{4\times 2} \sum_{i=1}^{4} A_i^2 = \frac{1}{4\times 2}(122.8^2 + 245.1^2 + 178.0^2 + 272.4^2)$$
$$= 181\,039.01 \div 8 = 22\,262.95$$

8. Find the sum of the squared row totals divided by the number of observations in a row

$$SS_3 = \frac{1}{4\times 2} \sum_{j=1}^{4} B_j^2 = \frac{1}{4\times 2}(166.1^2 + 122.1^2 + 212.3^2 + 317.8^2)$$
$$= 188\,565.75 \div 8 = 23\,570.72$$

9. Find the correction term defined as the grand total squared and divided by the total number of observations

$$SS_4 = \frac{\left(\sum_{i=1}^{4} \sum_{j=1}^{4} \sum_{u=1}^{2} y_{iju}\right)^2}{32} = 818.3^2 \div 32 = 669\,617.89 \div 32 = 20\,925.47$$

10. Find the sums of squares of deviations for factors A and B:

$$SS_A = SS_2 - SS_4 = 22\,262.95 - 20\,925.47 = 1\,704.48$$
$$SS_B = SS_3 - SS_4 = 23\,570.72 - 20\,925.47 = 2\,645.25$$

11. Find the sum of squares for error variance

$$SS_e = SS_1 - \frac{1}{n} \sum_{i=1}^{k} \sum_{j=1}^{m} y_{ij}^2 = SS_1 - \frac{1}{2} \sum_{i=1}^{4} \sum_{j=1}^{4} y_{ij}^2$$
$$= 32\,916.43 - 65\,724.61 \div 2 = 54.13$$

12. Find the total sum of squares

$$SS_{\text{tot}} = SS_1 - SS_4 = 32\,916.43 - 20\,295.47 = 11\,990.96$$

13. Find the sum of squares for the interaction term

$$SS_{AB} = SS_{\text{tot}} - SS_A - SS_B - SS_e$$
$$= 11\,990.96 - 1\,704.48 - 2\,645.25 - 54.13 = 7\,587.11$$

14. Find the respective variances:

$$s_A^2 = SS_A/(k-1) = 1\,704.48/(4-1) = 568.16$$
$$s_B^2 = SS_B/(m-1) = 2\,645.25/(4-1) = 881.75$$
$$s_e^2 = SS_e/mk(n-1) = 54.13/[4\times 4(2-1)] = 3.38$$
$$s_{AB}^2 = SS_{AB}/(k-1)(m-1) = 7\,587.11/[(4-1)(4-1)] = 843.01$$

The results are presented in the ANOVA table that follows.

Source of variation	Degrees of freedom	Sum of squares	Mean square
A	3	1 704.48	568.16
B	3	2 645.25	881.75
AB	9	7 587.11	843.01
Error	16	54.13	3.38
Total	31	11 990.97	

Evaluate the significance of the linear effects A and B and the interaction effect by applying a Fisher (F) test. The variance ratio for effect A is

$$F = s_A^2/s_e^2 = 568.16/3.38 = 168.09$$

and for effect B

$$F = s_B^2/s_e^2 = 881.75/3.38 = 260.87$$

The tabulated value of the F test statistic for a significance level $\alpha = 0.05$ and for degrees of freedom $\nu_1 = 3$ and $\nu_2 = 16$ is $F_{0.95}$ (3, 16) = 3.2. Since the computed variance ratios are larger than the tabulated value, factors A and B are both significant. In other words, the polymer yield depends in a significant manner on the type of solvent and alkyl used.

To evaluate the significance of the interaction term, we set up a variance ratio

$$F = s_{AB}^2/s_e^2 = 843.01/3.38 = 249.41$$

The tabulated value for the F statistic at $\alpha = 0.05$ and degrees of freedom $\nu_1 = 9$ and $\nu_2 = 16$ is $F_{0.95}$ (9, 16) = 2.65. Therefore,

$$s_{AB}^2/s_e^2 > F_{\text{tabulated}}$$

Hence, the interaction effect should be taken as significant.

To sum up, the degree to which the type of solvent affects the process of polymerization depends on which alkyl is used. By the same token, the effect of an alkyl varies from solvent to solvent.

3.4. Experimental Design for ANOVA. Latin and Hyper-Graeco-Latin Squares

In evaluating the effect of two factors on a process, we have found that the required number N of observations (without replication) is equal to the product of the factors investigated. If the factors all have the same number n of levels, the size of experiment for a two-factor ANOVA is $N = n^2$. With this number of observations, the experiment will cover all likely combinations of factors. Such an experiment is referred to as a *factorial experiment*. An experiment in which some of level combinations are omitted is referred to as a *fractional factorial experiment*.

Omission of some level combinations always leads to the loss of some information. Therefore, it is important to design a fractional factorial experiment so that the information lost is the least significant in a given case. Most frequently, use is made of fractional factorial experiments in which the information lost is related to the interaction of the factors. This is legitimate in cases where the interaction effect is known in advance to be nil or so insignificant that it may be ignored.

Consider a three-way ANOVA with the same number n of levels for each factor. If we were to try all likely combinations of factor levels, the number of observations we would have to make would be:

$$N = n^3 \tag{3.82}$$

The number of observations can be reduced, if we use a fractional factorial design called a *Latin Square design* proposed for the first time by Fisher. An $n \times n$ Latin Square is a square table composed of n elements (numbers or letters) in such a way that each element appears once and only once in each row (position) and once and only once in each column. Three elements form a 3×3 Latin Square:

$$\begin{matrix} A & B & C \\ B & C & A \\ C & A & B \end{matrix} \tag{3.83}$$

Four elements form a 4×4 Latin Square:

$$\begin{matrix} A & B & C & D \\ B & C & D & A \\ C & D & A & B \\ D & A & B & C \end{matrix} \tag{3.84}$$

Latin Squares in which the first row and the first column are arranged in alphabetic order (in the case of letters) or as a natural series (in the case of numbers) are called *standard* or *canonical*. For example, the Latin Squares defined by (3.83) and (3.84) are standard. They are obtained by single-step cyclic permutation in which the second row is obtained by placing the first element of the first row at the end of the row, the third row is obtained by placing the first element of the second row at the end of the second row, etc. Single-step cyclic permutation is the simplest way to construct a Latin Square. In the general case, an $n \times n$ Latin Square can be constructed by $n-1$ single-step cyclic permutations. The number of likely Latin Squares depends on the size of the square, and for $n > 3$ the number is very large. For example we can construct 576 Latin Squares of 4×4 size, and 161 280 Latin Squares of 5×5 size.

Latin Square design is resorted to in evaluating the effect of three factors A, B and C on a given process. Factors A and B may be related to the investigation itself, and the third factor C may be, say, the inhomogeneity of the material.

In a Latin Square design, all the three factors have the same number of levels: a_i, b_i, and c_i. For example, in the design illustrated in Table 3.7 each factor has two levels.

Table 3.7 illustrates a 2^2 factorial experiment on which a 2×2 Latin Square is superimposed. The design matrix, that is, the experimental design corresponding to Table 3.7, consisting of three columns and four rows, is shown in Table 3.8.

Table 3.7. 2×2 Latin Square

A	B	
	b_1	b_2
a_1	c_1	c_2
a_2	c_2	c_1

Table 3.8. Experimental Design ($n = 2, N = 4$)

Position	A	B	C	y
1	a_1	b_1	c_1	y_1
2	a_1	b_2	c_2	y_2
3	a_2	b_1	c_2	y_3
4	a_2	b_2	c_1	y_4

A Latin Square is part of an experimental design—a Latin Square design involves a third factor, *C*. Yet, it is customary to apply the name "Latin Square" to a design like that shown in Table 3.7. Since here each element appears once and only once in each row and column, so, whatever the disturbing properties of a square element, they will equally tell on the means for the columns and rows.

The design shown in Table 3.8 is half of the full two-level factorial design for three factors, 2^3 (Table 3.9.). Quite appropriately it is called the *half-replicate* of a full fractional design. The observations included in the half-replicate are marked with an asterisk.

Table 3.9. Full Two-Level Design for Three Factors (2^3 Factorial Design)

	a_1		a_2	
	b_1	b_2	b_1	b_2
c_1	*			*
c_2		*	*	

The model for a full factorial design is

$$y_{ijq} = \mu+\alpha_i+\beta_j+\gamma_q+\alpha_i\beta_j+\alpha_i\gamma_q+\beta_j\gamma_q+\alpha_i\beta_j\gamma_q+\varepsilon_{ijq} \tag{3.85}$$

In addition to linear effects, the model defined by Eq. (3.85) contains three pairwise interaction effects and one triple interaction effect. The reduction of observations in a fractional replicate (see Table 3.7) results in that the linear effects are *confounded* with the interaction effects:

— effect A is confounded with BC interaction
— effect B is confounded with AC interaction
— effect C is confounded with AB interaction.

In using a Latin Square, it is usual to presume that the interaction effects are negligible. Then the following linear model can be written for the experiment:

$$y_{ijq} = \mu+\alpha_i+\beta_j+\gamma_q+\varepsilon_{ijq} \tag{3.86}$$

An example of a 3×3 Latin Square design is shown in Table 3.10.

Table 3.10. 3×3 Latin Square

A	B_j			Row totals
	b_1	b_2	b_3	
a_1	c_1 y_1	c_2 y_2	c_3 y_3	A_1
a_2	c_2 y_4	c_3 y_5	c_1 y_6	A_2
a_3	c_3 y_7	c_1 y_8	c_2 y_9	A_3
Column totals	B_1	B_2	B_3	

From a structural point of view, a 3×3 Latin Square may be regarded as a one-third replicate of a full 3^3 factorial design. In the general case, an $n\times n$ Latin Square may be treated as a $1/n$th replicate of an n^3 full factorial design.

In applying the ANOVA to a Latin Square experiment without replications, it is convenient to proceed by the following sequence:

(1) Find the totals for rows (A_i), for columns (B_j), and for C_q. As an example, for the 3×3 Latin Square illustrated in Table 3.10 the row totals are:

$$A_1 = y_1+y_2+y_3$$

$$A_2 = y_4+y_5+y_6$$

$$A_3 = y_7+y_8+y_9$$

the column totals are:

$$B_1 = y_1+y_4+y_7$$
$$B_2 = y_2+y_5+y_8$$
$$B_3 = y_3+y_6+y_9$$

and the totals for C_q are:

$$C_1 = y_1+y_6+y_8$$
$$C_2 = y_2+y_4+y_9$$
$$C_3 = y_3+y_5+y_7$$

(2) find the sum of squares of all observations

$$SS_1 = \sum_{i=1}^{n} \sum_{j=1}^{n} y_{ij}^2 \tag{3.87}$$

(3) find the sum of squares for the row totals, divided by the number of observations per row (the row mean square):

$$SS_2 = \frac{1}{n} \sum_{i=1}^{n} A_i^2 \tag{3.88}$$

(4) find the sum of squares for the column totals, divided by the number of observations per column (the column mean square):

$$SS_3 = \frac{1}{n} \sum_{j=1}^{n} B_j^2 \tag{3.89}$$

(5) find the sum of squares for the C totals, divided by the number of observations per C:

$$SS_4 = \frac{1}{n} \sum_{q=1}^{n} C_q^2 \tag{3.90}$$

(6) find the correction term defined as the grand total squared and divided by the number of all observations

$$SS_5 = \frac{1}{n^2}\left(\sum_{i=1}^{n} A_i\right)^2 = \frac{1}{n^2}\left(\sum_{j=1}^{n} B_j\right)^2 = \frac{1}{n^2}\left(\sum_{q=1}^{n} C_q\right)^2 \tag{3.91}$$

(7) find the sum of squares for a row

$$SS_A = SS_2-SS_5 \tag{3.92}$$

(8) find the sum of squares for a column

$$SS_B = SS_3-SS_5 \tag{3.93}$$

(9) find the sum of squares for a C

$$SS_C = SS_4-SS_5 \tag{3.94}$$

(10) find the total sum of squares defined as the difference between the sum of squares of all observations and the correction term

$$SS_{\text{tot}} = SS_1 - SS_5 \tag{3.95}$$

(11) find the residual sum of squares

$$\begin{aligned} SS_{\text{res}} &= SS_{\text{tot}} - SS_A - SS_B - SS_C \\ &= SS_1 - SS_5 - SS_2 + SS_5 - SS_3 + SS_5 - SS_4 + SS_5 \\ &= SS_1 - SS_2 - SS_3 - SS_4 + 2SS_5 \end{aligned} \tag{3.96}$$

The residual sum of squares comprises the variance due to the measurement error and the variance due to the factor interactions if there are any:

(12) find s_A^2

$$s_A^2 = SS_A/(n-1) \tag{3.97}$$

(13) find s_B^2

$$s_B^2 = SS_B/(n-1) \tag{3.98}$$

(14) find s_C^2

$$s_C^2 = SS_C(n-1) \tag{3.99}$$

(15) find s_e^2

$$s_e^2 = SS_{\text{res}}/(n-1)(n-2) \tag{3.100}$$

These data are then presented in an ANOVA table, such as Table 3.11.

Table 3.11. Latin Square ANOVA (without Replications)

Source of variation	Degrees of freedom	Sum of squares	Mean square	Expected mean square
A	$n-1$	$SS_A = SS_2 - SS_5$	$s_A^2 = \dfrac{SS_A}{n-1}$	$n\sigma_A^2 + \sigma_e^2$
B	$n-1$	$SS_B = SS_3 - SS_5$	$s_B^2 = \dfrac{SS_B}{n-1}$	$n\sigma_B^2 + \sigma_e^2$
C	$n-1$	$SS_C = SS_4 - SS_5$	$s_C^2 = \dfrac{SS_C}{n-1}$	$n\sigma_C^2 + \sigma_e^2$
Residue (error)	$(n-1)(n-2)$	$SS_{res} = SS_1 - SS_2 - SS_3 - SS_4 + 2SS_5$	$s_e^2 = \dfrac{SS_{res}}{(n-1)(n-2)}$	σ_e^2
Totals	n^2-1	$SS_{tot} = SS_1 - SS_5$		

The significance of the linear effects is tested by applying the Fisher (F) test. If the variance ratios are such that

$$\begin{aligned} s_A^2/s_e^2 &< F_{1-\alpha}(\nu_1, \nu_2) \\ s_B^2/s_e^2 &< F_{1-\alpha}(\nu_1, \nu_2) \\ s_C^2/s_e^2 &< F_{1-\alpha}(\nu_1, \nu_2) \end{aligned} \tag{3.101}$$

where α is the significance level, $\nu_1 = n-1$, and $\nu_2 = (n-1)(n-2)$ then we accept the null hypotheses: $\alpha_i = 0$, $\beta_j = 0$, and $\gamma_q = 0$. If any of the above variance ratios exceeds the tabulated value of the F test statistic, the respective null hypothesis is rejected, and the effect of the factor is taken as significant. On accepting the hypothesis about the significance of a factor, that is, the hypothesis about the significance of difference in the means, the next step usually is to find out which means significantly differ between them by applying the Student (t) test or the Duncan multiple range test. If, on the other hand, one or two factors are, under the statement of the problem, sources of inhomogeneity whose effect is to be excluded in evaluating the principal effect (this is achieved by arranging a Latin Square design), the means over the sources are not determined and no statistic tests are applied to test the significance in their difference.

Example 3.3. A Latin Square design was used to evaluate the effect of three factors on an organic synthesis process, namely: (A) the type of an haloid alkyl at levels a_1, a_2, a_3, and a_4; (B) the type of solvent at levels b_1, b_2, b_3, b_4; and (C) the monomer-to-solvent ratio. The percent yields of the polymer are given in the accompanying table. The experiment was carried out without replications. Evaluate the effect of the three factors on the synthesis process.

A	B: b_1	b_2	b_3	b_4	Row totals
a_1	c_1 13.2	c_2 27	c_3 49.1	c_4 7.2	72.2
a_2	c_2 19.0	c_3 8.0	c_4 15.5	c_1 9.5	52.0
a_3	c_3 4.6	c_4 5.9	c_1 31.5	c_2 53.1	95.1
a_4	c_4 14.7	c_1 16.3	c_2 60.9	c_3 55.2	147.1
Column totals	51.5	32.9	157.0	125	

Solution. We shall follow the sequence outlined above, using Eqs. (3.88) through (3.100). The totals for rows A_1, A_2, A_3 and A_4, and also for columns B_1, B_2, B_3 and B_4 are presented in the above table. We proceed as follows.

(1) Find the total for Cs:

$$C_1 = 70.5, \quad C_2 = 135.7, \quad C_3 = 116.9, \quad C_4 = 43.3$$

(2) Find the sum of squares of all observations

$$SS_1 = \sum_{i=1}^{4} \sum_{j=1}^{4} y_{ij}^2 = 13.2^2 + 2.7^2 + \ldots + 55.2^2 = 14\,505.14$$

(3) Find the row mean square (defined as the sum of row totals squared and divided by the number of observations in a row)

$$SS_2 = \frac{1}{4}(72.2^2 + 52.0^2 + 95.1^2 + 147.1^2) = 9\,649.82$$

(4) Find the column mean square (defined as the sum of column totals squared and divided by the number of observations in a column)

$$SS_3 = \frac{1}{4}(51.5^2 + 32.9^2 + 157.0^2 + 125^2) = 11\,002.16$$

(5) Find the mean square for Cs (defined as the sum of totals for Cs squared and divided by the number of observations associated with each C)

$$SS_4 = \frac{1}{4}(70.5^2 + 135.7^2 + 116.9^2 + 43.3^2) = 9\,731.31$$

(6) Find the correction terms (defined as the grand total squared and divided by the total number of observations)

$$SS_5 = \frac{1}{4 \times 4}(72.2 + 52.0 + 95.1 + 147.1)^2 = (1/16)(51.5 + 32.9 + 157 + 125)^2$$
$$= (1/16)(70.5 + 135.7 + 116.9 + 43.3)^2 = 8\,390.56$$

(7) Find the sum of squares for a row

$$SS_A = SS_2 - SS_5 = 9\,649.82 - 8\,390.56 = 1\,259.26$$

(8) Find the sum of squares for a column

$$SS_B = SS_3 - SS_5 = 11\,002.16 - 8\,390.56 = 2\,611.60$$

(9) Find the sum of squares for C

$$SS_C = SS_4 - SS_5 = 9\,731.31 - 8\,390.56 = 1\,340.75$$

(10) Find the total sum of squares

$$SS_{tot} = SS_1 - SS_5 = 14\,505.14 - 8\,390.56 = 6\,114.58$$

(11) Find the residual sum of squares

$$SS_{res} = SS_{tot} - SS_A - SS_B - SS_C$$
$$= 6\,114.58 - 1\,259.26 - 2\,611.60 - 1\,340.75 = 902.97$$

(12) Find the variance for a row

$$s_A^2 = SS_A/3 = 1\,259.26 \div 3 = 419.75$$

(13) Find the variance for a column

$$s_B^2 = 2\,611.60 \div 3 = 870$$

(14) Find the variance for factor C

$$s_C^2 = 1\,340.75 \div 3 = 446.92$$

(15) Find the error variance

$$s_e^2 = 902.97 \div 6 = 150.5$$

The ANOVA table for the above data is

Source of variation	Degrees of freedom	Sum of squares	Mean square
A	3	1 259.28	419.75
B	3	2 611.60	870
C	3	1 340.75	446.92
Error	6	902.97	150.5
Totals	15	6 114.58	

The effects of factors A, B and C are tested for significance by applying the Fisher (F) test. The variance ratio for A is

$$F = s_A^2/s_e^2 = 419.75 \div 150.5 = 2.64$$

For B, it is

$$F = s_B^2/s_e^2 = 870 \div 150.5 = 5.62$$

For C, it is

$$F = s_C^2/s_e^2 = 446.92 \div 150.5 = 2.88$$

Entering a table of values for the F test statistic, we find that for $\alpha = 0.05$, $\nu_1 = 3$ and $\nu_3 = 6$, the tabulated value of the statistic is $F_{0.95}$ (3, 6) = 4.8. From a comparison with the computed values it follows that the effects of factors A and C should be taken as insignificant. The process is only affected by factor B, because

$$F = s_B^2/s_e^2 > F_{\text{tabul}}$$

Now let us compare the effects of factor B at the various levels by applying the Duncan multiple range test (see Sec. 2.14). The mean yields of the polymer for the various solvents are:

Solvent	b_1	b_2	b_3	b_4
Yield, $\bar{y}$	12.87	8.24	39.25	31.25

Arrange the means from low to high

b_2	b_1	b_4	b_3
$\bar{y}_2 = 8.24$	$\bar{y}_1 = 12.87$	$\bar{y}_4 = 31.25$	$\bar{y}_3 = 39.25$

Referring to the table above, the error variance for $\nu = 6$ is $s_{\bar{y}}^2 = 150.5$. The standard error of the mean is

$$s_{\bar{y}} = \sqrt{150.5/4} = 6.13$$

Entering Appendix Table 10 and using $\nu = 6$, and $p = 2, 3, 4$, list three significant ranges:

Ranges, r.......... 3.46 3.58 3.64

Multiply the above ranges by the standard error of the mean to form three least significant ranges:

$r \times s_{\bar{y}}$ 21.4 21.9 22.3

Test the observed ranges between means in order to evaluate the signifiance of difference between the various solvents:

$\bar{y}_3-\bar{y}_2 = 39.25- \ 8.24 = 31.01 > 22.3$ a significant difference

$\bar{y}_3-\bar{y}_1 = 39.25-12.87 = 26.38 > 21.9$ a significant difference

$\bar{y}_3-\bar{y}_4 = 39.25-31.25 = \ 8.00 < 21.4$ an insignificant difference

$\bar{y}_4-\bar{y}_2 = 31.25- \ 8.24 = 23.01 > 21.9$ a significant difference

$\bar{y}_4-\bar{y}_1 = 31.25-12.87 = 18.38 < 21.4$ an insignificant difference

$\bar{y}_1-\bar{y}_2 = 12.87- \ 8.24 = \ 4.63 < 21.4$ an insignificant difference

The ANOVA set forth above holds for a linear model. However, the goodness of fit of the adopted model cannot be tested in the absence of replications. If we carry out the same number of replications in each cell of the Latin Square, we shall be able to evaluate the significance of interactions between the factors. In the circumstances, the replications are only used to evaluate the measurement error (the error within the treatments). If the interaction effects are insignificant (the linear model fits the data), the residual variance insignificantly differs from the variance of the uncertainty due to the measurement error. Also, the significance of the linear effects can readily be tested. If, in contrast, the linear model does not fit the data and there are interactions between the factors, we cannot evaluate the significance of the linear effects, because they are confounded with the interaction effects. A way out is to set up some further hypotheses about the insignificance of some interactions.

Latin Square designs apply to cases involving three factors. For cases involving four factors, the design that serves its purpose better is a *Graeco-Latin square*. The problem is to add a fourth factor of interest to the three already present without changing the total number of observations, n^2. This can be done by finding an arrangement for the levels of factors C and D when each row and each column contains all n levels of factor C and all n levels of factor D, but also no two levels of factors C, and D appear more than once anywhere in the table. This form of arrangement is referred to as a *second-order Latin Square*. It is produced by combining two *orthogonal Latin Squares*.

Consider two Latin Squares one composed of Latin and the other of Greek letters.

$$\begin{array}{ccccc} & & \text{I} & & \\ A & B & C & D & E \\ C & D & E & A & B \\ E & A & B & C & D \\ B & C & D & E & A \\ D & E & A & B & C \end{array} \qquad \begin{array}{ccccc} & & \text{II} & & \\ \alpha & \beta & \gamma & \delta & \varepsilon \\ \delta & \varepsilon & \alpha & \beta & \gamma \\ \beta & \gamma & \delta & \varepsilon & \alpha \\ \varepsilon & \alpha & \beta & \gamma & \delta \\ \gamma & \delta & \varepsilon & \alpha & \beta \end{array} \tag{3.102}$$

If we superimpose one square on the other so as to form a third square in which each cell contains one Latin and one Greek letter from the respective cells of the original squares, we shall obtain

$$\begin{array}{ccccc} A\alpha & B\beta & C\gamma & D\delta & E\varepsilon \\ C\delta & D\varepsilon & E\alpha & A\beta & B\gamma \\ E\beta & A\gamma & B\delta & C\varepsilon & D\alpha \\ B\varepsilon & C\alpha & D\beta & E\gamma & A\delta \\ D\gamma & E\delta & A\varepsilon & B\alpha & C\beta \end{array} \tag{3.103}$$

In this square, each letter from one original square is connected once and only once with each letter from the other original square. Such two Latin Squares are called *orthogonal*. Another name for the resultant second-order square is a Graeco-Latin square. The problem of finding orthogonal Latin squares has not yet been solved in combinatorial mathematics. It has been proved that orthogonal Latin squares do exist for $n = 3, 4, 5, 7, 8$, and 9. It is known that they do not exist for $n = 6$. Therefore, for $n = 6$ we can construct an ordinary Latin square but we cannot construct a second-order square. The Latin square for $n = 10$ has not been investigated. If we have $k = n-1$ pairwise orthogonal Latin squares, they form what is called a *complete system of orthogonal Latin squares*. It has been shown that complete systems of orthogonal Latin squares exist for $n = p$ (where p is a prime number) and $n = p^a$ (that is, the power of the prime number). A complete system of orthogonal Latin squares for $n = p$ (where p is a prime number) can be constructed, using Galois fields. Let us, for example, construct a Galois field for modulo 5 residues. Two integers a and b are congruent modulo 5 if $a-b = \lambda 5$, where λ is any integer. Symbolically,

$$a \equiv b \pmod 5 \tag{3.104}$$

The congruence defined by Eq. (3.104) defines a field. This field contains five distinct elements 0, 1, 2, 3, and 4. Let us write a table of addition and a table of multiplication for this field:

$$\begin{array}{ccccc} \multicolumn{5}{c}{\textit{Addition}} \\ 0 & 1 & 2 & 3 & 4 \\ 1 & 2 & 3 & 4 & 0 \\ 2 & 3 & 4 & 0 & 1 \\ 3 & 4 & 0 & 1 & 2 \\ 4 & 0 & 1 & 2 & 3 \end{array} \qquad \begin{array}{cccc} \multicolumn{4}{c}{\textit{Multiplication}} \\ 1 & 2 & 3 & 4 \\ 2 & 4 & 1 & 3 \\ 3 & 1 & 4 & 2 \\ 4 & 3 & 2 & 1 \\ & & & \end{array} \tag{3.105}$$

Consider the square formed by the table of addition. If, in this square, we replace the pth row beginning with the element p ($p = 0, 1, 2, 3, 4$) by a row obtained by adding modulo 5 the number $p \times 2$ to the elements of the first row in the first square, we shall obtain a second square

$$\begin{array}{lcccccc} & \multicolumn{6}{c}{k = 2} \\ 2\times 0 = 0 & 0 & 0 & 1 & 2 & 3 & 4 \\ 2\times 1 = 2 & 2 & 2 & 3 & 4 & 0 & 1 \\ 2\times 2 = 4 & 4 & 4 & 0 & 1 & 2 & 3 \\ 2\times 3 = 1 & 1 & 1 & 2 & 3 & 4 & 0 \\ 2\times 4 = 3 & 3 & 3 & 4 & 0 & 1 & 2 \end{array} \tag{3.106}$$

In order to obtain the pth row in the third and fourth Latin squares we should add (mod. 5) the numbers $p\times 3$ and $p\times 4$, respectively, to the elements of the first row in the first square:

$$k = 3$$

$$\begin{array}{lc|ccccc} 3\times 0 = 0 & 0 & 0 & 1 & 2 & 3 & 4 \\ 3\times 1 = 3 & 3 & 3 & 4 & 0 & 1 & 2 \\ 3\times 2 = 1 & 1 & 1 & 2 & 3 & 4 & 0 \\ 3\times 3 = 4 & 4 & 4 & 0 & 1 & 2 & 3 \\ 3\times 4 = 2 & 2 & 2 & 3 & 4 & 0 & 1 \end{array} \tag{3.107}$$

$$k = 4$$

$$\begin{array}{lc|ccccc} 4\times 0 = 0 & 0 & 0 & 1 & 2 & 3 & 4 \\ 4\times 1 = 4 & 4 & 4 & 0 & 1 & 2 & 3 \\ 4\times 2 = 3 & 3 & 3 & 4 & 0 & 1 & 2 \\ 4\times 3 = 2 & 2 & 2 & 3 & 4 & 0 & 1 \\ 4\times 4 = 1 & 1 & 1 & 2 & 3 & 4 & 0 \end{array} \tag{3.108}$$

Thus, we have obtained a complete system of orthogonal Latin squares.

A Graeco-Latin square design is applicable to experiments involving four factors. The number of levels must be the same for all the four factors. Table 3.12 shows Graeco-Latin squares of sizes 3×3, 4×4, and 5×5.

A Graeco-Latin square is part of a four-factorial experimental design—the design is extended to include factors C and D. In the 5×5 design of Table 3.12, the levels of factor C correspond to the Latin letters and the levels of factor D, to the Greek letters of the Graeco-Latin square in (3.103), namely: $A-c_1$, $B-c_2$, $C-c_3$, $D-c_4$, $E-c_5$, and $\alpha-d_1$, $\beta-d_2$, $\gamma-d_3$, $\delta-d_4$, $\varepsilon-d_5$. However, it is customary to call the entire four-factorial design a Graeco-Latin square design. The design matrix corresponding to a 3×3 Graeco-Latin square is shown in Table 3.13.

In a Graeco-Latin square design, there are n^2 various combinations of factor levels instead of n^4 combinations in a full factorial design. Therefore, a Graeco-Latin square design is a $1/n^2$th replicate of a full factorial design. For example, the 3×3 Graeco-Latin square in Table 3.12 is a 1/9th replicate of a 3^4 full factorial design ($N = 81$), the 4×4 Graeco-Latin square in the same table is a 1/16th replicate of a 4^4 full factorial design ($N = 256$), and the 5×5 Graeco-Latin square is a 1/25th replicate of a 5^4 full factorial design ($N = 625$).

The ANOVA is applied to a Graeco-Latin square design in the same manner as to an ordinary Latin square, except that there is the fourth factor, D, to be taken into consideration. The sum of squares for factor D has $n-1$ degrees of freedom. The number of degrees of freedom for the residual sum of squares, defined as before as the difference between

Table 3.12. Graeco-Latin Squares

3×3

A \ B	b_1	b_2	b_3
a_1	c_1 d_1	c_2 d_2	c_3 d_3
a_2	c_2 d_3	c_3 d_1	c_1 d_2
a_3	c_3 d_2	c_1 d_3	c_2 d_1

4×4

A \ B	b_1	b_2	b_3	b_4
a_1	c_1 d_1	c_2 d_2	c_3 d_3	c_4 d_4
a_2	c_2 d_3	c_1 d_4	c_4 d_1	c_3 d_2
a_3	c_3 d_4	c_4 d_3	c_1 d_2	c_2 d_1
a_4	c_4 d_2	c_3 d_1	c_2 d_4	c_1 d_3

5×5

A \ B	b_1	b_2	b_3	b_4	b_5
a_1	c_1 d_1	c_2 d_2	c_3 d_3	c_4 d_4	c_5 d_5
a_2	c_3 d_4	c_4 d_5	c_5 d_1	c_1 d_2	c_2 d_3
a_3	c_5 d_2	c_1 d_3	c_2 d_4	c_3 d_5	c_4 d_1
a_4	c_2 d_5	c_3 d_1	c_4 d_2	c_5 d_3	c_1 d_4
a_5	c_4 d_3	c_5 d_4	c_1 d_5	c_2 d_1	c_3 d_2

the total sum of squares and the sums of squares of all factors, is $(n-1) \times (n-3)$. If we superimpose on one another three orthogonal Latin squares, we shall obtain a 3rd-order Latin square. If we superimpose n orthogonal squares, we shall obtain an nth order Latin square. Another name for such squares is hyper-Graeco-Latin squares.

Table 3.13. Experimental Design ($n = 3, N = 9$)

Observation No.	A	B	C	D	y	Observation No.	A	B	C	D	y
1	a_1	b_1	c_1	d_1	y_1	6	a_2	b_3	c_1	d_2	y_6
2	a_1	b_2	c_2	d_2	y_2	7	a_3	b_1	c_3	d_2	y_7
3	a_1	b_3	c_3	d_3	y_3	8	a_3	b_2	c_1	d_3	y_8
4	a_2	b_1	c_2	d_3	y_4	9	a_3	b_3	c_2	d_1	y_9
5	a_2	b_2	c_3	d_1	y_5						

At n levels, the number of factors that can be introduced in an experimental design is $n+1$. Then the residual sum of squares will have zero degrees of freedom. Such designs are called saturated. Let us construct a saturated design for $n = 5$. To do this, we superimpose the four 5×5 orthogonal Latin squares [see Eqs. (3.105) through (3.108)] that form a complete series

Table 3.14. A 4th-order Hyper-Graeco-Latin Square

A	B = 0	1	2	3	4
0	$C = 0$ $D = 0$ $E = 0$ $F = 0$	$C = 1$ $D = 1$ $E = 1$ $F = 1$	$C = 2$ $D = 2$ $E = 2$ $F = 2$	$C = 3$ $D = 3$ $E = 3$ $F = 3$	$C = 4$ $D = 4$ $E = 4$ $F = 4$
1	$C = 1$ $D = 2$ $E = 3$ $F = 4$	$C = 2$ $D = 3$ $E = 4$ $F = 0$	$C = 3$ $D = 4$ $E = 0$ $F = 1$	$C = 4$ $D = 0$ $E = 1$ $F = 2$	$C = 0$ $D = 1$ $E = 2$ $F = 3$
2	$C = 2$ $D = 4$ $E = 1$ $F = 3$	$C = 3$ $D = 0$ $E = 2$ $F = 4$	$C = 4$ $D = 1$ $E = 3$ $F = 0$	$C = 0$ $D = 2$ $E = 4$ $F = 1$	$C = 1$ $D = 3$ $E = 0$ $F = 2$
3	$C = 3$ $D = 1$ $E = 4$ $F = 2$	$C = 4$ $D = 2$ $E = 0$ $F = 3$	$C = 0$ $D = 3$ $E = 1$ $F = 4$	$C = 1$ $D = 4$ $E = 2$ $F = 1$	$C = 2$ $D = 0$ $E = 3$ $F = 2$
4	$C = 4$ $D = 3$ $E = 2$ $F = 1$	$C = 0$ $D = 4$ $E = 3$ $F = 2$	$C = 1$ $D = 0$ $E = 4$ $F = 3$	$C = 2$ $D = 1$ $E = 0$ $F = 4$	$C = 3$ $D = 2$ $E = 1$ $F = 0$

of 5×5 orthogonal Latin squares (Table 3.14). The original Latin square, (3.105), defines the levels of factor *C*, the second square (3.106) defines the levels of factor *D*, etc. The factor levels are designated by numerals. The corresponding experimental design for six factors is given in Table 3.15.

Table 3.15. 5^5 Experimental Design ($n = 5, N = 25$)

Observation No.	*A*	*B*	*C*	*D*	*E*	*F*	Observation No.	*A*	*B*	*C*	*D*	*E*	*F*
1	0	0	0	0	0	0	14	2	3	0	2	4	1
2	0	1	1	1	1	1	15	2	4	1	3	0	2
3	0	2	2	2	2	2	16	3	0	3	1	4	2
4	0	3	3	3	3	3	17	3	1	4	2	0	3
5	0	4	4	4	4	4	18	3	2	0	3	1	4
6	1	0	1	2	3	4	19	3	3	1	4	2	0
7	1	1	2	3	4	0	20	3	4	2	0	3	1
8	1	2	3	4	0	1	21	4	0	4	3	2	1
9	1	3	4	0	1	2	22	4	1	0	4	3	2
10	1	4	0	1	2	3	23	4	2	1	0	4	3
11	2	0	2	4	1	3	24	4	3	2	1	0	4
12	2	1	3	0	2	4	25	4	4	3	2	1	0
13	2	2	4	1	3	0							

This is a saturated design, because the residual sum of squares has zero degrees of freedom: $v = (n-1)(n-k+1) = 0$, where *k* is the number of factors involved. It also a 1/625th replicate design of a 5^6 full factorial design. Such designs are ordinarily used at the early stages of process investigation when a multiplicity of qualitative factors has to be tried in order to select the most promising combinations for further study and to discard the unacceptable ones. Thus, the use of Graeco-Latin and hyper-Graeco-Latin squares as experimental designs leads to both a reduction in the number of observations and a simplification in the accompanying calculations.

The basic assumption underlying the use of Graeco-Latin and higher squares is the absence of interaction between the factors. As with Latin squares, the adopted linear model can be tested for goodness of fit only if replicate observations have been made.

3.5. Latin Cubes

A full factorial design for three factors, that is an n^3 design (where $n > 2$), can be represented by a cubic arrangement of *n* elements in n^3 positions. Three edges of the cube represent factors *A*, *B* and *C* at levels 0, 1, 2, ..., $n-1$ (Fig. 3.1). If we add a fourth factor, *D*, and arrange its levels (0, 1,

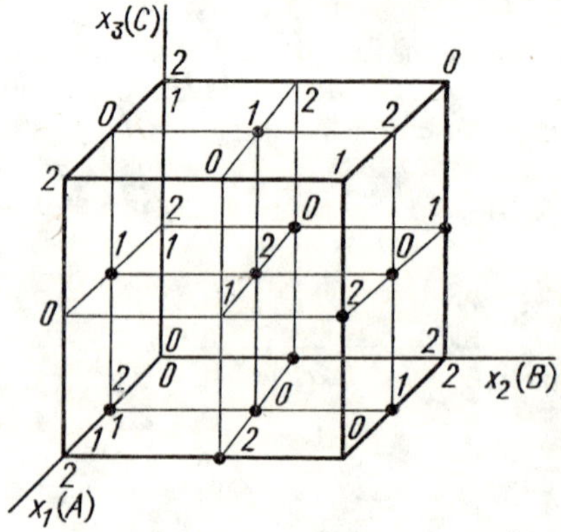

Fig. 3.1. First order Latin cube

2, ..., $n-1$) at the corresponding points of the cubic arrangement, the result will be a first-order Latin cube of size n.

A first-order Latin cube of size n can be represented by a cubic table of n elements arranged in n^3 positions such that each element appears in the table n^2 times and occurs in each of the $3n$ planes parallel with the coordinate planes x_1Ox_2, x_1Ox_3, and x_2Ox_3, n times, the same for all elements. To prove, the levels of the additional factor D (the elements of the Latin cube) occur in the design the same number of times, n^2_i and appear in each of the $3n$ coordinate planes (that is, with the levels of factors A, B and C) the

Table 3.16. First-Order 3×3×3 Latin Cube

	A	*B* = 0	*B* = 1	*B* = 2
C = 0	0	0	1	2
	1	2	0	1
	2	1	2	0
C = 1	0	2	0	1
	1	1	2	0
	2	0	1	2
C = 2	0	1	2	0
	1	0	1	2
	2	2	0	1

same number of times, n (Table 3.16). The corresponding design matrix for a Latin cube of sizes $n = 3$ and $r = 1$ is given in Table 3.17.

An experimental design based on a first-order Latin cube can cover four factors (A, B, C, and D). In contrast to a Graeco-Latin square design which also permits four factors to be studied, three factors (A, B, C) in a Latin cube are taken to be principal and the fourth factor (D) forms an eliminating

Table 3.17. 3^3 Experimental Design ($n = 3$, $N = 27$)

Observation No.	A	B	C	D	y	Observation No.	A	B	C	D	y
1	0	0	0	0	y_1	15	1	2	1	0	y_{15}
2	0	1	0	1	y_2	16	2	0	1	0	y_{16}
3	0	2	0	2	y_3	17	2	1	1	1	y_{17}
4	1	0	0	0	y_4	18	2	2	1	2	y_{18}
5	1	1	0	1	y_5	19	0	0	2	1	y_{19}
6	1	2	0	1	y_6	20	0	1	2	2	y_{20}
7	2	0	0	2	y_7	21	0	2	2	0	y_{21}
8	2	1	0	2	y_8	22	1	0	2	0	y_{22}
9	2	2	0	0	y_9	23	1	1	2	1	y_{23}
10	0	0	1	2	y_{10}	24	1	2	2	2	y_{24}
11	0	1	1	0	y_{11}	25	2	0	2	2	y_{25}
12	0	2	1	1	y_{12}	26	2	1	2	0	y_{26}
13	1	0	1	1	y_{13}	27	2	2	2	1	y_{27}
14	1	1	1	2	y_{14}						

group. In a Graeco-Latin square design, two factors, A and B, are principal, and the remaining two, C and D, form a double eliminating group. The number of observations in a Latin cube design is n times the number of observations in a Graeco-Latin square design.

A Latin cube design without replications is used, assuming that the process of interest can be described by a linear model:

$$y_{ijql} = \mu + \alpha_i + \beta_j + \gamma_q + \delta_l + \varepsilon_{ijql} \tag{3.108}$$

where μ is an overall mean, α_i is the effect due to factor A at ith level ($i = 0, 1, 2, \ldots, n-1$), β_j is the effect due to factor B at the jth level ($j = 0, 1, 2, \ldots, n-1$), γ_q is the effect due to factor C at the qth level ($q = 0, 1, 2, \ldots, n-1$), δ_l is the effect due to factor D at the lth level, and ε_{ijql} is the random error of experiment (within each set of observations).

A first-order Latin cube design can conveniently be statistically analyzed in the following sequence.

(1) Find the totals for all factors at each level:

$$A_i (i = 0, 1, 2, \ldots, n-1)$$
$$B_j (j = 0, 1, 2, \ldots, n-1)$$
$$C_q (q = 0, 1, 2, \ldots, n-1)$$
$$D_l (l = 0, 1, 2, \ldots, n-1)$$

With reference to the design illustrated in Table 3.17 we would have

$$A_0 = y_1 + y_2 + y_3 + y_{10} + y_{11} + y_{12} + y_{19} + y_{20} + y_{21}$$
$$A_1 = y_4 + y_5 + y_6 + y_{13} + y_{14} + y_{15} + y_{22} + y_{23} + y_{24}$$

$$A_2 = y_7+y_8+y_9+y_{16}+y_{17}+y_{18}+y_{25}+y_{26}+y_{27}$$
$$B_0 = y_1+y_4+y_7+y_{10}+y_{13}+y_{16}+y_{19}+y_{22}+y_{25}$$
$$B_1 = y_2+y_5+y_8+y_{11}+y_{14}+y_{17}+y_{20}+y_{23}+y_{26}$$
$$B_2 = y_3+y_6+y_9+y_{12}+y_{15}+y_{18}+y_{21}+y_{24}+y_{27}$$
$$C_0 = y_1+y_2+y_3+y_4+y_5+y_6+y_7+y_8+y_9$$
$$C_1 = y_{10}+y_{11}+y_{12}+y_{13}+y_{14}+y_{15}+y_{16}+y_{17}+y_{18}$$
$$C_2 = y_{19}+y_{20}+y_{21}+y_{22}+y_{23}+y_{24}+y_{25}+y_{26}+y_{27}$$
$$D_0 = y_1+y_5+y_9+y_{11}+y_{15}+y_{16}+y_{21}+y_{22}+y_{26}$$
$$D_1 = y_2+y_6+y_7+y_{12}+y_{13}+y_{17}+y_{19}+y_{23}+y_{27}$$
$$D_2 = y_3+y_4+y_8+y_{10}+y_{14}+y_{18}+y_{20}+y_{24}+y_{25}$$

(2) Find the sum of squares of all observations

$$SS_1 = \sum_{i=0}^{n-1} \sum_{j=0}^{n-1} \sum_{q=0}^{n-1} (y_{ijql})^2 \tag{3.109}$$

(3) Find the sum of squares of totals for factor A divided by n^2

$$SS_2 = \frac{1}{n^2} \sum_{i=0}^{n-1} A_i^2 \tag{3.110}$$

(4) Find the sum of squares of totals for factors B divided by n^2

$$SS_3 = \frac{1}{n^2} \sum_{j=0}^{n-1} B_j^2 \tag{3.111}$$

(5) Find the sum of totals for factor C squared and divided by n^2

$$SS_4 = \frac{1}{n^2} \sum_{q=0}^{n-1} C_q^2 \tag{3.112}$$

(6) Find the sum of totals for factor D, squared and divided by n^2

$$SS_5 = \frac{1}{n^2} \sum_{l=0}^{n-1} D_l^2 \tag{3.113}$$

(7) Find the correction term defined as the grand total squared and divided by the number of all observations

$$SS_6 = \frac{1}{n^3} \left(\sum_{i=0}^{n-1} A_i\right)^2 = \frac{1}{n^3} \left(\sum_{j=0}^{n-1} B_j\right)^2 = \frac{1}{n^3} \left(\sum_{q=0}^{n-1} C_q\right)^2$$
$$= \frac{1}{n^3} \left(\sum_{l=0}^{n-1} D_l\right)^2 \tag{3.114}$$

(8) Find the total sum of squares defined as the difference between the sum of squares of all observations and the correction term

$$SS_{tot} = SS_1 - SS_6 \tag{3.115}$$

(9) Find the sum of squares for factor A

$$SS_A = SS_2 - SS_6 \tag{3.116}$$

(10) Find the sum of squares for factor B

$$SS_B = SS_3 - SS_6 \tag{3.117}$$

(11) Find the sum of squares for factor C

$$SS_C = SS_4 - SS_6 \tag{3.118}$$

(12) Find the sum of squares for factor D

$$SS_D = SS_5 - SS_6 \tag{3.119}$$

(13) Find the residual sum of squares

$$SS_{\text{res}} = SS_{\text{tot}} - (SS_A + SS_B + SS_C + SS_D) \tag{3.120}$$

The ANOVA table for the above data is given in Table 3.18.

Table 3.18. ANOVA Table for a First-order Latin Cube Design (without Replications)

Source of variation	Degrees of freedom	Sum of squares	Mean square	Expected mean square
Factor A	$n-1$	SS_A	$\frac{SS_A}{n-1}$	$n\sigma_A^2+\sigma_e^2$
Factor B	$n-1$	SS_B	$\frac{SS_B}{n-1}$	$n\sigma_B^2+\sigma_e^2$
Factor C	$n-1$	SS_C	$\frac{SS_C}{n-1}$	$n\sigma_C^2+\sigma_e^2$
Factor D	$n-1$	SS_D	$\frac{SS_D}{n-1}$	$n\sigma_D^2+\sigma_e^2$
Residue	n^3-4n+3	SS_{res}	$\frac{SS_{\text{res}}}{n^3-4n+3}$	σ_e^2
Totals	n^3-1	SS_{tot}		

Two first-order Latin cubes of size n are orthogonal, if, on superimposition, each element of one cube appears with each element of the other cube n times. Two such orthogonal cubes superimposed on each other constitute a first-order Graeco-Latin cube of size n.

A first-order Graeco-Latin cube design permits an experiment to involve five factors. If we combine three and more orthogonal Latin cubes, we shall obtain a hyper-Graeco-Latin cube. A complete system of first-order orthogonal Latin cubes of size n that make up a completely orthogonal hyper-Graeco-Latin cube cannot contain more than n^2+n-2 cubes. The existence of such systems has been proved for n which is a prime number or an integer positive power of a prime number [11].

In first-order Latin cubes, all factors are set up at an equal number of levels, n, which is also the size of a cube, and all linear effects are determined with the same accuracy which is maximal for a given number of observa-

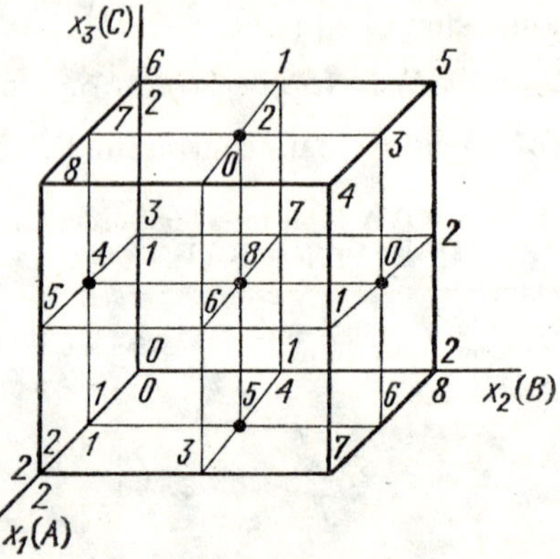

Fig. 3.2. Second order Latin cube

tions. In a second-order Latin cube, one factor is set up at n^2 levels, and all the remaining factors at n levels. A second-order Latin cube of size $n = 3$ is shown in Fig. 3.2. Factors A, B and C have each three levels: 0, 1 and 2; factor D has nine levels: 0, 1, 2, 3, 4, 5, 6, 7, and 8, arranged in a Latin cube. In tabular form, it is shown in Table 3.19.

Table 3.19. Second-order Latin Cube ($n = 3$, $r = 2$)

$C = 0$

A	B = 0	B = 1	B = 2
0	0	4	8
1	1	5	6
2	2	3	7

$C = 1$

A	B = 0	B = 1	B = 2
0	3	7	2
1	4	8	0
2	5	6	1

$C = 2$

A	B = 0	B = 1	B = 2
0	6	1	5
1	7	2	3
2	8	0	4

Latin cube designs can be very useful at the early stages of experimentation as a way of selecting an optimal combination of qualitative factors [11, 15, 16].

Example 3.4. A second-order Latin cube design was used in developing the compositions of a new high-density polyethylene-based polymer having an increased rigidity and ability to be thermally moulded. The experiment was concerned with a three-component system: high-density polyethylene, filler, and elastifier. The properties of the compositions were investigated with reference to three forms of elastifier systems and nine type of fillers, in which the amount of additive and the amount of filler was varied at three levels each. Type of additive, x_1: (1) No. 1, (2) No. 2 and (3) No. 3. Quantity of additive, x_2: (1) 3%, (2) 5% and (3) 10%. Quantity of filler, x_3: (1) 5%, (2) 10% and (3) 15%. Type of filler, x_4: talc, T(0), aeroseal, A (1), mica, M (2). T : A = 1 : 1 (3), T : A = 1 : 0.5 (4), T : A = 0.5 : 1 (5), A : M = 1 : 1 (6), A : M = 1 : 0.5 (7), A : M = 0.5 : 1 (8).

The experiment was run under laboratory conditions. The commercial value of the material was evaluated in terms of four responses: modulus of elasticity in bending y_1 in kgf cm^{-2}, ultimate stress in tension y_2 in kgf cm^{-2}, percent elongation in tension y_3, and a generalized dimensionless quality response (a generalized desirability function) D.

Solution. The experiment design and the results of specimen testing are given in Table 3.20 (see also Table 3.19).

Table 3.20. Second-order Latin Cube Design

Nos.	x_1	x_2	x_3	x_4	y_1, kgf cm^{-2}	y_2, kgf cm^{-2}	y_3, %	D
1	0	0	0	0	364	117	483	0.645
2	0	1	0	4	365	118	504	0.647
3	0	2	0	8	367	99	447	0.610
4	1	0	0	1	470	134	447	0.810
5	1	1	0	5	352	122	480	0.650
6	1	2	0	6	324	112	452	0.550
7	2	0	0	2	434	96	456	0.686
8	2	1	0	3	355	126	493	0.638
9	2	2	0	7	354	127	477	0.638
10	0	0	1	3	476	112	376	0.759
11	0	1	1	7	464	96	241	0.650
12	0	2	1	2	484	66	90	0.381
13	1	0	1	4	502	99	386	0.732
14	1	1	1	8	547	104	69	0.440
15	1	2	1	0	320	89	407	0.491
16	2	0	1	5	464	126	361	0.773
17	2	1	1	6	535	110	104	0.248
18	2	2	1	1	431	143	402	0.768
19	0	0	2	6	615	101	69	0.210
20	0	1	2	1	615	129	28	0.350
21	0	2	2	5	572	94	205	0.668
22	1	0	2	7	593	114	25	0.430
23	1	1	2	2	501	85	36	0.304
24	1	2	2	3	482	114	108	0.530
25	2	0	2	8	610	96	36	0.340
26	2	1	2	0	605	102	69	0.445
27	2	2	2	4	558	120	231	0.743

To identify the factors significantly affecting the quality responses, an ANOVA is applied, assuming a linear model, Eq. (3.108). The ANOVA sequence is as follows, with each operation being carried for each of the responses y_1, y_2, y_3, and D.

(1) Find the totals for each factor at all levels (Table 3.21) X_{1i} ($i = 0, 1, 2$), X_{2j} ($j = 0, 1, 2$), X_{3q} ($q = 0, 1, 2$), X_{4l} ($l = 0, 1, 2, \ldots, 8$).

(2) Find the sum of squares of all observations

$$SS_1 = \sum_{i=0}^{2} \sum_{j=0}^{2} \sum_{q=0}^{2} (y_{ijql})^2$$

For the modulus of elasticity in bending, y_1, it is

$$SS_1 = 364^2 + 365^2 + \ldots + 558^2 = 6\,275\,327$$

(3) Find the sum of totals for factor x_1 squared and divided by n^2

$$SS_2 = \frac{1}{n^2} \sum_{i=0}^{2} X_{1i}^2$$

For y_1 (see Table 3.21), this is

$$SS_2 = \frac{4\,322^2 + 4\,091^2 + 4\,346^2}{9} = 6\,033\,742$$

(4) Find the sum of totals for factor x_2, squared and divided by n^2

$$SS_3 = \frac{1}{n^2} \sum_{j=0}^{2} X_{2j}^2$$

For y_1 (see Table 3.21), it is

$$SS_3 = \frac{4\,528^2 + 4\,339^2 + 3\,892^2}{9} = 6\,053\,041$$

(5) Find the sum of totals for factor x_3, squared and divided by n^2

$$SS_4 = \frac{1}{n^2} \sum_{q=0}^{2} X_{3q}^2$$

For y_1 (see Table 3.21), it is

$$SS_4 = \frac{3\,385^2 + 4\,223^2 + 5\,151^2}{9} = 6\,202\,750$$

(6) Similarly, for factor x_4

$$SS_5 = \frac{1}{3} \sum_{l=0}^{8} X_{4l}^2$$

For y_1 (see Table 3.21), it is

$$SS_5 = \frac{1\,289^2 + 1\,516^2 + 1\,419^2 + \ldots + 1\,524^2}{3} = 6\,046\,886$$

(7) Find the correction term defined as the grand total squared and divided by the number of observations

$$SS_6 = \frac{1}{n^3} \left(\sum_{i=0}^{3} X_{1i}\right)^2 = \frac{1}{n^3} \left(\sum_{j=0}^{2} X_{2j}\right)^2 = \frac{1}{n^3} \left(\sum_{q=0}^{3} X_{3q}\right)^2 = \frac{1}{n^3} \left(\sum_{l=0}^{8} X_{4l}\right)^2$$

For y_1 (see Table 3.21), it is

$$SS_6 = \frac{(4\,322 + 4\,091 + 4\,346)^2}{27} = \frac{(4\,528 + 4\,339 + 3\,892)^2}{27}$$

$$= \frac{(1\,289 + \ldots + 1\,524)^2}{27} = 6\,029\,336$$

Table 3.21. Totals for Various Factor Levels

Responses	Additives, x_1			Qnty of an additive, x_2			Qnty of filler, x_3			Fillers, x_4								
	0	1	2	0	1	2	0	1	2	0	1	2	3	4	5	6	7	8
y_1	4 322	4 091	4 346	4 528	4 339	3 892	3 385	4 223	5 151	1 289	1 516	1 419	1 313	1 425	1 388	1 474	1 411	1 524
y_2	932	973	1 046	995	992	964	1 050	945	955	308	406	247	352	337	342	323	337	299
y_3	2 443	2 410	2 629	2 639	2 024	2 819	4 239	2 436	807	959	877	582	977	1 121	1 046	625	743	552
D	4.937	4.920	5.279	5.382	4.320	4.986	3.627	5.238	5.877	1.581	1.929	1.371	1.926	1.725	1.911	1.008	1.719	1.389

(8) Find the sum of squares for the first source of variation

$$SS_{x_1} = SS_2 - SS_6$$

For y_1, it is

$$SS_{x_1} = 6\,033\,742 - 6\,029\,336 = 4\,406$$

(9) Find the sum of squares for the second source

$$SS_{x_2} = SS_3 - SS_6$$

For y_1, it is

$$SS_{x_2} = 6\,053\,041 - 6\,029\,336 = 23\,705$$

(10) Find the sum of squares for the third source

$$SS_{x_3} = SS_4 - SS_6$$

For y_1, it is

$$SS_{x_3} = 6\,202\,750 - 6\,029\,336 = 173\,414$$

(11) Find the sum of squares for the fourth source

$$SS_{x_4} = SS_5 - SS_6$$

For y_1, it is

$$SS_{x_4} = 6\,046\,886 - 6\,029\,336 = 17\,547$$

(12) Find the total sum of squares defined as the difference between the sum of squares for all observations and the correction term

$$SS_{\text{tot}} = SS_1 - SS_6$$

For y_1, it is

$$SS_{\text{tot}} = 6\,275\,327 - 6\,029\,336 = 245\,991$$

(13) Find the residual sum of squares (the estimate of the experimental error)

$$SS_{\text{res}} = SS_{\text{tot}} - (SS_{x_1} + SS_{x_2} + SS_{x_3} + SS_{x_4})$$

For y_1, it is

$$SS_{\text{res}} = 245\,991 - (4\,406 + 23\,705 + 17\,341 + 17\,547) = 26\,919$$

The ANOVA table of the above data for all the four quality responses is given below.

In order to select an optimal composition, the effects of the factors at the various levels are compared by applying the Duncan multiple ranges test. Since the type of additive, x_1, affects significantly only the response y_2 (see the Table), we select the additive ensuring a maximum bending strength. The standard error of the mean for y_2 is

$$s_{\bar{y}_2} = \sqrt{s_e^2/9} = \sqrt{80 \div 9} = 2.99$$

Means of y_2 for various levels of x_1

$$\bar{y}_2^{(0)} = 104 \qquad \bar{y}_2^{(1)} = 108 \qquad \bar{y}_2^{(2)} = 116$$

Ranges, r (see Appendix Table 10)	3.08	3.23
$r \times s_{\bar{y}_2}$	9.2	9.7

$\bar{y}_2^{(2)} - \bar{y}_2^{(0)} = 12 > 9.7$ a significant difference

$\bar{y}_2^{(2)} - \bar{y}_2^{(1)} = 8 < 9.2$ an insignificant difference

$\bar{y}_2^{(1)} - \bar{y}_2^{(0)} = 4 < 9.2$ an insignificant difference

Response	Source of variation	Degrees of freedom	Sum of squares	Mean square	Significance test
	x_1	2	4 406	2 203	$F_1 = s^2_{x_1}/s^2_e < 1$
	x_2	2	23 705	11 853	$F_2 = s^2_{x_2}/s^2_e = 5.284$
y_1	x_3	2	173 414	86 707	$F_3 = s^2_{x_3}/s^2_e = 38.657$
	x_4	8	17 547	2 193	$F_4 = s^2_{x_4}/s^2_e < 1$
	Error	12	26 919	2 243	
	Totals	26	245 991		
	x_1	2	741	370	$F_1 = s^2_{x_1}/s^2_e = 4.625$
	x_2	2	65	33	$F_2 = s^2_{x_2}/s^2_e < 1$
y_2	x_3	2	761	381	$F_3 = s^2_{x_3}/s^2_e = 4.763$
	x_4	8	4 949	619	$F_4 = s^2_{x_4}/s^2_e = 7.730$
	Error	12	964	80	
	Totals	26	7 480		
	x_1	2	3 098	1 549	$F_1 = s^2_{x_1}/s^2_e < 1$
	x_2	2	38 617	19 309	$F_2 = s^2_{x_2}/s^2_e = 3.838$
y_3	x_3	2	654 929	327 465	$F_3 = s^2_{x_3}/s^2_e = 65.089$
	x_4	8	120 054	15 007	$F_4 = s^2_{x_4}/s^2_e = 2.983$
	Error	12	60 375	5 031	
	Totals	26	877 073		
	x_1	2	0	0	$F_1 < 1$
	x_2	2	0.058	0.029	$F_2 = 2.636$
D	x_3	2	0.299	0.150	$F_3 = 13.636$
	x_4	8	0.307	0.154	$F_4 = 14.00$
	Error	12	0.130	0.011	
	Totals	26	0.794		

We choose additive No. 3—it differs significantly from additive No. 1 and insignificantly from additive No. 2. Because factors x_3, x_2 and x_4 affect the quality differently (see Table 3.21), we choose the optimal composition on the basis of a factorial analysis of the generalized desirability function D. To this end, we find the standard error of the mean or response D:

— for factors x_2 and x_3

$$s_{\bar{D}} = \sqrt{0.011/9} = 0.035$$

— for factor x_4

$$s_{\bar{D}} = \sqrt{0.011/3} = 0.061$$

Levels of x_2	1	2	0
Means of D	0.480	0.554	0.598
Ranges, r		3.08	3.23
$r \times s_{\bar{D}}$		0.107	0.113

$\bar{D}^{(0)} - \bar{D}^{(1)} = 0.118 > 0.113$ a significant difference

$\bar{D}^{(0)} - \bar{D}^{(2)} = 0.044 < 0.107$ an insignificant difference

$\bar{D}^{(2)} - \bar{D}^{(1)} = 0.074 < 0.107$ an insignificant difference

Levels of x_3	0	1	2
Means of D	0.403	0.582	0.653

$\bar{D}^{(2)} - \bar{D}^{(0)} = 0.250 > 0.113$ a significant difference
$\bar{D}^{(2)} - \bar{D}^{(1)} = 0.071 < 0.107$ an insignificant difference
$\bar{D}^{(1)} - \bar{D}^{(0)} = 0.179 > 0.107$ a significant difference

Levels of x_4	6	2	8	0	7	4	3	1	5
Means of D	0.336	0.457	0.463	0.527	0.573	0.575	0.642	0.643	0.697
Ranges, r		3.08	3.23	3.33	3.36	3.40	3.42	3.44	3.44
$r \times s_{\bar{D}}$		0.187	0.197	0.201	0.205	0.208	0.209	0.210	0.220

$\bar{D}^{(5)} - \bar{D}^{(6)} = 0.331 < 0.210$ a significant difference
$\bar{D}^{(5)} - \bar{D}^{(2)} = 0.240 > 0.210$ a significant difference
$\bar{D}^{(5)} - \bar{D}^{(8)} = 0.234 > 0.209$ a significant difference
$\bar{D}^{(5)} - \bar{D}^{(0)} = 0.170 < 0.208$ an insignificant difference
$\bar{D}^{(5)} - \bar{D}^{(7)} = 0.124 < 0.205$ an insignificant difference
$\bar{D}^{(5)} - \bar{D}^{(4)} = 0.122 < 0.201$ an insignificant difference
$\bar{D}^{(5)} - \bar{D}^{(3)} = 0.055 < 0.197$ an insignificant difference
$\bar{D}^{(5)} - \bar{D}^{(1)} = 0.054 < 0.187$ an insignificant difference
$\bar{D}^{(1)} - \bar{D}^{(6)} = 0.307 > 0.210$ a significant difference
$\bar{D}^{(1)} - \bar{D}^{(2)} = 0.186 < 0.209$ an insignificant difference
$\bar{D}^{(1)} - \bar{D}^{(8)} = 0.180 < 0.208$ an insignificant difference
$\bar{D}^{(1)} - \bar{D}^{(0)} = 0.116 < 0.205$ an insignificant difference
$\bar{D}^{(1)} - \bar{D}^{(7)} = 0.070 < 0.201$ an insignificant difference
$\bar{D}^{(1)} - \bar{D}^{(4)} = 0.068 < 0.197$ an insignificant difference
$\bar{D}^{(1)} - \bar{D}^{(3)} = 0.001 < 0.187$ an insignificant difference
$\bar{D}^{(3)} - \bar{D}^{(6)} = 0.306 > 0.209$ a significant difference
$\bar{D}^{(3)} - \bar{D}^{(2)} = 0.185 < 0.208$ an insignificant difference
$\bar{D}^{(3)} - \bar{D}^{(8)} = 0.179 < 0.205$ an insignificant difference
$\bar{D}^{(3)} - \bar{D}^{(0)} = 0.115 < 0.201$ an insignificant difference
$\bar{D}^{(3)} - \bar{D}^{(7)} = 0.069 < 0.197$ an insignificant difference
$\bar{D}^{(3)} - \bar{D}^{(4)} = 0.067 < 0.187$ an insignificant difference
$\bar{D}^{(4)} - \bar{D}^{(6)} = 0.239 > 0.208$ a significant difference
$\bar{D}^{(4)} - \bar{D}^{(2)} = 0.118 < 0.205$ an insignificant difference
$\bar{D}^{(4)} - \bar{D}^{(8)} = 0.112 < 0.201$ an insignificant difference
$\bar{D}^{(4)} - \bar{D}^{(0)} = 0.048 < 0.197$ an insignificant difference
$\bar{D}^{(4)} - \bar{D}^{(7)} = 0.002 < 0.187$ an insignificant difference
$\bar{D}^{(7)} - \bar{D}^{(6)} = 0.237 > 0.205$ a significant difference
$\bar{D}^{(7)} - \bar{D}^{(2)} = 0.116 < 0.201$ an insignificant difference

$\bar{D}^{(7)} - \bar{D}^{(8)} = 0.110 < 0.197$ an insignificant difference

$\bar{D}^{(7)} - \bar{D}^{(0)} = 0.046 < 0.187$ an insignificant difference

$\bar{D}^{(0)} - \bar{D}^{(6)} = 0.191 < 0.201$ an insignificant difference

$\bar{D}^{(0)} - \bar{D}^{(2)} = 0.070 < 0.197$ an insignificant difference

$\bar{D}^{(0)} - \bar{D}^{(8)} = 0.064 < 0.187$ an insignificant difference

$\bar{D}^{(8)} - \bar{D}^{(6)} = 0.127 < 0.197$ an insignificant difference

$\bar{D}^{(8)} - \bar{D}^{(2)} = 0.006 < 0.187$ an insignificant difference

$\bar{D}^{(2)} - \bar{D}^{(6)} = 0.121 < 0.187$ an insignificant difference

On the basis of variance and factorial analysis, the following compositions were chosen:
(1) high-density polyethylene +10% (T : A = 1 : 1) +10% additive No. 3
(2) high-density polyethylene +15% (T : A = 1 : 1) +10% additive No. 3
(3) high-density polyethylene +10% (T : A = 1 : 0.5) +10% additive No.3

The properties of the optimal compositions are given in the following table.

Composition No.	y_1, kgf cm^{-2}	y_2, kgf cm^{-2}	y_3, %	Mouldability
1	545	135	450	Excellent
2	576	125	400	Good
3	498	130	470	Excellent

Chapter Four
CORRELATION AND REGRESSION ANALYSIS

4.1. Stochastic Relationship

Random variables are usually related in such a way that a change in one entails a change in the distribution of the other. This is known as stochastic relationship. The change brought about in a random variable Y by a change in another random variable X will generally contain two components, stochastic (connected to the dependence of Y on X) and random. If the stochastic component is absent, the two variables Y and X are independent of each other. If the random component is absent, Y and X are connected by a functional relationship. If both components are present, the relationship between them defines the *strength* (closeness) *of association*. Mathematical statistics has not yet defined a general index of stochastic relationship, but there are a number of indices that aid in describing various aspects of stochastic relationship. The most important among them is the *correlation coefficient*.

If two random variables are independent, the variance of their sum is equal to the sum of their variances:

$$\mathrm{Var}\,\{X+Y\} = \mathrm{Var}\,\{X\}+\mathrm{Var}\,\{Y\} \tag{4.1}$$

From the definition of variance and expected value it follows that

$$\begin{aligned}\mathrm{Var}\,\{X+Y\} &= E[X+Y-E(X+Y)]^2\\ &= E[X-E(X)]^2+2E\{[X-E(X)][Y-E(Y)]\}+E[Y-E(Y)]^2\\ &= \mathrm{Var}\,\{X\}+2E\{[X-E(X)]\,[Y-E(Y)]\}+\mathrm{Var}\,\{Y\}\end{aligned} \tag{4.2}$$

The relationship between X and Y can be deduced from the inequality

$$E[(X-\mu_x)\,(Y-\mu_y)] \neq 0 \tag{4.3}$$

The quantity defined by Eq. (4.3) is called the *cross covariance function*, or the *covariance*, of the random variables X and Y, frequently abbreviated Cov $\{XY\}$ or Covar_{xy}. The dimensionless quantity

$$r = \frac{E[(X-\mu_x)\,(Y-\mu_y)]}{\sigma_x\sigma_y} \tag{4.4}$$

is the *correlation coefficient*. If the two variables fluctuate in absolute independence, the value of r will be zero. In fact, it may also have the value zero for some dependent variables which are then referred to as *uncorrelated*. For normally distributed random variables, a correlation coefficient of zero indicates that there is no relationship between the variables.

The correlation coefficient will remain unchanged, if we add any nonrandom terms to X and Y, or if we multiply X and Y by a positive number. If we multiply one of the two variables by -1 while leaving the other variable unchanged, the correlation coefficient will likewise be multiplied by -1.

The correlation coefficient characterizes a linear, rather than just any relationship. The linear probabilistic relationship between random variables consists in that an increase in one causes the other to show the tendency to increase (or decrease) linearly.

As already noted, the correlation coefficient gives a measure of the degree of relationship between two variables. If the random variables X and Y are connected by a rigorously linear functional relationship such that $y = b_0+b_1x$, then $r_{xy} = \pm 1$, the sign being the same as that of the constant b_1. In the general case, when X and Y are connected by an arbitrary stochastic relationship the limits to the value of r are $+1$ and -1, that is,

$$-1 < r_{xy} < 1 \tag{4.5}$$

At $r_{xy} > 0$, the correlation is positive: at $r_{xy} < 0$, it is negative.

The correlation coefficient is equally responsive to a high degree of randomness and to a marked nonlinearity in the relationship. The correlation coefficient may remain less than unity, although the relationship between X and Y is a rigorously functional one.

Qualitatively, the existence or absence of association between two random variables can be ascertained from the appearance of the *correlation field.* Positive correlation between two random variables is illustrated in Fig. 4.1*a*. A more pronounced positive correlation close to a linear functional

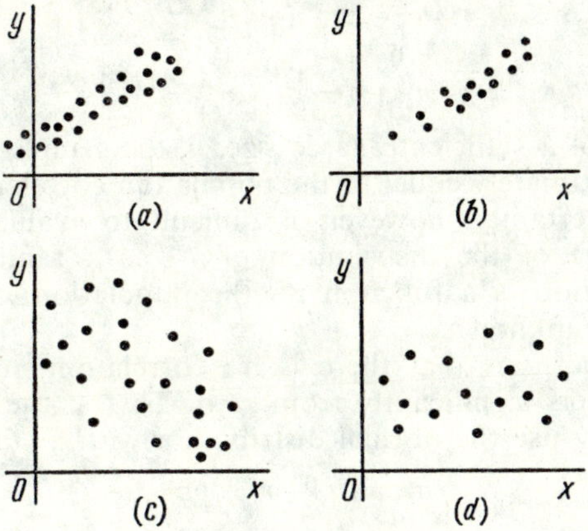

Fig. 4.1. Scatter diagrams

one is shown in Fig. 4.1*b*. An example of a relatively weak (or loose) negative correlation is shown in Fig. 4.1*c*, and an example of practically uncorrelated variables appears in Fig. 4.1*d*.

4.2. Sample Correlation Coefficient

The *sample correlation coefficient*, r^*, is defined in about the same manner as the population correlation coefficient, r, except that one uses the sample means and the sample variances.

Suppose that an experimentor has carried out n observations and noted in each the values of two random variables. Denoting the means as $\bar{x}$ and $\bar{y}$ and $\sum_{i=1}^{n} = \sum$, we may write

$$\bar{x} = \frac{1}{n}\sum x_i \quad \text{and} \quad \bar{y} = \frac{1}{n}\sum y_i$$

Hence, the sample correlation coefficient will be

$$r^* = \frac{\sum (x_i - \bar{x})(y_i - \bar{y})}{(n-1)s_x s_y} \tag{4.6}$$

where s_x and s_y are such that the respective sample variances are defined as

$$s_x^2 = \frac{1}{n-1}\sum (x_i - \bar{x})^2$$

$$s_y^2 = \frac{1}{n-1}\sum (y_i - \bar{y})^2$$

In calculation, it is convenient to use the following formulae:

$$\sum (x_i-\bar{x})(y_i-\bar{y}) = \sum x_i y_i - \frac{1}{n}\sum x_i \sum y_i$$

$$s_x^2 = \frac{1}{n-1}\left[\sum x_i^2 - \frac{1}{n}\left(\sum x_i\right)^2\right]$$

$$s_y^2 = \frac{1}{n-1}\left[\sum y_i^2 - \frac{1}{n}\left(\sum y_i\right)^2\right] \quad (4.7)$$

With a sample of a sufficiently large size, n, the sample correlation coefficient r^* is approximately equal to the population correlation coefficient r. The resultant uncertainty, however, is difficult to evaluate, because this involves knowledge of the distribution of r^* as a random variable. In turn, this distribution is a function of the population correlation coefficient, r, which is unknown.

To test the hypothesis that there is no correlation, it is necessary to ascertain if r^* differs significantly from zero. To test the null hypothesis, H_0: $r = 0$, we may use the normal distribution with a standard deviation

$$\sigma_{r*} \approx (1-r^{*2})/\sqrt{n} \quad (4.8)$$

Assuming a confidence coefficient of $1-\alpha = 0.95$, the correlation coefficient will lie between the following confidence limits:

$$r^* - \frac{1.96\,(1-r^{*2})}{\sqrt{n}} \leqslant r \leqslant r^* + \frac{1.96\,(1-r^{*2})}{\sqrt{n}} \quad (4.9)$$

We may state with probability 0.95 that there is a relationship between the two random variables if 0 lies beyond the confidence interval, that is, if

$$|r^*| - \frac{1.96\,(1-r^{*2})}{\sqrt{n}} > 0 \quad (4.10)$$

With a small-size sample and a relatively close correlation, the distribution of the correlation coefficient substantially differs from normal (Fig. 4.2*a*). The confidence interval can be constructed by using the Fisher transformation:

$$r^* = \tanh z = (e^{2z}-1)/(e^{2z}+1) \quad (4.11)$$

Hence,

$$z = \frac{1}{2}\ln\frac{1+r^*}{1-r^*} \quad (4.12)$$

The z-distribution remains practically unchanged in shape as r^* and n are changed, and rapidly approaches with an increase in n the normal distribution (Fig. 4.2*b*) with a mean

$$\mu_z = \frac{1}{2}\ln\frac{1+r}{1-r} \quad (4.13)$$

and a standard deviation

$$\sigma_z \approx 1/\sqrt{n-3} \quad (4.14)$$

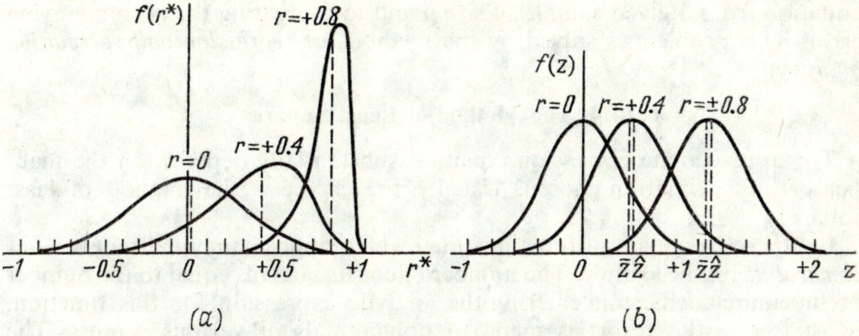

Fig. 4.2. Probability density of sample correlation coefficient

Then, with a confidence coefficient $1-\alpha$, the value of the unknown μ_z lies between the limits

$$z-\frac{u_{1-\alpha}}{\sqrt{n-3}} \leqslant \mu_z \leqslant z+\frac{u_{1-\alpha}}{\sqrt{n-3}} \tag{4.15}$$

where $u_{1-\alpha}$ is the quantile of the normal distribution. At $1-\alpha = 0.95$, $u_{1-\alpha} = 1.96$; hence

$$z-\frac{1.96}{\sqrt{n-3}} \leqslant \mu_z \leqslant z+\frac{1.96}{\sqrt{n-3}} \tag{4.16}$$

On finding the confidence limits for μ_z

$$z_1 = z-\frac{1.96}{\sqrt{n-3}} \quad \text{and} \quad z_2 = z+\frac{1.96}{\sqrt{n-3}} \tag{4.17}$$

we can determine the confidence limits for the population correlation coefficient by substituting z_1 and z_2 in Eq. (4.11).

4.3. Regression

The relationship between random variables is completely defined by the conditional distribution function. For a system of two random variables, the condition distribution function $F(x, y)$ is a function of two variables, etc. In practice, however, it is difficult to use conditional distribution functions; instead, recourse is had to conditional means, μ_c, and conditional variances, σ_c^2. The dependence of σ_c^2 on the parameter x is called the *scedastic relation*, but it is used relatively seldom. Practically, if the variances are homogeneous, they are averaged. The dependence of the conditional mean μ_c on x is called the *regression* of μ_c on x.

In processing experimental data, one usually finds an approximate regression equation and evaluates the magnitude and probability of the uncertainty. Thus, the problem reduces to developing an approximate regression equation from a given sample of size n and to evaluating the accompanying error. This problem is solved by the methods of *regression and correlation analysis*.

4.4. The Method of Least Squares

The approximate regression equation substantially depends on the manner of approximation chosen. Usually, the choice is the method of least squares.

Let there be a class of functions $f(x)$ which impose an equal number l of *constraints* on the sample. The number of constraints l is equal to the number of indefinite coefficients entering the analytic expression for this function. Most frequently, resort is made to polynomials of various powers. The best approximate regression equation is yielded by that function out of the class in question for which the sum of squares

$$\Phi = \sum [y_i - f(x_i)]^2 \tag{4.18}$$

yields the lowest value.

When the random variables involved are normally distributed, the method of least squares is treated in probability theory as a special case of the maximum likelihood principle. In the circumstances, we may invoke *sufficient statistics*, that is, functions of observations (estimates of population parameters) with the aid of which we can extract all information about the population parameters from the observations.

The problem of determining the regression equation parameters practically reduces to finding the maximum of a function of many variables. If the function

$$\hat{y} = f(x, b_0, b_1, b_2, \ldots) \tag{4.19}$$

is differentiable and we are to choose $b_0, b_1, b_2, \ldots$ such that

$$\Phi = \sum [y_i - f(x_i, b_0, b_1, b_2, \ldots)]^2 = \min \tag{4.20}$$

a necessary condition for $\Phi(b_0, b_1, b_2, \ldots)$ to be a minimum is to satisfy the following equalities

$$\partial\Phi/\partial b_0 = 0, \quad \partial\Phi/\partial b_1 = 0, \ldots \tag{4.21}$$

or

$$\sum 2[y_i - f(x_i, b_0, b_1, b_2, \ldots)] \frac{\partial f(x_i)}{\partial b_0} = 0$$

$$\sum 2[y_i - f(x_i, b_0, b_1, b_2, \ldots)] \frac{\partial f(x_i)}{\partial b_1} = 0 \tag{4.22}$$

On re-arranging,

$$\sum y_i \frac{\partial f(x_i)}{\partial b_0} - \sum f(x_i, b_0, b_1, b_2, \ldots) \frac{\partial f(x_i)}{\partial b_0} = 0$$

$$\sum y_i \frac{\partial f(x_i)}{\partial b_1} - \sum f(x_i, b_0, b_1, b_2, \ldots) \frac{\partial f(x_i)}{\partial b_1} = 0 \tag{4.23}$$

$$\ldots\ldots\ldots\ldots\ldots\ldots\ldots\ldots\ldots\ldots$$

Equations (4.23) are as many as there are unknown coefficients $b_0, b_1, b_2, \ldots$ in the regression equation; in mathematical statistics, they are known as a set of *normal equations*.

At any values of $b_0, b_1, b_2, \ldots$, the quantity Φ is greater than or equal to zero, so it must have at least one minimum. Therefore, if the set of normal equations has a unique solution, this solution is the minimum for Φ. Equations (4.23) cannot be solved in the general form. For their solution, we must specify a particular form for the function f.

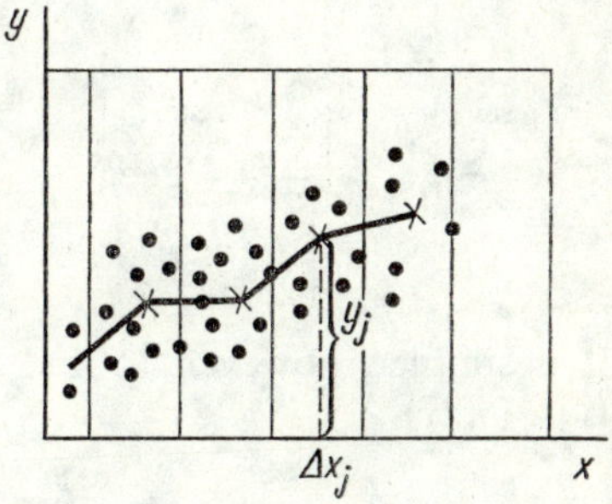

Fig. 4.3. Drawing an empirical regression line

In cases involving the dependence on one variable, it is useful to plot an *empirical line of regression* and to use it in order to determine the form of the equation of regression. For this purpose, the entire range of values of the variable x on the correlation field (Fig. 4.3) is broken down into equal intervals, or increments, Δx. All points falling within a given interval Δx_j are referenced to its centre, x_j. This is done by finding the mean $\bar{y}_j$ for each of the intervals

$$\bar{y}_j = \frac{1}{n_j} \sum_{i=1}^{n_j} y_{ji} \tag{4.24}$$

where n_j is the number of points in the interval Δx_j, such that

$$\sum_{j=1}^{k} n_j = n \tag{4.25}$$

where n is the sample size and k is the number of sample intervals.

Then the points $(x_j, \bar{y}_j)$ are consecutively joined by straight-line segments. The result is the empirical line of regression of y on x. Finally, the equation of regression $\hat{y} = f(x)$ is chosen that fits the empirical line of regression best.

4.5. The Linear Regression Equation

Using the method of least squares, we seek to estimate the coefficients of a linear equation of regression

$$\hat{y} = b_0 + b_1 x \tag{4.26}$$

from a sample of size n.

The set of normal equations for the above case will be $\left(\sum_{i=1}^{n} = \sum\right)$

$$\sum y_i - \sum (b_0 + b_1 x_i) = 0$$
$$\sum y_i x_i - \sum (b_0 + b_1 x_i) x_i = 0$$

or

$$\begin{aligned} n b_0 + b_1 \sum x_i &= \sum y_i \\ b_0 \sum x_i + b_1 \sum x_i^2 &= \sum x_i y_i \end{aligned} \tag{4.27}$$

The coefficients b_0 and b_1 can readily be found, using the determinants

$$b_0 = \frac{\begin{vmatrix} \sum y_i & \sum x_i \\ \sum x_i y_i & \sum x_i^2 \end{vmatrix}}{\begin{vmatrix} n & \sum x_i \\ \sum x_i & \sum x_i^2 \end{vmatrix}} = \frac{\sum y_i \sum x_i^2 - \sum x_i \sum x_i y_i}{n \sum x_i^2 - (\sum x_i)^2} \tag{4.28}$$

$$b_1 = \frac{\begin{vmatrix} n & \sum y_i \\ \sum x_i & \sum x_i y_i \end{vmatrix}}{\begin{vmatrix} n & \sum x_i \\ \sum x_i & \sum x_i^2 \end{vmatrix}} = \frac{n \sum x_i y_i - \sum x_i \sum y_i}{n \sum x_i^2 - (\sum x_i)^2} = \frac{\sum (x_i - \bar{x})(y_i - \bar{y})}{\sum (x_i - \bar{x})^2} \tag{4.29}$$

It is simpler, however, to find b_0, once b_1 is known, from the first line of Eqs. (4.27)

$$b_0 = \bar{y} - b_1 \bar{x} \tag{4.30}$$

where $\bar{y}$ and $\bar{x}$ are the sample means of y and x.

Among other things, Eq. (4.30) suggests the existence of a correlation between b_0 and b_1. To evaluate the strength of correlation, Eq. (4.26), we should find the sample coefficient of correlation r^*:

$$r^* = \frac{\sum (x_i - \bar{x})(y_1 - \bar{y})}{(n-1) s_x s_y} \tag{4.31}$$

where s_x and s_y are the sample standard deviations.

From Eqs. (4.29) and (4.31), we have

$$r^* = \frac{b_1 s_x}{s_y} = b_1 \sqrt{\frac{n \sum x_i^2 - (\sum x_i)^2}{n \sum y_i^2 - (\sum y_i)^2}} \tag{4.32}$$

Example 4.1. Determine the dependence of the solubility, y, of barium chloride in water in the presence of a percentage, x, of calcium chloride at 70 °C. Assume that the sample size is $n = 6$. The experimental data are as follows.

x, %	0	5	8	10	15	20
y, %	32	25	20	17	11	5

Solution. Find the coefficients of the regression equation of the form

$$\hat{y} = b_0 + b_1 x$$

The coefficient b_1 can be found from Eq. (4.29), and the coefficient b_0 from Eq. (4.30). For this purpose, we tabulate the experimental data and the results of calculation as follows.

A check on the results can be made, using the equation

$$\sum (x_i + y_i)^2 = \sum x_i^2 + 2 \sum x_i y_i + \sum y_i^2 \tag{4.33}$$

In our example,

$$4\,738 = 814 + 2 \times 720 + 2\,484$$

which indicates that the calculations have been carrried out correctly.

Experiment No.	x	y	x^2	xy	y^2	$x+y$	$(x+y)^2$
1	0	32	0	0	1024	32	1024
2	5	25	25	125	625	30	900
3	8	20	64	160	400	28	784
4	10	17	100	170	289	27	720
5	15	11	225	165	121	26	676
6	20	5	400	100	25	25	625
$\sum$	58	110	814	720	2484	168	4738

Referring to the table above, we find the coefficients b_0 and b_1 to be:

$$b_1 = \frac{(6\times720-58\times110)}{6\times814-58^2} = -1.355$$

$$b_0 = \frac{110+1.355\times58}{6} = 31.43$$

Using Eq. (4.32), we find the sample coefficient of correlation to be

$$r^* = -1.355\sqrt{\frac{6\times814-3\,364}{6\times2\,484-12\,100}} = -1.355\sqrt{\frac{1\,520}{2\,804}} = -0.99$$

As is seen, the correlation coefficient is unity very nearly, so the dependence of y on x in the range of values in question is practically linear and has the form

$$\hat{y} = 31.43-1.355x$$

After the regression equation has been found, it must be subjected to statistical analysis. This analysis consists in the following: all regression coefficients are tested for significance in comparison with the sample standard deviation (replication error), and the equation itself is tested for the goodness of fit. This constitutes regression analysis. In conducting regression analysis, it is customarily assumed that:

(1) the input variable x has been measured with a negligible error in comparison with the error in y. The greater error in y is attributable to the fact that any process involves unobservable variables that are not included in the equation of regression;

(2) the observed values of the output variable, $y_1, y_2, \ldots, y_n$, are independent, normally distributed random variables;

(3) with the sample size being n and with each run (series of observations) containing m_i observations (where $i = 1, 2, \ldots, n$), the sample variances $s_1^2, s_2^2, \ldots, s_n^2$, are homogeneous. The sample variances are tested for homogeneity by applying Cochran's test in the case of an equal number of degrees of freedom, or Bartlet's test, in the case of different degrees of freedom. Knowledge of the error mean square, s_e^2, found from the replicate observations is essential in testing the coefficients of the regression equation for

significance and the equation itself for the goodness of fit to the experimental data.

The coefficients of the regression equation are tested for significance by applying the Student t-test

$$t_j = |b_j|/s_{b_j} \tag{4.34}$$

where b_j is the jth estimated coefficient of the regression equation and s_{bj} is the sample standard deviation of the jth coefficient.

If t_j exceeds the tabulated value of $t_\alpha(\nu)$ for the selected significance level α and degrees of freedom ν, the estimated coefficient b_j is substantially nonzero; s_{b_j} is found by the error accumulation law, Eq. (2.36):

$$s_{b_j} = \sqrt{\sum_{i=1}^{n} \left(\frac{\partial b_j}{\partial y_i}\right)^2 s_i^2} \tag{4.35}$$

If the sample variances $s_1^2, s_2^2, \ldots, s_n^2$ are homogeneous, then $(\sum_{i=1}^{n} = \sum)$:

$$s_{b_0} = \sqrt{\frac{s_e^2 \sum x_i^2}{n\sum x_i^2 - (\sum x_i)^2}} \tag{4.36}$$

$$s_{b_1} = \sqrt{\frac{s_e^2 n}{n\sum x_i^2 - (\sum x_i)^2}} \tag{4.37}$$

The nonsignificant estimated coefficients are dropped from the regression equation, and the remaining estimated coefficients are recalculated, because they are interrelated. The equation is tested for adequacy (goodness of fit) by applying Fisher's test (on the basis of Fisher's variance ratio):

$$F = s_g^2/s_e^2 \tag{4.38}$$

where s_e^2 is the error mean square and s_g^2 is the goodness-of-fit mean square defined as

$$s_g^2 = SS_g/\nu_g \tag{4.39}$$

where SS_g is the sum of the goodness-of-fit squares defined as

$$SS_g = SS_{res} - SS_e \tag{4.40}$$

ν_g is the degrees of freedom of the goodness-of-fit variance

$$\nu_g = \nu_{res} - \nu_e = n - l \tag{4.41}$$

where l is the number of coefficients in the regression equation. SS_e is the sum of squares related to the error mean square s_e^2 as follows:

$$SS_e = \sum_{i=1}^{n} \sum_{u=1}^{m_i} (y_{iu} - \bar{y}_i)^2 \tag{4.42}$$

$$\bar{y}_i = \frac{1}{m_i} \sum_{u=1}^{m_i} y_{iu} \tag{4.43}$$

$$s_e^2 = SS_e/\nu_e \tag{4.44}$$

ν_e is the degrees of freedom of the error mean square (replication variance) s_e^2:

$$\nu_e = \sum_{i=1}^{n}(m_i-1) \tag{4.45}$$

$$s_e^2 = \frac{\sum_{i=1}^{n}\sum_{u=1}^{m_i}(y_{iu}-\bar{y}_i)^2}{\sum_{i=1}^{n}(m_i-1)} \tag{4.46}$$

SS_{res} is the residual sum of squares:

$$SS_{\text{res}} = \sum_{i=1}^{n}\sum_{u=1}^{m_i}(y_{iu}-\hat{y}_i)^2 \tag{4.47}$$

$$\nu_{\text{res}} = \sum_{i=1}^{n} m_i - l \tag{4.48}$$

and ν_{res} is the degrees of freedom of the residual variance, s_{res}^2, defined as

$$s_{\text{res}}^2 = \frac{SS_{\text{res}}}{\nu_{\text{res}}} = \frac{\sum_{i=1}^{n}\sum_{u=1}^{m_i}(y_{iu}-\hat{y}_i)^2}{\sum_{i=1}^{n} m_i - l} \tag{4.49}$$

If Fisher's variance ratio,

$$F = s_g^2/s_e^2 \tag{4.50}$$

proves smaller than the tabulated value of $F_{1-\alpha}(\nu_1, \nu_2)$ for the selected significance level α and degrees of freedom $\nu_1 = \nu_g$ and $\nu_2 = \nu_e$, the equation is a good fit.

If the same number of observations is made in each run, so that $m_1 = m_2 = \ldots = m_i = \ldots = m_n = m$, the calculations are simplified:

$$s_g^2 = \frac{m}{n-l}\sum_{i=1}^{n}(\bar{y}_i-\hat{y}_i)^2 \tag{4.51}$$

$$s_e^2 = \frac{1}{n(m-1)}\sum_{i=1}^{n}\sum_{u=1}^{m}(y_{iu}-\bar{y}_i)^2 \tag{4.52}$$

If no replicate observations are made and the error mean square is found from a separate run of m observations, then

$$s_g^2 = s_{\text{res}}^2 = \frac{1}{n-l}\sum_{i=1}^{n}(y_i-\hat{y}_i)^2 \tag{4.53}$$

$$s_e^2 = \frac{1}{m-1}\sum_{u=1}^{m}(y_u^0-\bar{y}^0)^2 \tag{4.54}$$

$$\bar{y}^0 = \frac{1}{m}\sum_{u=1}^{m} y_u^0 \tag{4.55}$$

If no replicate observations are made and no error mean square is available, the goodness of fit for the adopted equation can be tested by comparing the residual mean square, s_{res}^2, and the estimated variance about the mean, s_y^2:

$$s_y^2 = \frac{1}{n-1}\sum_{i=1}^{n}(y_i-\bar{y})^2 \tag{4.56}$$

on the basis of the Fisher F-test

$$F = s_y^2(\nu_1)/s_{\text{res}}^2(\nu_2) \tag{4.57}$$

In this case, the F-ratio is the factor by which the dispersion about the regression line is reduced relative to the dispersion about the mean. The greater the value of F in comparison with its tabulated value, $F_{1-\alpha}(\nu_1, \nu_2)$, for the selected significance level α and degrees of freedom $\nu_1 = n-1$ and $\nu_2 = n-l$, the more efficient the estimated regression equation is.

4.6. Parabolic Regression

If an estimated regression equation is a polynomial of some order, in the method of least squares the coefficients of this polynomial are found by solving a set of linear equations. For example, it is desired to estimate by the method of least squares the coefficients of a quadratic equation

$$\hat{y} = b_0+b_1x+b_2x^2 \tag{4.58}$$

which describes a second-order parabola, such that

$$\partial f(x)/\partial b_0 = 1 \quad \partial f(x)/\partial b_1 = x, \text{ and } \partial f(x)/\partial b_2 = x^2$$

and the set of normal equations has the form $(\sum_{i=1}^{n} = \sum)$

$$\begin{aligned} b_0 n+b_1\sum x_i+b_2\sum x_i^2 &= \sum y_i \\ b_0\sum x_i+b_1\sum x_i^2+b_2\sum x_i^3 &= \sum x_i y_i \\ b_0\sum x_i^2+b_1\sum x_i^3+b_2\sum x_i^4 &= \sum x_i^2 y_i \end{aligned} \tag{4.59}$$

Equations similar in structure will also define the coefficients of a parabola of any order.

The goodness of fit of the equation to experimental data is improved by raising the order (or degree) of the polynomial. In the circumstances, owing to correlation among the coefficients, all regression coefficients must be calculated anew. In passing from the kth to the $k+1$st order, the right-hand side of the regression equation is extended to include a complete $k+1$st-order polynomial containing $(k+2)$ new unknown coefficients, rather than a single term of the form $b_{k+1}x^{k+1}$. The test is carried out relative to the residual variance

$$s_{\text{res}}^2 = \frac{\sum (y_i-\hat{y}_i)^2}{n-l}$$

Once $s_{(k+1),\,\text{res}}^2$ ceases to be significantly smaller than $s_{k,\,\text{res}}^2$, the order k of the polynomial should not be raised any more. The significance of the discrepancy between s_k^2 and s_{k+1}^2 is tested by applying Fisher's F-test:

$$F = s_k^2/s_{k+1}^2$$

If the F-ratio thus obtained proves smaller the tabulated value of $F_{1-\alpha}(\nu_1, \nu_2)$ for the selected significance level and degrees of freedom $\nu_1 = \nu_k$ and $\nu_2 = \nu_{k+1}$, the order k of the polynomial should not be raised any more.

4.7. Chebyshev's Polynomials

The regression equation expressed in terms of Chebyshev's polynomials has the form

$$\hat{y} = b_0 P_0(x) + b_1 P_1(x) + \dots + b_k P_k(x) \tag{4.60}$$

where $P_0(x)$, $P_1(x)$, ..., $P_k(x)$ are orthogonal Chebyshev's polynomials on a set of points $x_1, x_2, \dots, x_n$. This means that for all $u \neq j$, the condition

$$\textstyle\sum_{i=1}^{n} P_u(x_i) P_j(x_i) = 0 \tag{4.61}$$

where $P_{k+1}(x)$ depends solely on the sample size n, is satisfied. If the Chebyshev polynomials $P_{k+1}(x)$ are known, only the coefficient b_{k+1} needs to be calculated each time the degree of the regression equation is incremented. Chebyshev's polynomials are evaluated, using the equations

$$P_0(x) = 1, \tag{4.62}$$

$$P_1(x) = x - \frac{n+1}{2}, \tag{4.63}$$

....

$$P_{k+1}(x) = P_1(x) P_k(x) - \frac{k^2(n^2-k^2)}{4(4k^2-1)} P_{k-1}(x). \tag{4.64}$$

For example,

$$P_2(x) = x^2 - (n+1)x + \frac{(n+1)(n+2)}{6} \tag{4.65}$$

$$P_3(x) = x^3 - \frac{3(n+1)}{2} x^2 + \frac{6n^2+15n+11}{10} x - \frac{(n+1)(n+2)(n+3)}{20} \tag{4.66}$$

$$P_4(x) = x^4 - 2(n+1)x^3 + \frac{9n^2+21n+4}{7} x^2 - \frac{(n+1)(2n^2+7n+10)}{7} x + \frac{(n+1)(n+2)(n+3)(n+4)}{80} \tag{4.67}$$

Using the least squares method to evaluate the coefficients $b_0, b_1, \dots, b_k$ of the regression equation, Eq. (4.60), we obtain $(\sum_{i=1}^{n} = \sum)$

$$b_0 = \frac{1}{n} \sum y_i$$

$$b_1 = \frac{\sum y_i P_1(x_i)}{\sum P_1^2(x_i)} \tag{4.68}$$

....

$$b_k = \frac{\sum y_i P_k(x_i)}{\sum P_k^2(x_i)}$$

The coefficients b_j found by Eq. (4.68) are independent of the degree of the regression equation involved. In finding the regression equation by the method of successive approximations, use is made of all the b_j's found previously. Raising the degree of the regression equation by one leads to finding only one coefficient. This procedure results in equations convenient to use in determining the residual variance for the regression equation of the kth order:

$$s^2_{k,\,\text{res}} = SS_k/(n-k-1) \qquad (4.69)$$

where the sums of squares of deviations, SS_k, are found, using the recurrence equation

$$SS_k = SS_{k-1} - b_k^2 \sum P_k^2(x_i) \qquad (4.70)$$

The only pre-requisite is to evaluate SS_0:

$$SS_0 = \sum [y_i - b_0 P_0(x_i)]^2 = \sum (y_i - \bar{y})^2 = \sum y_i^2 - \frac{1}{n}\left(\sum y_i\right)^2 \qquad (4.71)$$

The evaluation of the coefficients is simplified, if the argument takes on equidistant values

$$x_2 = x_1 + h$$

$$x_3 = x_1 + 2h$$

$$\dots\dots\dots\dots\dots$$

$$x_n = x_1 + (n-1)h$$

where h is the step of interpolation. On changing the variables

$$z = \frac{x - x_1}{h} + 1 \qquad (4.72)$$

each value of x_i can be replaced by its ordinal number, that is, $z_i = i$. Let us seek a regression equation of the form

$$\hat{y} = a_0 P_0(z) + a_1 P_1(z) + \dots + a_k P_k(z) \qquad (4.73)$$

where

$$a_0 = \frac{\sum y_i}{n} \qquad (4.74)$$

$$a_1 = \frac{\sum y_i P_1(i)}{\sum P_1^2(i)} \qquad (4.75)$$

$$\dots\dots\dots\dots\dots$$

$$a_k = \frac{\sum y_i P_k(i)}{\sum P_k^2(i)} \qquad (4.76)$$

The sums $\sum_{i=1}^{n} = \sum$ in the denominator can be found, using an abbreviated equation

$$\sum P_k^2(i) = \frac{(k\,!)^2\, n(n^2-1)\,(n^2-4)\dots(n^2-k^2)}{[(2k-1)!\,!]^2\, 2^{2k}(2k+1)} \qquad (4.77)$$

where $(2k-1)!!$ is the product of all odd numbers from 1 to $2k-1$ inclusive. In particular,

$$\sum P_1^2(i) = \frac{n(n^2-1)}{12} \tag{4.78}$$

$$\sum P_2^2(i) = \frac{n(n^2-1)(n^2-4)}{180} \tag{4.79}$$

$$\sum P_3^2(i) = \frac{n(n^2-1)(n^2-4)(n^2-9)}{2\,800} \tag{4.80}$$

$$\sum P_4^2(i) = \frac{n(n^2-1)(n^2-4)(n^2-9)(n^2-16)}{44\,100} \tag{4.81}$$

These sums are also used to evaluate SS_k necessary to determine the residual variance:

$$SS_k = SS_{k-1} - a_k^2 \sum_{i=1}^{n} P_k^2(i) \tag{4.82}$$

Once the regression equation, Eq. (4.73), is found, the variable z is changed back to the original variable x.

Example 4.2. Determine how the degree of dissociation, d, of hydrogen iodide depends on temperature t. The experimental data are tabulated below.

t, °C	280	300	320	340	360	380	400	420	440	460	480
d	.178	.182	.186	.191	.196	.202	.207	.213	.220	.228	.236

The sample size is $n = 11$. The temperature was varied in steps of 20 deg. C ($h = 20$).

Solution. An equation of regression of the degree of dissociation with temperature can be developed, using the Chebyshev method. To begin with, we change the variable:

$$z = (t-260)/20$$

Also, to make the calculations more convenient, we change from d to y:

$$y = 1\,000(d-0.178)$$

The values of z_i, y_i and Chebyshev's polynomials obtained by Eqs. (4.63), (4.65) and (4.66) in which $i = 1, 2, \ldots, 11$ are substituted for x are tabulated below.

$z_i = i$	y_i	$P_1(i)$	$y_iP_1(i)$	$P_2(i)$	$y_iP_2(i)$	$P_3(i)$	$y_iP_3(i)$
1	0	−5	0	15	0	−36	0
2	4	−4	−16	6	24	7.2	28.3
3	8	−3	−24	−1	−8	26.4	24.2
4	13	−2	−26	−6	−78	27.5	358.8
5	18	−1	−18	−9	−162	16.8	302.4
6	24	0	0	−10	−240	0	0
7	29	1	29	−9	−261	−16.8	−487.2
8	35	2	70	−6	−210	−27.6	−96.6
9	42	3	126	−1	−42	−26.4	−1108.8
10	50	4	200	6	300	−7.2	−360
11	58	5	290	15	870	36	2088
$\sum$	281		631		193		67.2

Now, we find the coefficients of the equation

$$\hat{y} = a_0 P_0(z) + a_1 P_1(z)$$

where

$$P_0(z) = 1 \quad \text{and} \quad P_1(z) = z - 6$$

Using Eq. (4.74), we obtain

$$a_0 = 281/11 = 25.6$$

By the abbreviated equation (4.78), we find $\sum_{i=1}^{n} P_1^2(i)$ to be

$$\sum_{i=1}^{n} P_1^2(i) = 11(121-1)/12 = 110$$

From Eq. (4.75), the coefficient a_1 is

$$a_1 = 631/110 = 5.73$$

The first-order regression equation has the form

$$\hat{y} = 25.6 + 5.73(z-6)$$

Determine the residual variance, s^2_{res}. To this end, we use Eqs. (4.71) and (4.82) to find the respective sums of squares:

$$SS_0 = \sum_{i=1}^{11} y_i^2 - \frac{1}{11}\left(\sum_{i=1}^{11} y_i\right)^2 = 10\,843 - \frac{78\,961}{11} = 3\,665$$

$$SS_1 = SS_0 - a_1^2 \sum_{i=1}^{11} P_1^2(i) = 3\,665 - 5.73^2 \times 110 = 53$$

Hence,

$$s^2_{\text{res},\,1} = SS_1/(n-2) = 53/9 = 5.9$$

Now we evaluate a second-order equation of regression:

$$\hat{y} = a_0 P_0(z) + a_1 P_1(z) + a_2 P_2(z)$$

From Eq. (4.65), we find $P_2(z)$ to be

$$P_2(z) = z^2 - 12z + 26$$

Using Eq. (4.79), we obtain

$$\sum_{i=1}^{11} P_2^2(i) = \frac{11(121-1)(121-9)}{180} = 858$$

and

$$a_2 = 193/858 = 0.225$$

Hence, the second-order regression equation is found to be

$$\hat{y} = 25.6 + 5.73(z-6) + 0.225(z^2 - 12z + 26)$$

Determine the residual variance $s^2_{\text{res},\,2}$:

$$s^2_{\text{res},\,2} = \frac{SS_2}{n-3} = \frac{SS_1 - a_2^2 \sum P_2^2(i)}{n-3} = \frac{53 - 0.225^2 \times 858}{8} = \frac{9.6}{8} = 1.2$$

Ascertain the significance of discrepancy between $s^2_{\text{res},\,1}$ and $s^2_{\text{res},\,2}$ by applying the Fisher (F) test:

$$F = s^2_{\text{res},\,1}/s^2_{\text{res},\,2} = 5.9/1.2 = 4.9$$

$$F_{0.95}(9,\,8) = 3.4$$

$$F > F_{1-\alpha}(\nu_1,\,\nu_2)$$

As is seen, the second-order regression equation is a marked improvement upon the first-order equation. Now evaluate the third-order regression equation:

$$\hat{y} = a_0P_0(z) + a_1P_1(z) + a_2P_2(z) + a_3P_3(z)$$

On substituting $n = 11$ in Eq. (4.66), we obtain $P_3(z) = z^3 - 18z^2 + 90.2z - 191.1$. Whence by Eq. (4.80)

$$\sum_{i=1}^{11} P_3^2(i) = \frac{11(121-1)(121-4)(121-9)}{2\,800} = 6\,177.7$$

Hence,

$$a_3 = 67.2/177.6 = 0.010\,9$$

So, finally, the third-order regression equation has the form

$$\hat{y} = 25.6 + 5.73(z-6) + 0.225(z^2 - 12z + 26) + 0.0109(z^3 + 18z^2 + 90z - 191.1)$$

Now we check to see if the use of a third-order regression equation instead of the second-order one can improve the accuracy. To this end, we find

$$s_{3,\,res}^2 = \frac{SS_3}{n-4} = \frac{SS_2 - a_3^2 \sum_{i=1}^{n} P_3^2(i)}{n-4} = \frac{9.6 - 0.0109^2 \times 6\,177.6}{7} = \frac{8.9}{7} = 1.27$$

$$s_{3,\,res}^2 > s_{2,\,res}^2$$

Thus, it will be well-advised to stop on the second-order regression. On changing from y and z back to d and t, we obtain

$$1000(d - 0.178) = 25.6 + 5.73\left(\frac{t-260}{20} - 6\right) + 0.225\left[\left(\frac{t-260}{20}\right)^2 - 12\frac{t-260}{20} + 26\right]$$

and finally,

$$d = 0.1731 - 0.000139t + 0.00000056t^2$$

This is the final equation in the polynomial class. Now we evaluate the strength of correlation, using the index of correlation to be defined later, Eq. (4.89):

$$\theta = \sqrt{1-\xi}$$

where

$$\xi = SS_2/SS_0 = 9.6/3\,665 = 0.002\,72$$

and

$$\theta = \sqrt{1 - 0.002\,72} = 0.986$$

The index of correlation θ is close to unity, so the relationship thus established is a functional one very nearly.

4.8. Transcendental Regression

With a small sample size n, an increase in the order of the polynomial might lead to an increase in the residual variance (residual mean square). One way to reduce the number of coefficients to be evaluated is to use *transcendental regression.* The evaluation of transcendental regression coefficients may prove a tedious task because one has to solve a system of non-linear equations. The task may be simplified by using a transformation of variables. For example, exponential and fractional-power relations such as

$$\hat{y} = b_0 b_1^x \tag{4.83}$$

$$\hat{y} = b_0 x_1^b \tag{4.84}$$

can be linearized by taking logarithms

$$\log_{10} \hat{y} = \log_{10} b_0 + x \log_{10} b_1 \quad (4.85)$$

$$\log_{10} \hat{y} = \log_{10} b_0 + b_1 \log_{10} x \quad (4.86)$$

On setting

$$\log_{10} \hat{y} = \hat{z},\ \log_{10} b_0 = a,\ \log_{10} b_1 = a_1, \text{ and } \log_{10} x = t$$

we obtain linear equations in the new variables

$$\hat{z} = a_0 + a_1 x \qquad z = a_0 + b_1 t$$

The coefficients are estimated by the least squares method. Once a_0 and a_1 are found, one can estimate b_0 and b_1. It should be remembered, however, that the coefficients of regression equations (4.83) and (4.84) thus obtained are *biased estimators* for the respective general coefficients.

4.9. Estimate of the Strength of Correlation

If we assume that a regression equation has been found with sufficient accuracy, the residual mean square will be only due to the error mean square, that is,

$$s_{\text{res}}^2 \approx s_e^2 \quad (4.87)$$

The smaller the contribution of $s_{\text{res}}^2 \approx s_e^2$ to the overall estimated variance s_y^2, the stronger the correlation between Y and X, because randomness in this correlation is reduced in proportion. Accordingly, the strength of correlation may be defined as

$$\xi = (n-l) s_{\text{res}}^2 / (n-1) s_y^2 \quad (4.88)$$

where

$$s_{\text{res}}^2 = \frac{1}{n-l} \sum_{i=1}^{n} (y_i - \hat{y}_i)^2$$

$$s_y^2 = \frac{1}{n-1} \sum_{i=1}^{n} (y_i - \bar{y})^2$$

and

$$\bar{y} = \frac{1}{n} \sum_{i=1}^{n} y_i$$

The value of ξ decreases with increasing strength of correlation. The quantity

$$\sqrt{1-\xi} = \theta \quad (4.89)$$

is termed the *index of correlation*. The greater the value of θ, the stronger the relationship between the two variables

$$0 \leqslant \theta \leqslant 1 \quad (4.90)$$

If $\theta = 1$, a functional relationship exists between the two variables. At $\theta = 0$, however, the variables Y and X cannot be deemed independent. Without affecting the respective mean squares and variances, the dependence between them may manifest itself in moments of a higher order. It is only in the case of normal distribution that a correlation index of zero may be regarded as an indication that there is no functional relationship between the variables. Like the coefficient of correlation in linear regression, the index of correlation gives a measure of the degree of relationship between random variables. Analysis of correlation strength on the basis of θ constitutes *correlation analysis*.

With linear regression, the index of correlation is equal to the coefficient of correlation:

$$\theta = \sqrt{1-[(n-2)/(n-1)]\,(s_{\text{res}}^2/s_y^2)} = |r^*| \tag{4.91}$$

4.10. Multiple Correlation Method

If one is to analyse correlation between many variables, one uses a multiple regression equation of the form

$$\hat{y} = b_0+b_1x_1+b_2x_2+\ldots+b_kx_k \tag{4.92}$$

Instead of a regression line, one has to deal with a regression surface already at $k = 2$ and with a regression hypersurface at $k > 2$. In the general case, as already stated, this is a response surface.

In constructing a response surface, the numerical values of independent variables (factors) are laid off on the coordinate axes of a factor space. The statistical material used for the purpose is presented in tabular form like Table 4.1.

Table 4.1. Raw Statistical Data on a Natural Scale

Experiment No.	x_1	x_2	x_3	...	x_k	y
1	x_{11}	x_{21}	x_{31}	...	x_{k1}	y_1
2	x_{12}	x_{22}	x_{32}	...	x_{k2}	y_2
3	x_{13}	x_{23}	x_{33}	...	x_{k3}	y_3
.	.	.	.	...	...	.
n	x_{1n}	x_{2n}	x_{3n}	...	x_{kn}	y_n

The first step is to transform the raw data to a new scale. This is done by normalizing all random variables by the use of equations

$$y_i^0 = \frac{y_i-\bar{y}}{s_y}, \qquad x_{ji}^0 = \frac{x_{ji}-\bar{x}_j}{s_{x_j}} \tag{4.93}$$

$$i = 1, 2, \ldots, n \qquad j = 1, 2, \ldots, k$$

where y_i^0 and x_{ji}^0 are the normalized values of the respective factors, $\bar{y}$ and $\bar{x}_j$ are the sample means of the factors, s_y and s_{x_j} are the sample standard deviations of the factors defined as

$$s_y = \sqrt{\frac{1}{n-1}\sum_{i=1}^{n}(y_i-\bar{y})^2}, \qquad s_{x_j} = \sqrt{\frac{1}{n-1}\sum_{i=1}^{n}(x_{ji}-\bar{x}_j)^2}$$

The statistical material after normalization is presented in Table 4.2

Table 4.2. Raw Statistical Data on a Non-Dimensional Scale

Experiment No.	x_1^0	x_2^0	x_3^0	...	x_k^0	y^0
1	x_{11}^0	x_{21}^0	x_{31}^0		x_{k1}^0	y_1^0
2	x_{12}^0	x_{22}^0	x_{32}^0		x_{k2}^0	y_2^0
3	x_{13}^0	x_{23}^0	x_{33}^0		x_{k3}^0	y_3^0
.						
.						
.						
n	x_{1n}^0	x_{2n}^0	x_{3n}^0		x_{kn}^0	y_n^0

On the new scale we have

$$\bar{x}_j^0 = 0, \quad \bar{y}^0 = 0, \quad s_{x_j}^0 = 1, \quad \text{and} \quad s_{y^0} = 1 \tag{4.94}$$

Then the sample correlation coefficient is

$$r^*_{y^0x_j^0} = \frac{1}{n-1}\sum_{i=1}^{n} y_i^0\, x_{ji}^0$$

$$r^*_{x_l^0x_m^0} = \frac{1}{n-1}\sum_{i=1}^{n} x_{l_i}^0\, x_{m_i}^0 \qquad l, m = 1, 2, \ldots, k \quad l > m \tag{4.95}$$

The sample correlation coefficient found from Eq. (4.95) is equal to the correlation coefficient for the variables taken on a natural scale, $r^*_{yx_j}$ and $r^*_{x_lx_m}$.

The estimated regression equation for normalized variables has no free term and takes the form

$$\hat{y}^0 = a_1x_1^0 + a_2x_2^0 + \ldots + a_kx_k^0 \tag{4.96}$$

The coefficients of Eq. (4.96) are found from the condition

$$S = \sum_{i=1}^{n}(y_i^0 - \hat{y}_i^0)^2 = \text{minimum}$$

The condition for the minimum of the function S is defined in the same manner as in the case of a function of a single variable

$$\partial S/\partial a_1 = 0, \quad \partial S/\partial a_2 = 0, \ \ldots, \ \partial S/\partial a_k = 0 \tag{4.97}$$

and the system of normal equations takes the form ($\sum = \sum_{i=1}^{n}$)

$$\begin{aligned} a_1 \sum (x_{1i}^0)^2 + a_2 \sum x_{1i}^0 x_{2i}^0 + \ldots + a_k \sum x_{1i}^0 x_{ki}^0 &= \sum x_{1i}^0 y_i^0 \\ a_1 \sum x_{2i}^0 x_{1i}^0 + a_2 \sum (x_{2i}^0)^2 + \ldots + a_k \sum x_{2i}^0 x_{ki}^0 &= \sum x_{2i}^0 y_i^0 \\ \cdots\cdots\cdots\cdots\cdots\cdots & \\ \cdots\cdots\cdots\cdots\cdots\cdots & \\ a_1 \sum x_{ki}^0 x_{1i}^0 + a_2 \sum x_{ki}^0 x_{2i}^0 + \ldots + a_k \sum (x_{ki}^0)^2 &= \sum x_{ki}^0 y_i^0 \end{aligned} \tag{4.98}$$

Multiplying the left- and right-hand sides of the equation by $1/(n-1)$, we get a sample correlation coefficient, r^*, according to Eq. (4.95), for each coefficient a_j. Noting that

$$\frac{1}{n-1} \sum_{i=1}^{n} (x_{ji}^0)^2 = s_{x_j^0}^2 = 1,$$

we have a system of normal equations of the form

$$\begin{aligned} a_1 + a_2 r_{x_1 x_2}^* + a_3 r_{x_1 x_3}^* + \ldots + a_k r_{x_1 x_k}^* &= r_{yx_1}^* \\ a_1 r_{x_2 x_1}^* + a_2 + a_3 r_{x_2 x_3}^* + \ldots + a_k r_{x_2 x_k}^* &= r_{yx_2}^* \\ \cdots\cdots\cdots\cdots\cdots\cdots & \\ \cdots\cdots\cdots\cdots\cdots\cdots & \\ a_1 r_{x_k x_1}^* + a_2 r_{x_k x_2}^* + a_3 r_{x_k x_3}^* + \ldots + a_k &= r_{yx_k}^* \end{aligned} \tag{4.99}$$

It should be remembered that $r_{x_l x_m}^* = r_{x_m x_l}^*$. For multivariable processes, Eq. (4.99) has a very high order, and it can be solved only on a computer.

On solving Eq. (4.99), one finds the coefficient of multiple correlation, R:

$$R = \sqrt{a_1 r_{yx_1}^* + a_2 r_{yx_2}^* + \ldots + a_k r_{yx_k}^*} \tag{4.100}$$

The coefficient of multiple correlation is a measure of the strength of correlation in the case of multiple regression:

$$0 \leqslant R \leqslant 1 \tag{4.101}$$

With samples of small size, the value of R must be corrected for the systematic error. The fewer the degrees of freedom, $v = n-l$, the more the degree of correlation given by the multiple correlation coefficient is overestimated. The correction formula is defined as

$$R' = \sqrt{1-(1-R^2)(n-1)/(n-l)} \tag{4.102}$$

where R' is the corrected multiple correlation coefficient, l is the number of coefficients in the regression equation. For Eq. (4.92), $l = k+1$.

Before Eq. (4.96) can be used for practical purposes, one should change back to the natural scale, using the equations

$$\begin{aligned} b_j &= a_j s_y / s_{x_j} \qquad j = 1, 2, \ldots, k \qquad j \neq 0, \\ b_0 &= \bar{y} - \sum_{j=1}^{k} b_j \bar{x}_j \end{aligned} \tag{4.103}$$

With replicate observations, we can find the error mean square (the replication variance) and analyse the equation of regression statistically.

Example 4.4. Determine how the degree of extraction (y) of sulphuric acid from pickling liquor depends on the following factors: the concentration of sulphuric acid in the fresh liquor, x_1; the concentration of iron sulphate, x_2; and the volumetric ratio of alcohol to acid, x_3. The raw statistical data are available as a sample of 105 observations obtained in a passive experiment.

Solution. It is known that the degree of extraction of sulphuric acid is a linear function of the factors listed above in the region of interest. Therefore, the respective relationship can be described by a linear regression equation

$$\hat{y} = b_0 + b_1x_1 + b_2x_2 + b_3x_3$$

which can be solved by the multiple correlation method. Using Eqs. (4.93), we convert all experimental data to a standard scale. Then we find the sample correlation coefficients from Eqs. (4.95):

$$r^*_{yx_1} = 0.212, \quad r^*_{yx_2} = 0.043, \quad r^*_{yx_3} = 0.903,$$

$$r^*_{x_1x_2} = -0.417, \quad r^*_{x_1x_3} = -0.128, \quad r^*_{x_2x_3} = 0.046$$

Next we substitute the above correlation coefficients into Eqs. (4.99) and obtain

$$a_1 - 0.417a_2 - 0.128a_3 = 0.212$$

$$-0.417a_1 + a_2 + 0.046a_3 = 0.043$$

$$-0.128a_1 + 0.046a_2 + a_3 = 0.903$$

On solving the set of equations, we get $a_1 = 0.397$, $a_2 = 0.166$, and $a_3 = 0.903$, and recast the regression equation on a standard scale:

$$\hat{y}^0 = 0.397x_1^0 + 0.166x_2^0 + 0.903x_3^0$$

or by (4.103) on the natural scale:

$$\bar{y} = -26.5 + 1.987x_1 + 1.17x_2 + 14.14x_3$$

The goodness of fit of the above equation can be ascertained by applying the Fisher (F-) test:

$$F = s^2_{\text{res}}/s^2_e$$

The error mean square (replication variance), s^2_e, can be found from three replicate observations

$$s^2_e = \frac{1}{2}\sum_{u=1}^{3}(y^0_u - \bar{y}^0) = 3.82$$

where $\bar{y}^0$ is the replicate mean.

The error mean square s^2_e has two degrees of freedom. The residual variance is found by Eq. (4.53):

$$s^2_{\text{res}} = \textstyle\sum_{i=1}^{105}(y_i - \hat{y}_i)^2/(105-4) = 36.03$$

The residual variance s^2_{res} has 101 degrees of freedom; the F-ratio is 9.4. The tabulated value of the F-ratio for the significance level $\alpha = 0.05$ and degrees of freedom $\nu_1 = 101$ and $\nu_2 = 2$ is

$$F_{1-\alpha}(\nu_1, \nu_2) = 19.5$$

Therefore, the regression equation obtained shows a good fit to the experimental data.

4.11. Regression Analysis in Matrix Form

This form of regression analysis is readily adaptable to computer solution. The first step is to estimate the coefficients of the regression equation by the least squares procedure on the basis of Table 4.2:

$$\hat{y} = b_0x_0+b_1x_1+b_2x_2+\ldots+b_kx_k$$

where x_0 is a dummy variable equal to unity.

Now the raw statistical material is presented in matrix form. The matrix

$$X = \begin{bmatrix} x_{01} & x_{11} & \ldots & x_{k1} \\ x_{02} & x_{12} & \ldots & x_{k2} \\ \cdot & \cdot & & \cdot \\ \cdot & \cdot & & \cdot \\ \cdot & \cdot & & \cdot \\ x_{0n} & x_{1n} & \ldots & x_{kn} \end{bmatrix} \tag{4.104}$$

will be called a matrix of independent variables, and the column vector

$$Y = \begin{bmatrix} y_1 \\ y_2 \\ \cdot \\ \cdot \\ \cdot \\ y_n \end{bmatrix} \tag{4.105}$$

will be called the observation vector. We also introduce a coefficient column of the form

$$B = \begin{bmatrix} b_0 \\ b_1 \\ \cdot \\ \cdot \\ \cdot \\ b_k \end{bmatrix}$$

and the transpose of matrix X

$$X^T = \begin{bmatrix} x_{01} & x_{02} & \ldots & x_{0n} \\ x_{11} & x_{12} & \ldots & x_{1n} \\ .. & .. & \ldots & .. \\ .. & .. & \ldots & .. \\ x_{k1} & x_{k2} & \ldots & x_{kn} \end{bmatrix} \tag{4.106}$$

The set of normal equations for estimating $b_0, b_1, \ldots, b_k$ has the form

$$\begin{aligned}
b_0 \sum_{i=1}^{n} x_{0i}^2 + b_1 \sum_{i=1}^{n} x_{0i}x_{1i} + \ldots + b_k \sum_{i=1}^{n} x_{0i}x_{ki} &= \sum_{i=1}^{n} x_{0i}y_i \\
b_0 \sum_{i=1}^{n} x_{1i}x_{0i} + b_1 \sum_{i=1}^{n} x_{1i}^2 + \ldots + b_k \sum_{i=1}^{n} x_{1i}x_{ki} &= \sum_{i=1}^{n} x_{1i}y_i \\
\cdots\cdots\cdots\cdots\cdots\cdots\cdots\cdots\cdots\cdots\cdots\cdots \\
\cdots\cdots\cdots\cdots\cdots\cdots\cdots\cdots\cdots\cdots\cdots\cdots \\
b_0 \sum_{i=1}^{n} x_{ki}x_{0i} + b_1 \sum_{i=1}^{n} x_{ki}x_{1i} + \ldots + b_k \sum_{i=1}^{n} x_{ki}^2 &= \sum_{i=1}^{n} x_{ki}y_i
\end{aligned} \tag{4.107}$$

In matrix form, they reduce to

$$X^T XB = X^T Y \tag{4.108}$$

because multiplying matrix X^T by matrix X gives

$$X^T X = \begin{bmatrix} \sum x_{0i}^2 & \sum x_{0i}x_{1i} & \ldots & \sum x_{0i}x_{ki} \\ \sum x_{1i}x_{0i} & \sum x_{1i}^2 & \ldots & \sum x_{1i}x_{ki} \\ \cdots & \cdots & \cdots & \cdots \\ \cdots & \cdots & \cdots & \cdots \\ \sum x_{ki}x_{0i} & \sum x_{ki}x_{1i} & \ldots & \sum x_{ki}^2 \end{bmatrix} \tag{4.109}$$

where $X^T X$ is the moment matrix. Multiplying matrix $X^T X$ by the B column vector gives a column vector

$$X^T XB = \begin{bmatrix} b_0 \sum x_{0i}^2 + b_1 \sum x_{0i}x_{1i} + \ldots + b_k \sum x_{01}x_{ki} \\ b_0 \sum x_{1i}x_{01} + b_1 \sum x_{1i}^2 + \ldots + b_k \sum x_{1i}x_{ki} \\ \cdots\cdots\cdots\cdots\cdots\cdots \\ \cdots\cdots\cdots\cdots\cdots\cdots \\ b_0 \sum x_{ki}x_{0i} + b_1 \sum x_{ki}x_{1i} + \ldots + b_k \sum x_{ki}^2 \end{bmatrix} \tag{4.110}$$

Multiplying matrix X^T by the observation vector, Y, we obtain

$$X^T Y = \begin{bmatrix} \sum_{i=1}^{n} x_{0i}y_i \\ \sum_{i=1}^{n} x_{1i}y_i \\ \cdot \\ \cdot \\ \cdot \\ \sum_{i=1}^{n} x_{ki}y_i \end{bmatrix} \tag{4.111}$$

From Eq. (4.108), the column vector $\boldsymbol{B}$ of coefficients is defined as

$$\boldsymbol{B} = (\boldsymbol{X}^T\boldsymbol{X})^{-1}\boldsymbol{X}^T\boldsymbol{Y} \qquad (4.112)$$

where $(\boldsymbol{X}^T\boldsymbol{X})^{-1}$ is an inverse of matrix $(\boldsymbol{X}^T\boldsymbol{X})$:

$$(\boldsymbol{X}^T\boldsymbol{X})^{-1} = \begin{bmatrix} c_{00} & c_{01} & \dots & c_{0k} \\ c_{10} & c_{11} & \dots & c_{1k} \\ \dots & \dots & \dots & \dots \\ \dots & \dots & \dots & \dots \\ c_{k0} & c_{k1} & \dots & c_{kk} \end{bmatrix} \qquad (4.113)$$

The elements of the inverse matrix are defined by the relation

$$c_{ju} = \frac{1}{\Delta}\left(\sum_{i=1}^{n} x_{ui}\, x_{ji}\right)' \qquad (4.114)$$

where Δ is the determinant of product matrix $\boldsymbol{X}^T\boldsymbol{X}$, $(\sum_{i=1}^{n} x_{ui}\, x_{ji})'$ is the cofactor of the element $(\sum_{i=1}^{n} x_{ui}\, x_{ji})$ in matrix $\boldsymbol{X}^T\boldsymbol{X}$.

For an inverse matrix to exist, matrix $\boldsymbol{X}^T\boldsymbol{X}$ must be nonsingular. Therefore, in using this method of calculation, it is important that the variables x_1, x_2, ..., x_k be linearly independent. Then the elements of one column in the matrix of independent variables will not be a linear combination of the respective elements in other columns.

To find the residual mean square, one should evaluate the $\hat{\boldsymbol{Y}}$ column vector

$$\hat{\boldsymbol{Y}} = \begin{bmatrix} \hat{y}_1 \\ \hat{y}_2 \\ \cdot \\ \cdot \\ \cdot \\ \hat{y}_n \end{bmatrix} = \boldsymbol{X}\boldsymbol{B} \qquad (4.115)$$

The numerator of the residual mean square is found by multiplying the matrices together:

$$[\boldsymbol{Y}-\hat{\boldsymbol{Y}}]^T[\boldsymbol{Y}-\hat{\boldsymbol{Y}}] = \textstyle\sum_{i=1}^{n} (y_i-\hat{y}_i)^2 \qquad (4.116)$$

Let β designate the column vector of true regression coefficients and let the expected value of $\boldsymbol{B}$ be $E(\boldsymbol{B}) = \beta$. Then

$$E\{[\boldsymbol{B}-\beta]\,[\boldsymbol{B}-\beta]^T\} = \begin{bmatrix} \sigma^2_{b_0} & \mathrm{Cov}_{b_0 b_1} & \dots & \mathrm{Cov}_{b_0 b_k} \\ \mathrm{Cov}_{b_1 b_0} & \sigma^2_{b_1} & \dots & \mathrm{Cov}_{b_1 b_k} \\ \dots & \dots & \dots & \dots \\ \dots & \dots & \dots & \dots \\ \mathrm{Cov}_{b_k b_0} & \mathrm{Cov}_{b_k b_1} & \dots & \sigma^2_{b_k} \end{bmatrix} \qquad (4.117)$$

where $\sigma^2_{b_j}$ is the variance of b_j, $\text{Cov}_{b_j b_k}$ is the covariance, or the correlation moment between b_j and b_k.

Thus, the diagonal terms of the matrix are the coefficient variances required for significance testing, while the nondiagonal terms are the covariances of the respective regression estimators determining the statistical relationship between the coefficients.

Let the matrix $E\{[\boldsymbol{B}-\beta][\boldsymbol{B}-\beta]^T\}$ be expressed in terms of observations, noting that $\boldsymbol{B} = (\boldsymbol{X}^T\boldsymbol{X})^{-1}\boldsymbol{X}^T\boldsymbol{Y}$. Hence:

$$E\{[\boldsymbol{B}-\beta][\boldsymbol{B}-\beta]^T\} = E\{(\boldsymbol{X}^T\boldsymbol{X})^{-1}\boldsymbol{X}^T\boldsymbol{Y}^0[(\boldsymbol{X}^T\boldsymbol{X})^{-1}\boldsymbol{X}^T\boldsymbol{Y}^0]^T\}$$

where $\boldsymbol{Y}^0$ is a random normal vector with independent elements having a variance of σ_y^2:

$$\boldsymbol{Y}^0 = \boldsymbol{Y} - E(\boldsymbol{Y}) = \begin{bmatrix} y_1 - m(y_1) \\ y_2 - m(y_2) \\ \cdot \\ \cdot \\ \cdot \\ y_n - m(y_n) \end{bmatrix}$$

The coefficient matrix of normal equations is symmetrical, and so

$$[(\boldsymbol{X}^T\boldsymbol{X})^{-1}]^T = (\boldsymbol{X}^T\boldsymbol{X})^{-1}$$

Putting $\sigma^2_{y_1} = \sigma^2_{y_2} = \ldots = \sigma^2_{y_n} = \sigma^2_y$ and recalling that errors are statistically independent, we get

$$E\{\boldsymbol{Y}^0\boldsymbol{Y}^{0T}\} = \begin{bmatrix} \sigma^2_{y_1} & & & & 0 \\ & \sigma^2_{y_2} & & & \\ & & \cdot & & \\ & & & \cdot & \\ & & & & \cdot \\ 0 & & & & \sigma^2_{y_n} \end{bmatrix} = \boldsymbol{C}\sigma^2_y \qquad (4.118)$$

Thus, we have

$$E[(\boldsymbol{B}-\beta)(\boldsymbol{B}-\beta)^T] = (\boldsymbol{X}^T\boldsymbol{X})^{-1}\sigma^2_y \qquad (4.119)$$

whence

$$\sigma^2_{b_j} = c_{jj}\sigma^2_y, \quad \text{Cov}_{b_j b_u} = c_{ju}\sigma^2_y \qquad (4.120)$$

The matrix $(\boldsymbol{X}^T\boldsymbol{X})^{-1}$ is called an *error matrix*, or a *variance-(covariance) matrix*.

Since the variance matrix is non-diagonal and, as a consequence, all regression coefficients are interrelated, each coefficient cannot be tested for significance separately. Therefore, the relation

$$t_j = \frac{|b_j|}{s_y\sqrt{c_{jj}}} \qquad (4.121)$$

can only be looked upon as a means for ranking the factors. A procedure has been proposed for consecutively eliminating insignificant factors: a factor for which t_j is the least is dropped and the calculation is repeated, as long as the residual mean square decreases. This improves the interpolating properties of the regression equation, but the resultant coefficients are biased estimates. With a larger number of factors, the calculation of multiple regression calls for the use of a digital computer.

4.12. Development of Multiple Regression Equations by Brandon's Method [17]

In this method, the estimate regression equation is given the form

$$\hat{y} = af_1(x_1)\, f_2(x_2)\ \ldots f_j(x_j) \ldots f_k(x_k) \tag{4.122}$$

where $f_j(x_j)$ is any function of x_j.

The arrangement of the factors $x_1, x_2, \ldots, x_k$ in Eq. (4.122) has a direct bearing on the accuracy in the reduction of observations: the greater the effect that x_j has on y, the smaller the j-index should be. The form of the function f_j is chosen on the basis of graphical construction. To begin with, a correlation field (scatter diagram) and an empirical regression line, $y - x_1$, are plotted on the basis of samples of quantities $y_1, x_1, x_2, \ldots, x_k$. Thus, one finds that $\hat{y}_{x1} = f_1(x_1)$ and, using the least squares procedure, estimates the coefficients of this regression equation. Now a new sample is built up

$$y_1 = y/f_1(x_1) \tag{4.123}$$

This quantity is no longer dependent on x_1, and solely determined by the parameters $x_2, x_3, \ldots, x_k$. Therefore, one may write

$$\hat{y}_1 = af_2(x_2)f_3(x_3) \ldots f_k(x_k) \tag{4.124}$$

Using new samples of y_1 and x_2, another correlation field and another empirical regression line depicting the dependence of y_1 on x_2 are constructed such that

$$\hat{y}_{x_2} = f_2(x_2)$$

The coefficients are again estimated, and a sample of the new quantity is built up

$$y_2 = y_1/f_2(x_2) = y/f_1(x_1)f_2(x_2) \tag{4.125}$$

This quantity is independent of two factors, x_1 and x_2, and can be found from the following regression equation:

$$\hat{y}_2 = af_3(x_3) \ldots f_k(x_k) \tag{4.126}$$

This procedure is iterated until the sample of y_k is obtained

$$y_k = \frac{y_{k-1}}{f(x_k)} = \frac{y}{f_1(x_1)f_2(x_2) \ldots f_k(x_k)} \tag{4.127}$$

This quantity is independent of all factors, $x_1, \ldots, x_k$, and is defined by the coefficient of the original equation

$$\hat{y}_k = a = \frac{1}{n}\sum_{i=1}^{n} y_{ki}, \tag{4.128}$$

where n is the sample size.

Example 4.5. [18]. Determine how the output of a high-density polyethylene tubular reactor depends on the process variables. The factors affecting the reactor output (y) are as follows: reactor pressure, x_1; reactor temperature, x_2; oxygen concentration in the reaction feed, x_3; and flow-rate of the feed gas, x_4. The raw statistical data are a sample of 200 observations carried out under normal process conditions.

Solution. According to Eq. (4.122), the relationship between the reactor output and the selected factors can be written as

$$\hat{y} = af_1(x_1)f_2(x_2)f_3(x_3)f_4(x_4)$$

Let us evaluate the unknown functions $f_j(x_j)$ and the coefficient a by Brandon's method. To begin with, we plot y as a function of reactor pressure, x_1 (Fig. 4.4a). The empirical regression line shows that the function $f_1(x_1)$ should preferably be described by the equation of a second-order parabola:

$$f_1(x_1) = b_0 + b_1x_1 + b_{11}x_1^2$$

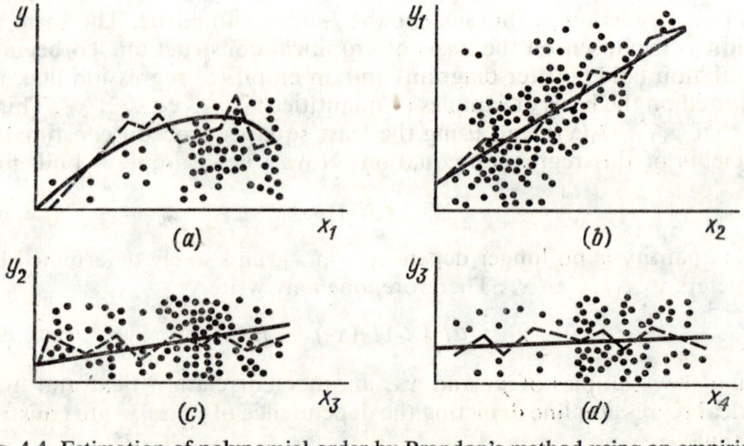

Fig. 4.4. Estimation of polynomial order by Brandon's method using an empirical line regression

Evaluating the coefficients b_0, b_1, and b_{11} by the method of least squares, we get

$$f_1(x_1) = -211 + 0.033x_1 - 1.16\times10^{-4}x_1^2$$

Then using Eq. (4.123), we determine the sample size for y_1 and plot the correlation field and the empirical regression line $y_1 - x_2$ (Fig 4.4b). It is well approximated by the following linear equation of regression:

$$\hat{y}_{x_2} = f_2(x_2) = -1.46 + 0.013x_2$$

Following a similar procedure for the remaining factors (Fig. 4.4c and d), we finally evaluate the relationship between the reactor output and the selected process variables:

$$\begin{aligned}\hat{y} = {} & 1.02(-211 + 0.33x_1 - 1.16\times10^{-4}x_1^2)\\ & \times(0.013x_2 - 1.46)\,(0.0077x_3 + 0.42)\\ & \times(0.00127x_4 + 0.747)\end{aligned}$$

Part Two

DESIGN OF EXPERIMENTS

Chapter Five

DESIGN OF EXPERIMENTS WITH EXTREMES

Many experiments in chemistry and chemical engineering belong to the extremum problem type, e.g. determination of the optimum conditions of a process, finding of the optimum constituents of a compound, *etc.* In such cases to effectively tackle the problem a specific design of an experiment is selected depending on the goals to be achieved in experimentation and the features of the object being investigated. The investigation scheme is usually broken down into some individual steps to evaluate the strategy for further experimentation by using the experimental results obtained from each previous step. Thus the experimenter is able to conduct the experiment in an optimum way. With a designed experiment, it is possible to vary all factors simultaneously and obtain quantitative estimates of the main effects and factor interactions. The designed experiment ensures a lesser error in determining the effects of interest than any orthodox method could afford. Arranging the points in experimental, or factor, space in an optimal way and using a linear transformation of coordinates the experimenter can avoid some disadvantages of the classical regression analysis, in particular, the computation of the correlation coefficients for a regression equation. All in all when the experimenter plans his experiments, he achieves wanted results in a much more economical manner.

5.1. Full-Factorial Design

A full factorial design is one in which all possible combinations of the factors at all levels involved in the experiment are used. The number N of all possible combinations to be performed during the experiment is given by

$$N = n^k$$

where n is the number of levels and k is the number of factors.

In the case of a two-level factorial experiment in which every factor is fixed at two levels and all combinations of k factors are used, we deal with a

2^k full factorial design. The term level of the factor will refer to the boundary of the region to be searched for a given process variable.

Let one be interested in the effect of three factors on the product yield (y, %), namely temperature (z_1) in the range 100–200 °C, pressure (z_2) in the range 2×10^5 - 6×10^5 Pa, and residence time in the range 10 to 20 min. If the upper temperature level z_1^{max} is 200 °C and the lower level z_1^{min} is 100 °C, then we have

$$z_1^0 = \frac{z_1^{max}+z_1^{min}}{2} = 150°$$

$$\Delta z_1 = \frac{z_1^{max}-z_1^{min}}{2} = 50°$$

Generally, for any factor z_j, we have

$$z_j^0 = \frac{z_j^{max}+z_j^{min}}{2}; \quad j = 1, 2, \ldots, k \tag{5.1}$$

$$\Delta z_j = \frac{z_j^{max}-z_j^{min}}{2} \tag{5.2}$$

The point with coordinates ($z_1^0, z_2^0, \ldots, z_k^0$) is called the *center point of the design*, or the *basic level*, Δz_j is the *unit* or *interval of variation* on the z_j-axis. It is usual to pass from the $z_1, z_2, \ldots, z_k$ coordinates to a new dimensionless system of coordinates $x_1, x_2, \ldots, x_k$ by the coding equation

$$x_j = \frac{z_j - z_j^0}{\Delta z_j}; \quad j = 1, 2, \ldots, k \tag{5.3}$$

In the dimensionless coordinate system, the upper and lower levels are at +1 and −1, respectively, the coordinates of the center point of the design are zero and coincide with the origin of coordinates. In our problem, $k = 3$. The number of all possible combinations, N, of the three factors each fixed at two levels is $N = 2^k = 2^3 = 8$. The design matrix may be reduced to a table form such as Table 5.1.

Table 5.1. 2^3 Full Factorial Design

	Factors on natural scale			Coded factors			Yield
observation No.	z_1	z_2	z_3	x_1	x_2	x_3	y
1	100	20	10	−1	−1	−1	2
2	200	20	10	+1	−1	−1	6
3	100	60	10	−1	+1	−1	4
4	200	60	10	+1	+1	−1	8
5	100	20	30	−1	−1	+1	10
6	200	20	30	+1	−1	+1	18
7	100	60	30	−1	+1	+1	8
8	200	60	30	+1	+1	+1	12

The coded design of Table 5.1 can be portrayed geometrically as a cube (Fig. 5.1) whose eight corners represent the eight experimental points. If we augment the design matrix for the 2^3 full factorial design with a column

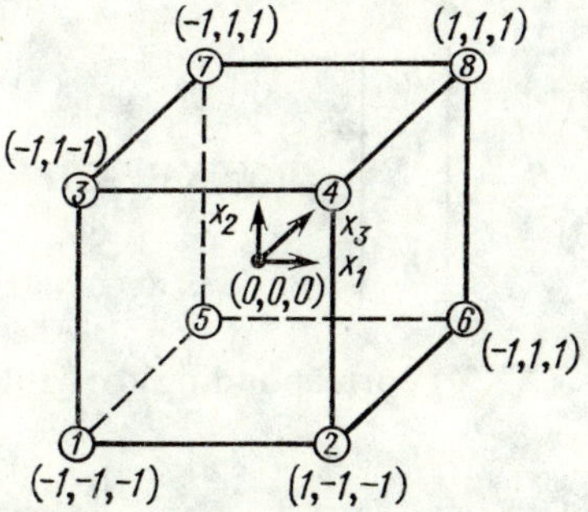

Fig. 5.1. Full factorial design 2^3

Table 5.2. Experimental Matrix

N	x_0	x_1	x_2	x_3	y
1	+1	−1	−1	−1	y_1
2	+1	+1	−1	−1	y_2
3	+1	−1	+1	−1	y_3
4	+1	+1	+1	−1	y_4
5	+1	−1	−1	+1	y_5
6	+1	+1	−1	+1	y_6
7	+1	−1	+1	+1	y_7
8	+1	+1	+1	+1	y_8

of 1's for a dummy variable ($x_0 = 1$), we have the experimental matrix shown in Table 5.2.

The experimental matrix of Table 5.2 has the following properties

$$\textstyle\sum_{i=1}^{N} x_{ui}x_{ji} = 0, \qquad u \neq j;\ u,\ j = 0,\ 1,\ \ldots,\ k \tag{5.4a}$$

$$\textstyle\sum_{i=1}^{N} x_{ji} = 0, \qquad j = 1,\ 2,\ \ldots,\ k;\ j \neq 0; \tag{5.4b}$$

$$\textstyle\sum_{i=1}^{N} x_{ji}^2 = N; \qquad j = 0,\ 1,\ \ldots,\ k \tag{5.4c}$$

where k is the number of factors and N is the number of experiments. The first property expressed by Eq. (5.4*a*) is called the property of *orthogonality*, that is, the equality of the scalar products of all column vectors to zero. This property materially reduces the difficulties in the estimation of coefficients for the regression equation, because the coefficient matrix $(\boldsymbol{X}^T\boldsymbol{X})$ of normal equations becomes diagonal, and its diagonal elements are equal to the number of experiments, N, in the design matrix.

Using the least squares method, we can determine coefficients of the regression equation in a way of matrix calculus

$$\boldsymbol{B} = \begin{bmatrix} b_0 \\ b_1 \\ b_2 \\ b_3 \end{bmatrix} = (\boldsymbol{X}^T\boldsymbol{X})^{-1}\boldsymbol{X}^T\boldsymbol{Y}$$

The moment matrix ($\boldsymbol{X}^T\boldsymbol{X}$) corresponding to Table 5.2, $k = 3$, takes the form ($\sum = \sum_{i=1}^{8}$):

$$\boldsymbol{X}^T\boldsymbol{X} = \begin{bmatrix} \sum x_{0i}^2 & \sum x_{0i}x_{1i} & \sum x_{0i}x_{2i} & \sum x_{0i}x_{3i} \\ \sum x_{1i}x_{0i} & \sum x_{1i}^2 & \sum x_{1i}x_{2i} & \sum x_{1i}x_{3i} \\ \sum x_{2i}x_{0i} & \sum x_{2i}x_{1i} & \sum x_{2i}^2 & \sum x_{2i}x_{3i} \\ \sum x_{3i}x_{0i} & \sum x_{3i}x_{1i} & \sum x_{3i}x_{2i} & \sum x_{3i}^2 \end{bmatrix}$$

In agreement with the properties described by Eqs. (5.4), a diagonal moment matrix results

$$\boldsymbol{X}^T\boldsymbol{X} = \begin{bmatrix} 8 & 0 & 0 & 0 \\ 0 & 8 & 0 & 0 \\ 0 & 0 & 8 & 0 \\ 0 & 0 & 0 & 8 \end{bmatrix} \tag{5.5}$$

The inverse of the matrix product $\boldsymbol{X}^T\boldsymbol{X}$ (the variance matrix) is

$$(\boldsymbol{X}^T\boldsymbol{X})^{-1} = \begin{bmatrix} 1/8 & 0 & 0 & 0 \\ & 1/8 & 0 & 0 \\ & & 1/8 & 0 \\ \text{sym} & & & 1/8 \end{bmatrix} \tag{5.6}$$

The last product for $\boldsymbol{B}$ is

$$\boldsymbol{X}^T\boldsymbol{Y} = \begin{bmatrix} \sum x_{0i}y_i \\ \sum x_{1i}y_i \\ \sum x_{2i}y_i \\ \sum x_{3i}y_i \end{bmatrix} \tag{5.7}$$

Thus we have

$$B = \begin{bmatrix} b_0 \\ b_1 \\ b_2 \\ b_3 \end{bmatrix} = \begin{bmatrix} 1/8 & & & 0 \\ & 1/8 & & \\ & & 1/8 & \\ 0 & & & 1/8 \end{bmatrix} \times \begin{bmatrix} \sum x_{0i} y_i \\ \sum x_{1i} y_i \\ \sum x_{2i} y_i \\ \sum x_{3i} y_i \end{bmatrix} \quad (5.8)$$

As a consequence, any coefficient b_j of the estimated regression equation is defined by the scalar product of a y column by the respective x_j column divided into the number of experiments N in the design matrix

$$b_j = \frac{1}{N} \sum_{i=1}^{N} x_{ji} y_i \quad (5.9)$$

Using the design presented in Table 5.1, we first find the coefficients of the estimated linear regression equation

$$\hat{y} = b_0 + b_1 x_1 + b_2 x_2 + b_3 x_3 \quad (5.10)$$

For example, b_1, can be determined by multiplying together the x_1 and y columns and adding together the products

$$\begin{array}{c} x_1 \\ \begin{bmatrix} -1 \\ +1 \\ -1 \\ +1 \\ -1 \\ +1 \\ -1 \\ +1 \end{bmatrix} \end{array} \times \begin{array}{c} y \\ \begin{bmatrix} 2 \\ 6 \\ 4 \\ 8 \\ 10 \\ 18 \\ 8 \\ 12 \end{bmatrix} \end{array} = \begin{bmatrix} -2 \\ +6 \\ -4 \\ +8 \\ -10 \\ +18 \\ -8 \\ +12 \end{bmatrix}$$

$$\sum_{i=1}^{8} x_{1i} y_i = 20$$

$$b_1 = \frac{\sum^{8} x_{1i} y_i}{8} = \frac{20}{8} = +2.5$$

Similarly,

$$b_0 = 8.5 \quad b_2 = -0.5 \quad b_3 = +3.5$$

In the case of a fuller regression equation containing two and three factor interaction coefficients

$$\hat{y} = b_0 + b_1 x_1 + b_2 x_2 + b_3 x_3 + b_{12} x_1 x_2 + b_{13} x_1 x_3 + b_{23} x_2 x_3 + b_{123} x_1 x_2 x_3 \quad (5.11)$$

where one has to determine b_{12}, b_{13} and b_{23} (the second order interaction terms) and b_{123} (the third order interaction term), the experimental matrix of Table 5.2 has an augmented form shown in Table 5.3.

Table 5.3. Experimental Matrix for 2^3 Full Experimental Design

N	x_0	x_1	x_2	x_3	x_1x_2	x_1x_3	x_2x_3	$x_1x_2x_3$	y
1	+1	−1	−1	−1	+1	+1	+1	−1	2
2	+1	+1	−1	−1	−1	−1	+1	+1	6
3	+1	−1	+1	−1	−1	+1	−1	+1	4
4	+1	+1	+1	−1	+1	−1	−1	−1	8
5	+1	−1	−1	+1	+1	−1	−1	+1	10
6	+1	+1	−1	+1	−1	+1	−1	−1	18
7	+1	−1	+1	+1	−1	−1	+1	−1	8
8	+1	+1	+1	+1	+1	+1	+1	+1	12

The interaction terms are determined in the same manner as the linear terms. To find b_{12}

$$\begin{array}{c} x_1x_2 \\ \begin{bmatrix} +1 \\ -1 \\ +1 \\ +1 \\ +1 \\ -1 \\ -1 \\ +1 \end{bmatrix} \end{array} \times \begin{array}{c} y \\ \begin{bmatrix} 2 \\ 6 \\ 4 \\ 8 \\ 10 \\ 18 \\ 8 \\ 12 \end{bmatrix} \end{array} = \begin{bmatrix} +2 \\ -6 \\ -4 \\ +8 \\ +10 \\ -18 \\ -8 \\ +12 \end{bmatrix}$$

$$\sum_{i=1}^{8} (x_1x_2)_i y_i = -4$$

$$b_{12} = \frac{1}{N} \sum_{i=1}^{N} (x_1x_2)_i y_i = \frac{-4}{8} = -0.5$$

The remaining coefficients are found in a similar manner

$$b_{13} = +0.5 \quad b_{23} = -1.5 \quad b_{123} = 0.11$$

By making replicate observations, it will be possible to determine s_e^2, to test the significance of the regression coefficients and, in the case of degrees of freedom, also the adequacy of the equation.

Since the variance (covariance) matrix $(\boldsymbol{X}^T\boldsymbol{X})^{-1}$ for a designed experiment is a diagonal matrix

$$(\boldsymbol{X}^T\boldsymbol{X})^{-1} = \begin{bmatrix} 1/N & & & & 0 \\ & 1/N & & & \\ & & \cdot & & \\ & & & \cdot & \\ & & & & \cdot \\ 0 & & & & 1/N \end{bmatrix} \tag{5.12}$$

the coefficients of the regression equation are uncorrelated. The significance of each coefficient can be tested on the basis of the Student t-test. Deletion of an insignificant coefficient from the regression equation, Eq. (5.11), will have no effect on the remaining coefficients. Thus the sample coefficients b_j prove to be *unbiased estimates* for the respective actual coefficients β_j

$$E(b_j) = \beta_j \quad \text{or} \quad b_j \to \beta_j \tag{5.13}$$

that is, the coefficients of the regression equation represent the contribution of each factor to y without confounding.

The diagonal elements of the variance matrix are equal to one another, and so all coefficients in Eqs. (5.11) and (5.12) can be determined with the same accuracy

$$s_{b_j} = s_e/\sqrt{N} \tag{5.14}$$

For example, three replicate observations have been made at the center point of the design to yield the following values: $y_1^0 = 8$, $y_2^0 = 9$, and $y_3^0 = 8.8$. Hence

$$\bar{y}^0 = \frac{1}{3} \sum_{u=1}^{3} y_u^0 = 8.6$$

$$s_e^2 = \frac{1}{2} \sum_{u=1}^{3} (y_u^0 - y^0)^2 = 0.28$$

$$s_e = 0.55$$

$$s_{b_j} = \frac{0.55}{\sqrt{8}} = 0.2$$

The significance of the coefficients can be tested, using the Student t-test

$$t_0 = \frac{|b_0|}{s_{b_0}} = \frac{8.5}{0.2} = 42.5 \qquad t_1 = \frac{|b_1|}{s_{b_1}} = \frac{2.6}{0.2} = 12.5$$

$$t_2 = \frac{|b_2|}{s_{b_2}} = 2.5 \qquad t_3 = \frac{|b_3|}{s_{b_3}} = 17.5$$

$$t_{12} = \frac{|b_{12}|}{s_{b_{12}}} = 2.5 \qquad t_{13} = \frac{|b_{13}|}{s_{b_{13}}} = 2.5$$

$$t_{23} = \frac{|b_{23}|}{s_{b_{23}}} = 7.5 \qquad t_{123} = \frac{|b_{123}|}{s_{b_{123}}} = 1.25$$

For a significance level of $\alpha = 0.05$ and $\nu = 2$ degrees of freedom, the tabulated value of the Student t-distribution is $t_\alpha(\nu) = 4.3$. Thus, the coefficients b_2, b_{12}, b_{13} and b_{123} are insignificant, and they must be deleted from the equation. After the deletion of the insignificant coefficients, the regression equation takes the form

$$\hat{y} = 8.5+2.5x_1+3.5x_3-1.5x_2x_3$$

Now the estimated regression equation is tested to see how it fits the observations, using Fisher's test, the variance ratio, $F = s^2_{\text{res}}/s^2_e$ where

$$s^2_{\text{res}} = \frac{\sum\limits^{8}(y_i-\hat{y}_i)^2}{N-l} = \frac{6}{4} = 1.5$$

$$s^2_e = 0.28$$

and $l = 4$ is the number of significant coefficients in the regression equation. Then

$$F = 1.5 \div 0.28 = 5.3$$

The tabulated value of Fisher's F for $\alpha = 0.05$, $\nu_1 = 4$ and $\nu_2 = 2$ is

$$F_{1-\alpha}(\nu_1, \nu_2) = 19.3$$

Thus, $F < F_{1-\alpha}(\nu_1, \nu_2)$, and the estimated regression equation above fits the experimental data adequately.

5.2. Fractional Replicates

If, in developing a regression equation, a linear approximation is adequate, the amount of experimentation needed may be cut down by using a so-called fractional factorial design, or a fractional replicate of a full factorial design (see Sec. 3.4). If a fractional replicate is to be an orthogonal design in itself, the nearest full factorial design should be taken as a replicate. The number of observations should then be greater than or equal to that of unknown coefficients in the regression equation. Suppose we seek a linear approximation for a small area on the response surface with three factors

$$\hat{y} = b_0+b_1x_1+b_2x_2+b_3x_3$$

Four observations may be sufficient to solve the problem, if the x_1x_2 column in the 2^2 full factorial design (Table 5.4) is used as a factor for x_3 (Table 5.5).

This is known as the half-replicate of a full two-level factorial design for three variables, 2^3. This design is satisfactory for estimating the intercept and the three coefficients of linear terms in the regression equation.

In practice, the equality of interaction terms to zero cannot be postulated *a priori*. However, there may be grounds for assuming that some of them are negligible in comparison with the linear terms. If the regression coeffi-

Table 5.4. 2^2 Full Factorial Design

N	x_0	x_1	x_2	x_1x_2
1	+1	+1	+1	+1
2	+1	−1	−1	+1
3	+1	−1	+1	−1
4	+1	+1	−1	−1

Table 5.5. Half-Replicate of 2^3 Full Factorial Design

N	x_0	x_1	x_2	x_3
1	+1	+1	+1	+1
2	+1	−1	−1	+1
3	+1	−1	+1	−1
4	+1	+1	−1	−1

cients of the paired products are non-zero, the estimated coefficients are *biased estimates* of the actual coefficients as

$$b_1 \rightarrow \beta_1+\beta_{23} \quad b_2 \rightarrow \beta_2+\beta_{13} \quad b_3 \rightarrow \beta_3+\beta_{12} \tag{5.15}$$

where the β's are the expected values of the respective coefficients.

The actual coefficients above cannot be estimated separately on the basis of a design envisaging as few as four trials (Table 5.5) because there is no way to distinguish between the columns for linear terms and paired products. Such factors are said to be *aliased* together. If, for example, in addition to the columns in Table 5.5, we evaluate one more column for the product x_1x_3, we shall see that the elements of this column are identically equal to the elements of the x_2-column. Thus, short of running the entire design, it is impossible to distinguish such aliases apart, which results in the appearance of biased estimates of the coefficients. In order to determine which actual effects are confounded, it is convenient to substitute x_3 for x_1x_2 (Table 5.5) to obtain

$$x_3 = x_1x_2 \tag{5.16}$$

known as the *defining relation*. Multiplying both sides of the defining relation by x_3 gives

$$x_3^2 = x_1x_2x_3$$

What we have got on the left-hand side is a unit column

$$\mathbf{1} = x_1x_2x_3 \tag{5.17}$$

The product, Eq. (5.17), is called the *defining contrast*. It readily identifies the columns containing like elements.

Multiplying the defining contrast in turn by x_1, x_2, x_3, we get

$$x_1 = x_1^2x_2x_3 = x_2x_3, \quad x_2 = x_1x_3, \quad x_3 = x_1x_2 \tag{5.18}$$

The above relations (5.18) correspond to a system of biased estimates (5.15).

When use is made of a fractional design, we must have a clear idea about the resolution of the fractional replicate, that is, know in advance which coefficients are unbiased estimates of the respective general coefficients.

Then, according to the objective sought, we can select a fractional replicate which will yield a maximum amount of information from the experiment.

For example, in the case of a fractional design for four factors $k = 4$, the defining relation may be one of the form

$$x_4 = x_1 x_2 x_3 \tag{5.19}$$

and any of the paired products, say,

$$x_4 = x_1 x_2 \tag{5.20}$$

The design matrix for the defining relation (5.19) is given in Table 5.6.

Table 5.6. Half-Replicate of 2^4 Full Factorial Design for Defining Relation (5.19)

N	x_0	x_1	x_2	x_3	x_4
1	+1	+1	+1	+1	+1
2	+1	−1	−1	+1	+1
3	+1	−1	+1	+1	−1
4	+1	+1	−1	+1	−1
5	+1	+1	+1	−1	−1
6	+1	−1	−1	−1	−1
7	+1	−1	+1	−1	+1
8	+1	+1	−1	−1	+1

Using the defining contrast $\mathbf{1} = x_1 x_2 x_3 x_4$, we get the following system of estimates for the regression coefficients

$$\begin{array}{llll} x_1 = x_2 x_3 x_4 & b_1 \to \beta_1 + \beta_{234} & x_4 = x_1 x_2 x_3 & b_4 \to \beta_4 + \beta_{123} \\ x_2 = x_1 x_3 x_4 & b_2 \to \beta_2 + \beta_{134} & x_1 x_2 = x_3 x_4 & b_{12} \to \beta_{12} + \beta_{34} \\ x_3 = x_1 x_2 x_4 & b_3 \to \beta_3 + \beta_{124} & x_1 x_3 = x_2 x_4 & b_{13} \to \beta_{13} + \beta_{24} \\ & & x_1 x_4 = x_2 x_3 & b_{14} \to \beta_{14} + \beta_{23} \end{array} \tag{5.21}$$

In practice, third-order interactions are equal to zero more often than second-order interactions. In other words, if we are interested more in the estimates of the linear terms, we should preferably use the defining relation $x_4 = x_1 x_2 x_3$.

With a defining relation of the form given by Eq. (5.20), the design will be as shown in Table 5.7.

Table 5.7. Half-replicate of 2^4 Full Fractional Design for Defining Relation (5.20)

N	x_0	x_1	x_2	x_3	x_4
1	+1	+1	+1	+1	+1
2	+1	−1	−1	+1	+1
3	+1	−1	+1	+1	−1
4	+1	+1	−1	−1	−1
5	+1	+1	+1	−1	+1
6	+1	−1	−1	−1	+1
7	+1	−1	+1	−1	−1
8	+1	+1	−1	−1	−1

The defining contrast is defined by $\mathbf{1} = x_1x_2x_4$. Thus, we get the following alias structure and system of estimates

$$\begin{aligned}
x_1 &= x_2x_4 & b_1 &\to \beta_1+\beta_{24}\\
x_2 &= x_1x_4 & b_2 &\to \beta_2+\beta_{14}\\
x_3 &= x_1x_2x_3x_4 & b_3 &\to \beta_3+\beta_{1234}\\
x_4 &= x_1x_2 & b_4 &\to \beta_4+\beta_{12}\\
x_1x_3 &= x_2x_3x_4 & b_{13} &\to \beta_{13}+\beta_{234}\\
x_2x_3 &= x_1x_3x_4 & b_{23} &\to \beta_{23}+\beta_{134}\\
x_3x_4 &= x_1x_2x_3 & b_{34} &\to \beta_{34}+\beta_{123}
\end{aligned} \tag{5.22}$$

Consequently, it will be advantageous to use a fractional replicate with the defining relation $x_4 = x_1x_2$, if our interest is mainly in the coefficients β_{13}, β_{23} and β_{34}.

We see that in the half-replicate of a 2^5 full factorial design all the linear effects and second-order interaction effects are confounded only with the third-order interactions and interactions of higher orders

$$\begin{aligned}
x_0 &= x_1x_2x_3x_4x_5 & b_0 &\to \beta_0+\beta_{12345}\\
x_1 &= x_2x_3x_4x_5 & b_1 &\to \beta_1+\beta_{2345}\\
x_2 &= x_1x_3x_4x_5 & b_2 &\to \beta_2+\beta_{1345}\\
x_3 &= x_1x_2x_4x_5 & b_3 &\to \beta_3+\beta_{1245}\\
x_4 &= x_1x_2x_3x_5 & b_4 &\to \beta_4+\beta_{1235}\\
x_5 &= x_1x_2x_3x_4 & b_5 &\to \beta_5+\beta_{1234}\\
x_1x_2 &= x_3x_4x_5 & b_{12} &\to \beta_{12}+\beta_{345}
\end{aligned}$$

$$
\begin{aligned}
x_1x_3 &= x_2x_4x_5 & b_{13} &\to \beta_{13}+\beta_{245} \\
x_1x_4 &= x_2x_3x_5 & b_{14} &\to \beta_{14}+\beta_{235} \\
x_1x_5 &= x_2x_3x_4 & b_{15} &\to \beta_{15}+\beta_{234} \\
x_2x_3 &= x_1x_4x_5 & b_{23} &\to \beta_{23}+\beta_{145} \\
x_2x_4 &= x_1x_3x_5 & b_{24} &\to \beta_{24}+\beta_{135} \\
x_2x_5 &= x_1x_3x_4 & b_{25} &\to \beta_{25}+\beta_{134} \\
x_3x_4 &= x_1x_2x_5 & b_{34} &\to \beta_{34}+\beta_{125} \\
x_3x_5 &= x_1x_2x_4 & b_{35} &\to \beta_{35}+\beta_{124} \\
x_4x_5 &= x_1x_2x_3 & b_{45} &\to \beta_{45}+\beta_{123}
\end{aligned}
\tag{5.23}
$$

In practice, if one neglects the third-order and higher interaction terms, one may assume that with $k \geqslant 5$ half-replicates of full factorial design provide for unbiased estimates of the linear effects and second-order interaction effects. Use is also made of replicates of higher fractional orders such as one-fourth replicates, one-eighth replicates, etc. Generally, a fractional replicate in which p linear terms are equated to interaction terms is designated 2^{k-p}. If we deal with a one-fourth replicate of a 2^{5-2} design ($k = 5$), the following defining relations can be used

$$x_4 = x_1x_2x_3 \qquad x_5 = x_1x_2$$

In this case, we have the defining contrasts as follows

$$\mathbf{1} = x_1x_2x_3x_4 \qquad \mathbf{1} = x_1x_2x_5 \tag{5.24}$$

Multiplying the defining contrasts together we get a *generalized defining contrast*. With the relations (5.24) taken into account, it completely specifies the resolution of higher fractional replicates

$$\mathbf{1} = x_1x_2x_3x_4 = x_1x_2x_5 = x_3x_4x_5 \tag{5.25}$$

The respective system of biased estimates is

$$
\begin{aligned}
x_0 &= x_1x_2x_3x_4 = x_1x_2x_5 = x_3x_4x_5 & b_0 &\to \beta_0+\beta_{1234}+\beta_{125}+\beta_{345} \\
x_1 &= x_2x_3x_4 = x_2x_5 = x_1x_3x_4x_5 & b_1 &\to \beta_1+\beta_{234}+\beta_{25}+\beta_{1345} \\
x_2 &= x_1x_3x_4 = x_1x_5 = x_2x_3x_4x_5 & b_2 &\to \beta_2+\beta_{134}+\beta_{15}+\beta_{2345} \\
x_3 &= x_1x_2x_4 = x_1x_2x_3x_5 = x_4x_5 & b_3 &\to \beta_3+\beta_{124}+\beta_{1235}+\beta_{45} \\
x_4 &= x_1x_2x_3 = x_1x_2x_4x_5 = x_3x_5 & b_4 &\to \beta_4+\beta_{123}+\beta_{1245}+\beta_{35} \\
x_5 &= x_1x_2x_3x_4x_5 = x_1x_2 = x_3x_4 & b_5 &\to \beta_5+\beta_{12345}+\beta_{12}+\beta_{34} \\
x_1x_3 &= x_2x_4 = x_2x_3x_5 = x_1x_4x_5 & b_{13} &\to \beta_{13}+\beta_{24}+\beta_{235}+\beta_{145} \\
x_1x_4 &= x_2x_3 = x_2x_4x_5 = x_1x_3x_5 & b_{14} &\to \beta_{14}+\beta_{23}+\beta_{245}+\beta_{135}
\end{aligned}
\tag{5.26}
$$

The design matrix of the experiment is given in Table 5.8.

Table 5.8. One-fourth Replicate of the 2^5 Full Factorial Design for the Defining Relations of Eq. (5.24)

N	x_1	x_2	x_3	x_4	x_5
1	+1	+1	+1	+1	+1
2	−1	−1	+1	+1	+1
3	−1	+1	+1	−1	−1
4	+1	−1	+1	−1	−1
5	+1	+1	−1	−1	+1
6	−1	−1	−1	−1	+1
7	−1	+1	−1	+1	−1
8	+1	−1	−1	+1	−1

The resolution of this one-fourth replicate is small, since all linear effects are confounded with the second-order interaction effects. A fractional factorial design can be transformed into a full factorial design by using the missing fractional replicates. In this case, the remaining three replicates have the following definition relations

$$\begin{aligned} x_4 &= x_1x_2x_3 & x_5 &= -x_1x_2 \\ x_4 &= -x_1x_2x_3 & x_5 &= x_1x_2 \\ x_4 &= -x_1x_2x_3 & x_5 &= -x_1x_2 \end{aligned} \tag{5.27}$$

The generalized defining contrasts take the form

$$\begin{aligned} I &= x_1x_2x_3x_4 = -x_1x_2x_5 = -x_3x_4x_5 \\ I &= -x_1x_2x_3x_4 = x_1x_2x_5 = -x_3x_4x_5 \\ I &= -x_1x_2x_3x_4 = -x_1x_2x_5 = x_3x_4x_5 \end{aligned} \tag{5.28}$$

If the three one-fourth replicates are used, we obtain the unbiased estimates of all theoretical coefficients.

Should the number N of observations in a fractional replicate be chosen to the inequality

$$k+1 \leq N < 2^k$$

where k is the number of factors, the unconfounded estimates of linear effects result. If the number N of observations is equal to $k+1$, which is the number of coefficients to be determined for a linear regression equation, the fractional replicate is a *linear saturated orthogonal design*. Table 5.9 shows a saturated orthogonal design for $k = 7$, which is a one-sixteenth replicate of a 2^7 full factorial design.

In Table 5.9, the factors x_4, x_5, x_6, x_7 are aliased to the factor interactions resulting in the following alias structure

$$x_4 = x_1x_2, \quad x_5 = x_1x_3, \quad x_6 = x_2x_3, \quad x_7 = x_1x_2x_3$$

Therefore, all linear effects will be confounded with interaction effects.

Table 5.9. Linear Saturated Design ($k = 7$)

N	x_0	x_1	x_2	x_3	x_4	x_5	x_6	x_7
1	+1	−1	−1	−1	+1	+1	+1	−1
2	+1	+1	+1	−1	+1	−1	−1	−1
3	+1	−1	+1	−1	−1	+1	−1	+1
4	+1	+1	−1	−1	−1	−1	+1	+1
5	+1	−1	−1	+1	+1	−1	−1	+1
6	+1	+1	+1	+1	+1	+1	+1	+1
7	+1	−1	+1	+1	−1	−1	+1	−1
8	+1	+1	−1	+1	−1	+1	−1	−1

As saturated designs have the number of observations equal to the number of estimated coefficients, the number of degrees of freedom of the residual mean square is zero. To test the estimated linear regression equation derived from the saturated design for goodness of fit, the experimenter must carry out an additional experiment.

The advantages of optimum 2^k and 2^{k-p} designs are as follows: the designs are orthogonal, and this makes all computations very simple; all coefficients are estimated independently of one another; each coefficient is estimated by using the results of all N observations. All regression coefficients are estimated to the same variance, which is minimal, i.e. for a given number N of observations the determinant of the variance matrix $(\boldsymbol{X}^T\boldsymbol{X})^{-1}$ is a minimum (D-optimality). It should also be stressed that the linear 2^k and 2^{k-p} designs have the property of *rotatability*. Since there is no correlation between the coefficients, according to the rule of addition of variances in the case of a linear equation in k factors, we have

$$s_{\hat{y}}^2 = s_{b_0}^2 + x_1^2 s_{b_1}^2 + \ldots + x_k^2 s_{b_k}^2 \tag{5.29}$$

Since $s_{b_j}^2 = s_e^2/N$, then

$$s_{\hat{y}}^2 = \frac{s_e^2}{N}(1 + x_1^2 + \ldots + x_k^2) = \frac{s_e^2}{N}(1 + \varrho^2) \tag{5.30}$$

$$\varrho^2 = \textstyle\sum_{j=1}^k x_j^2$$

where ϱ^2 is the square of the radius of the sphere in a k-dimensional space.

The reciprocal of $s_{\hat{y}}^2$ may be taken as a measure of the information contained in the regression equation.

In agreement with Eq. (5.30), the amount of information decreases in proportion to the square of the radius of the sphere, ϱ^2, and is the same for all equidistant points (Fig. 5.2). Designs having this property are called *rotatable designs*.

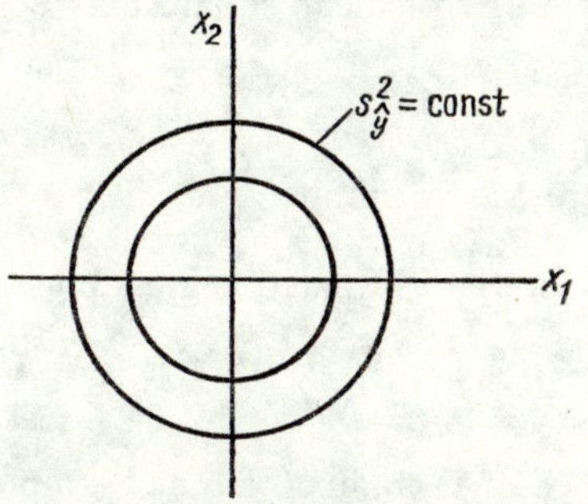

Fig. 5.2. Rotatability of the linear 2^2 design

Now sum up in general the steps involved in the analysis of variance and regression analysis of a designed experiment having a design matrix with N design points and every point (observation) replicated m times (Table 5.10).

Table 5.10. Design Matrix and Observation Results

Design point	x_0	x_1	x_2	x_k	y	$\bar{y}$	s^2
1	+1	+1	−1	... +1	$y_{11}, y_{12}, \ldots, y_{1m}$	$\bar{y}_1$	s_1^2
2	+1	−1	−1	... +1	$y_{21}, y_{22}, \ldots, y_{2m}$	$\bar{y}_2$	s_2^2
3	+1	+1	+1	... +1	.	.	.
.	.	.	.	.	.	.	.
.	.	.	.	.	.	.	.
.	.	.	.	.	.	.	.
N	+1	−1	+1	... −1	$y_{N1}, y_{N2}, \ldots, y_{Nm}$	$\bar{y}_N$	s_N^2

(1) For each row of the design matrix, the mean of m replicate observations is found

$$\bar{y}_i = \frac{1}{m}\sum_{u=1}^{m} y_{iu}, \qquad i = 1, 2, \ldots, N \tag{5.31}$$

and the sample variance is determined

$$s_i^2 = \frac{1}{m-1}\sum_{u=1}^{m} (y_{iu}-\bar{y}_i)^2 \qquad i = 1, 2, \ldots, N \tag{5.32}$$

(2) The sample variances are tested for homogeneity on the basis of Cochran's criterion. To this end, the ratio is formed

$$G = s^2_{\max}/\sum_{i=1}^{N} s_i^2$$

where s^2_{max} is the maximum value of the sample variance and $\sum_{i=1}^{N} s_i^2$ is the sum of all variances. The ratio G is compared with the tabulated value of Cochran's criterion, $G_{1-\alpha}(\nu_1, \nu_2)$, where α is the significance level, $\nu_1 = m-1$, and $\nu_2 = N$. If $G < G_{1-\alpha}(\nu_1, \nu_2)$, then the sample variances are homogeneous and the error mean square may be estimated

$$s_e^2 = \frac{1}{N}\sum_{i=1}^{N} s_i^2 \tag{5.33}$$

The error mean square has a number ν of degrees of freedom given by $\nu = N(m-1)$.

(3) The coefficients of the estimated regression equations are given by

$$b_j = \frac{1}{N}\sum_{i=1}^{N} x_{ji}\bar{y}_i \tag{5.34}$$

(4) Since the variance of $\bar{y}$ obtained from a sample of size m is m times less than the variance of single observation

$$s_{\bar{y}}^2 = s_e^2/m \tag{5.35}$$

then in the case under consideration (Table 5.10) the sample variance $s_{b_j}^2$ of the jth coefficient is given by

$$s_{b_j}^2 = s_e^2/Nm \tag{5.36}$$

(5) The coefficients of the regression equation are tested for significance on the basis of the Student t-test.

Under the null hypothesis H_0: $\beta_j = 0$, for all coefficients of the regression equation a t-ratio is formed

$$t_j = |b_j|/s_{b_j} \tag{5.37}$$

where b_i is the jth estimated coefficient of the regression equation and standard deviation of the jth coefficient.

The value of t_j is compared to the tabulated value of $t_\alpha(\nu)$ for the selected significance level α and the number of degrees of freedom $\nu = N(m-1)$. If $t_j < t_\alpha(\nu)$, then the hypothesis that the linear term of the regression equation, β_j, equals zero is accepted and the respective sample coefficient b_j is deleted from the estimated regression equation since it is not significant. Owing to the fact that the design matrix is orthogonal, the remaining coefficients do not require recalculation.

(6) The experimenter finds whether or not the estimated regression equation fits the experimental data using Fisher's test, a procedure analogous to that employed in handling the data of passive experiment. In the design matrix (Table 5.9), each observation is repeated m times. For the goodness-of-fit (adequacy) test, one forms Fisher's variance ratio

$$F = s_g^2/s_e^2$$

where s_g^2 is the adequacy mean square that is given by

$$s_g^2 = \frac{m \sum_{i=1}^{N} (\bar{y}_i - \hat{y}_i)^2}{N-l} \tag{5.38}$$

where l is the number of significant coefficients in the regression equation.

The equation is an adequate fit to the experiment if the F-ratio is smaller than the tabulated value of $F_{1-\alpha}(\nu_1, \nu_2)$

$$F < F_{1-\alpha}(\nu_1, \nu_2) \tag{5.39}$$

where α is the significance level, ν_1 is the number of degrees of freedom for the adequate variance

$$\nu_1 = N-l \tag{5.40a}$$

and ν_2 is the number of degrees of freedom of the replication mean square

$$\nu_2 = N(m-1) \tag{5.40b}$$

If one has

$$F > F_{1-\alpha}(\nu_1, \nu_2) \tag{5.41}$$

then an adequately approximating model is obtained by increasing the order of the approximating polynomial.

5.3. Optimization by Steepest Ascent on a Response Surface

The objective of optimization may be stated as follows: we seek to determine the coordinates of the extremum point, $x_{1,op}, x_{2,op}, \ldots, x_{k,op}$, on the response surface, $y = f(x_1, x_2, \ldots, x_k)$. We construct contours y = const at $k = 2$ (Fig. 5.3a). In conventional experiment it is customary to hold one of the factors, say x_1, constant and to move in the direction of the other factor, x_2, from point L. It is assumed that the coordinates of point L are known from previous observations. The motion in the x_2 direction continues until y increases no longer (Fig. 5.3b). At the point of the best yield (M), x_2 is held constant, and the experimenter moves in the x_1 direction. At

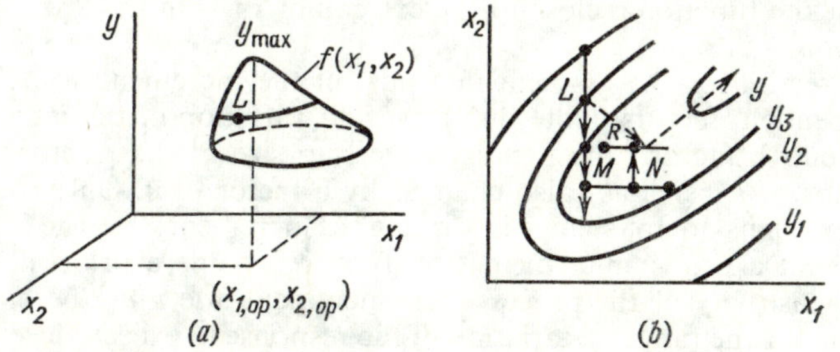

Fig. 5.3. Moving on a response surface: (a) toward an extremum in a single factor experiment, and (b) by the steepest ascent

point N, x_1 is again held constant, and the motion is resumed in the x_2 direction, etc. It is obvious that the broken line $LMNR$ is not the shortest path to the optimum. As will be recalled, the shortest path to the optimum is that of steepest ascent, that is in the direction of the gradient, at right angles to the contours of constant values of y (shown by the dashed arrows in Fig. 5.3b). If, in the general case, the response surface is described by an equation of the form

$$y = f(x_1, x_2, \ldots, x_k)$$

then the gradient of the function is

$$\text{grad } f = \frac{\partial f}{\partial x_1}\vec{i} + \frac{\partial f}{\partial x_2}\vec{j} + \ldots + \frac{\partial f}{\partial x_k}\vec{k} \tag{5.42}$$

where $\vec{i}, \vec{j}, \ldots, \vec{k}$ are the unit vectors in the direction of the coordinate axes.

It is assumed that the function f is continuous, univalued and has no singular points.

Box and Wilson [8] have proposed a step-by-step procedure for advance on the response surface. The experimenter starts working by selecting a local region for experimentation in the vicinity of point L and describes the response surface by an estimated linear regression equation of the form

$$\hat{y} = b_0 + b_1 x_1 + b_2 x_2 + \ldots + b_k x_k \tag{5.43}$$

Then he calculates the components of the gradient of y

$$\frac{\partial \hat{y}}{\partial x_1} = b_1 \quad \frac{\partial \hat{y}}{\partial x_2} = b_2 \ldots \frac{\partial \hat{y}}{\partial x_k} = b_k \tag{5.44}$$

and moves on the response surface in the direction of the linear approximation gradient.

The relative step size should be proportional to the product of the coefficient b_j by the increment size Δz_j. If one linear approximation proves insufficient, the experimenter selects a new local region for experimentation centred around a point where y is a maximum, and ascertains the direction of steepest ascent again. This step-by-step procedure is carried on until the experimenter finds that the function is close to the extremum, or is in the "nearly stationary" region.

The direction of the gradient depends on the increment size chosen for the independent factors. Note that an n-fold variation of the increment size for a given jth factor causes the relative step size of that factor to change by a factor of n^2, since b_j also changes by a factor of n. Only the gradient component signs do not change in varying the increment size. The experimenter can select a proper increment size if he has some a priori data on the variable sensitivity of the process. The increment size must be of sufficient magnitude for the range of variation of the response to exceed the error mean square by at least 3 or 4 times. On the other hand, a linear approximation of the response surface is deemed adequate in the case of a small increment size only. If no constraints are imposed on the increment size, then the latter is so selected that the regression equation is symmetric relative to its coefficients associated with the linear terms. The experimental data yielded by the steepest ascent should be given a thorough statistical analysis.

Example 5.1. We seek to determine an optimum composition of a photochrome glass belonging to the system $Li_2O-Al_2O_3-SiO_2$. The optical density of the glass in irradiated condition, y, serves as an optimization parameter. It is necessary to determine the glass composition and melting conditions that provide for a maximal optical density.

The factors to be studied are: z_1, the initial concentration of chlorine, in gram-atom per 100 g of the charge; z_2, the initial composition of bromine, in gram-atom per 100 g of the charge; z_3, the Ag/Cl ratio; z_4, the melting temperature, °C; z_5, the hold time, in hours; z_6, the Al_2O_3 content, in molar fractions; and z_7, the Li_2O/SiO_2 ratio.

The principal level and increment size are listed in a table below.

	z_1	z_2	z_3	z_4	z_5	z_6	z_7
Principal level z_j^0	0.0425	0.0187	0.0675	1325	1.75	0.1395	0.4165
Increment size Δz_j	0.0205	0.0093	0.0325	25	0.025	0.0125	0.0835
+1	0.063	0.028	0.1	1350	2	0.152	0.5
−1	0.022	0.0094	0.035	1300	1.5	0.127	0.333

Solution. In order to determine the coefficients of the estimated linear regression equation

$$\hat{y} = b_0 + b_1x_1 + b_2x_2 + b_3x_3 + b_4x_4 + b_5x_5 + b_6x_6 + b_7x_7$$

use is made of a one-sixteenth replicate of a 2^7 full factorial design with defining relations

$$x_4 = x_1x_2x_3, \quad x_5 = x_1x_2, \quad x_6 = x_1x_3, \quad x_7 = x_2x_3$$

Each observation in the design matrix (Table 5.11) is repeated twice.

Table 5.11

N	x_0	x_1	x_2	x_3	x_4	x_5	x_6	x_7	y_1	y_2	$\bar{y}$	$s^2 \cdot 10^4$	$\hat{y}$	$(\bar{y}-\hat{y})^2 \cdot 10^4$
1	+	−	+	−	+	−	+	−	0	0	0	0	0.017	2.9
2	+	+	+	−	−	+	−	−	0.108	0.15	0.129	8.82	0.136	0.49
3	+	−	−	−	−	+	+	+	0	0	0	0	−0.0088	0.77
4	+	+	−	−	+	−	−	+	0.194	0.16	0.177	5.78	0.1618	2.3
5	+	−	+	+	−	−	−	+	0.298	0.292	0.295	0.18	0.2896	0.29
6	+	+	+	+	+	+	+	+	0.400	0.408	0.404	0.32	0.4086	0.21
7	+	−	−	+	+	+	−	−	0.255	0.278	0.266	2.6	0.2638	0.073
8	+	+	−	+	−	−	+	−	0.453	0.408	0.431	10.1	0.4344	0.1156

The mean of the optical density, $\bar{y}$, is found from two repeated observations. The variances s_i^2, $i = 1, 2, \ldots, 8$ are tested for homogeneity on the basis of Cochran's criterion. The sum of the variances is

$$\sum_{i=1}^{8} s_i^2 = 27.80\times10^{-4}$$

and Cochran's criterion is

$$G = \frac{s_{max}^2}{\sum_{i=1}^{N} s_i^2} = \frac{10.1\times10^{-4}}{27.80\times10^{-4}} = 0.363$$

For the significance level $\alpha = 0.05$ and the numbers of degrees of freedom $\nu_1 = 1$, $\nu_2 = 8$, we look up the following value of Cochran's criterion

$$G_{0.95}(1,\ 8) = 0.6798$$

We see that $G < G_{0.95}(1, 8)$, which means that the variances are homogeneous. Thus the error mean square may be found as the mean

$$s_e^2 = \frac{1}{N}\sum_{i=1}^{N} s_i^2 = \frac{27.80\times10^{-4}}{8} = 3.5\times10^{-4}$$

The number of degrees of freedom of the error mean square is

$$\nu = N(m-1) = 8(2-1) = 8$$

The coefficients of the estimated regression equation are given by Eq. (5.34)

$$b_0 = 0.2128 \qquad b_1 = 0.0724$$
$$b_2 = -0.00575 \qquad b_3 = 0.1363$$
$$b_4 = -0.00088 \qquad b_5 = -0.0129$$
$$b_6 = -0.041 \qquad b_7 = 0.00625$$

Using the Student t-test, one can test the coefficients for significance. To this end, the value of s_{b_j} is found from Eq. (5.36)

$$s_{b_j} = \sqrt{\frac{3.5\times10^{-4}}{2\times8}} = 0.468\times10^{-2}$$

Then, one can form t-ratios for all regression coefficients

$$t_0 = \frac{0.2128}{0.468\times10^{-2}} = 45.5 \qquad t_1 = \frac{0.0724}{0.468\times10^{-2}} = 15.45$$

$$t_2 = \frac{0.00575}{0.468\times10^{-2}} = 1.23 \qquad t_3 = \frac{0.1363}{0.468\times10^{-2}} = 29.1$$

$$t_4 = \frac{0.00088}{0.468\times10^{-2}} = 0.188 \qquad t_5 = \frac{0.0129}{0.468\times10^{-2}} = 2.77$$

$$t_6 = \frac{0.0041}{0.468\times10^{-2}} = 0.875 \qquad t_7 = \frac{0.00625}{0.468\times10^{-2}} = 1.35$$

The tabulated value of the Student t-test is $t_{0.05}(8) = 2.31$. The coefficients b_2, b_4, b_6, b_7 are not significant since their t-ratios above are less than the tabulated value. With these coefficients deleted from the estimated regression equation, it takes the form

$$\hat{y} = 0.2128 + 0.0724x_1 + 0.1363x_3 - 0.0129x_5$$

Let us now determine whether the equation is fit to the experimental data using Fisher's test. The adequacy mean square is given by Eq. (5.38)

$$s_g^2 = \frac{2\sum^{8}(\bar{y}_i-\hat{y}_i)^2}{8-4} = \frac{2\times6.77\times10^{-4}}{4} = 3.385\times10^{-4}$$

Thus Fisher's variance ratio is

$$F = \frac{s_g^2}{s_e^2} = \frac{3.385\times10^{-4}}{3.35\times10^{-4}} = 1.01$$

With $\alpha = 0.05$, $\nu_1 = 4$, and $\nu_2 = 8$, the tabulated value of Fisher's F is $F_{0.95}(4, 8) = 3.8$. We see that $F < F_{0.95}(4, 8)$ and therefore the equation fits the experiment. Using this equation, the optical density can be increased by steepest ascent on the response surface. During the ascent, the nonsignificant parameters are fixed at zero level, as the holding time at the lower level is 1.5 hours. Subject to variation are therefore only the original concentration of chlorine (z_1) and the Ag/Cl ratio (z_3). First three observations in the steepest ascent (see Nos. 9, 10, 11 in the table below) are imaginary ones (not physically implemented).

	z_1	z_3	y
z_j^0	0.0425	0.0675	
Δz_j	0.0205	0.099	
b_j	0.0724	0.1363	
$b_j\,\Delta z_j$	0.00148	0.00443	
Size	0.0036	0.0111	
N			
9	0.0462	0.0787	—
10	0.0498	0.0897	—
11	0.0536	0.1008	—
12	0.0573	0.1119	0.552
13	0.0610	0.1230	0.500
14	0.0647	0.1341	0.476
15	0.0683	0.1452	0.436
16	0.0719	0.1563	0.426

The steps are selected to be 2.5 times more than the respective products $b_j\Delta z_j$. The best result corresponds to point 12. Further increase in z_1 and z_3 tends to deteriorate the photochrome property of the glass. Thus we decide to run points 10 and 11 which give

$$y_{10} = 0.496 \qquad y_{11} = 0.561$$

Thus, the result obtained at point 11 is considered an optimum.

5.4. Description of a Region Close to the Extremum. The Box-Wilson Composite Designs

The region close to the extremum is also known as a "nearly stationary" region. It is a substantially nonlinear area which can be described adequately by means of nonlinear polynomials. The second-order polynomials are in common use for the purpose because second-order designs are well devised. Besides, second-order response surfaces lend themselves to systematization. The latter property means that the extremum point can be detected with ease. Finally, the use of the approximating polynomial of higher order involves a considerable increase in the number of observations.

As a rule, the experiment designed to find the optimum conditions of a process is described adequately by a second-order polynomial. For this polynomial, the number N of observations included in the design should be not less than the number of estimated coefficients of the second-order regression equation for k factors:

$$\hat{y} = b_0+b_1x_1+b_2x_2+\ldots+b_kx_k+b_{12}x_1x_2+\ldots$$
$$+b_{k-1,k}x_{k-1}x_k+b_{11}x_1^2+\ldots+b_{kk}x_k^2 \tag{5.45}$$

The coefficients of Eq. (5.45) are estimates of the respective coefficients of the theoretical regression equation

$$m_y = \beta_0+\beta_1x_1+\ldots+\beta_kx_k+\beta_{12}x_1x_2+\ldots$$
$$+\beta_{k-1,k}x_{k-1}x_k+\beta_{11}x_1^2+\ldots+\beta_{kk}x_k^2 \tag{5.46}$$

The number l of coefficients in Eq. (5.45) can be given by

$$l = k+1+k+C_k^2 = 2k+1+\frac{k!}{2!(k-2)!} = \frac{(k+1)(k+2)}{2} \tag{5.47}$$

where C_k^2 is the number of combinations of two factors taken from a total of k factors, equal to the number of two factor (second-order) interaction terms in Eq. (5.45).

If a "nearly stationary" region can be described adequately by virtue of the theoretical second-order regression equation, Eq. (5.46), then the interaction and quadratic terms found during experimentation become significant. This allows the experimenter to visualize the fact that he is in a "nearly stationary" region. The closeness of the "nearly stationary" region can be ascertained by replicating the center point of the design ($x_1 = 0, x_2 = 0, \ldots x_k = 0$) and computing the mean, $\bar{y}^0$. The quantity $\bar{y}^0$ is an estimate of the free term (intercept) of the theoretical regression equation

$$\bar{y}^0 \rightarrow \beta_0 \tag{5.48}$$

while b_0 calculated in the factorial experiment by the equation

$$b_0 = \frac{1}{N}\sum_{i=1}^{N} x_{0i}y_i$$

is a biased estimate of the free term and the sum of the quadratic terms

$$b_0 \to \beta_0 + \sum_{j=1}^{k} \beta_{jj} \tag{5.49}$$

Therefore, the difference

$$b_0 - \bar{y}_0 \to \sum_{j=1}^{k} \beta_{jj} \tag{5.50}$$

may to a certain extent serve as a measure of the curvature of the surface.

In order to describe the response surface by second-order polynomials, the factors in experimental designs should take at least three different values. A three-level factorial experiment in which all possible combinations of k factors at all levels are used is called a 3^k full factorial design. Table 5.12 shows a design matrix of a 3^2 full factorial design.

Table 5.12. 3^2 Full Factorial Design

N	x_1	x_2	y
1	0	0	y_1
2	+1	0	y_2
3	−1	0	y_3
4	0	+1	y_4
5	+1	+1	y_5
6	−1	+1	y_6
7	0	−1	y_7
8	+1	−1	y_8
9	−1	−1	y_9

A 3^k full factorial design involves a very great number of observations that exceeds considerably the number l of the estimated coefficients even when $k > 2$.

Tabulated below are the number of observations involved in a 3^k design and the number l of the estimated coefficients of the second-order regression equation

k	2	3	4	5	6
3^k	9	27	81	243	729
l	6	10	15	21	28

The number of observations can be reduced through the use of *composite* or *sequential designs* proposed by Box and Wilson [8]. The kernel of such a design is a 2^k design for $k < 5$ or a half-replicate thereof at $k \geqslant 5$. Even in the case of the half-replicate above, the experimenter can obtain unconfounded estimates of the linear effects and second-order interaction effects (see Sec. 5.2).

If the estimated linear regression equation fails the adequacy-of-fit test, it is necessary:

(1) to add $2k$ star points positioned on the coordinate axes of factorial space $(\pm\alpha, 0, \ldots, 0)$, $(0, \pm\alpha, 0, \ldots, 0) \ldots (0, 0, \ldots, 0, \pm\alpha)$, where α is the distance from the center point of the design to a star point, or a *star arm*;

(2) to increase the number of runs, n_0, at the center point of the design.

To illustrate consider a composite design for $k = 2$ (Fig. 5.4). Points *1, 2, 3, 4* form a 2^2 full factorial design, while points *5, 6, 7* and *8* are star points with coordinates $(\pm\alpha, 0)$ and $(0, \pm\alpha)$, the observations at the center point, n_0, have zero coordinates (0, 0) (Table 5.13).

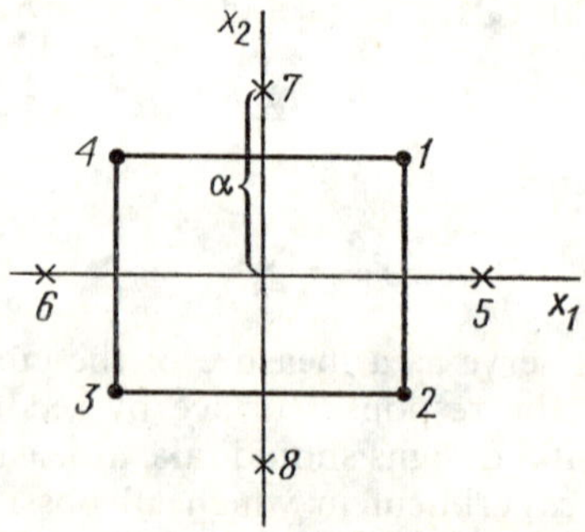

Fig. 5.4. Composite design at $k = 2$

Table 5.13. Second-order Composite Design for Two Factors

Design point	x_0	x_1	x_2	x_1x_2	x_1^2	x_2^2
1	+1	+1	+1	−1	+1	+1
2	+1	+1	−1	+1	+1	+1
3	+1	−1	−1	−1	+1	+1
4	+1	−1	+1	+1	+1	+1
5	+1	$+\alpha$	0	0	α^2	0
6	+1	$-\alpha$	0	0	α^2	0
7	+1	0	$+\alpha$	0	0	α^2
8	+1	0	$-\alpha$	0	0	α^2
9	+1	0	0	0	0	0
10	+1	0	0	0	0	0
.	.	.	.	.	.	.
.	.	.	.	.	.	.
N	+1	0	0	0	0	0

The information or moments matrix of a second-order composite design for $k = 2$ takes the form

$$X^TX = \begin{array}{c} \\ b_0 \\ b_1 \\ b_2 \\ b_{12} \\ b_{11} \\ b_{22} \end{array}\begin{array}{c} \begin{array}{cccccc} b_0 & b_1 & b_2 & b_{12} & b_{11} & b_{22} \end{array} \\ \left[\begin{array}{cccccc} \sum x_{0i}^2 & 0 & 0 & 0 & \sum x_{0i}x_{1i}^2 & \sum x_{0i}x_{2i}^2 \\ 0 & \sum x_{1i}^2 & 0 & 0 & 0 & 0 \\ 0 & 0 & \sum x_{2i}^2 & 0 & 0 & 0 \\ 0 & 0 & 0 & \sum x_{1i}^2x_{2i}^2 & 0 & 0 \\ \sum x_{1i}^2x_{0i} & 0 & 0 & 0 & \sum x_{1i}^4 & \sum x_{1i}^2x_{2i}^2 \\ \sum x_{2i}^2x_{0i} & 0 & 0 & 0 & \sum x_{2i}^2x_{1i}^2 & \sum x_{1i}^4 \end{array}\right] \end{array} \quad (5.51)$$

where

$$\sum_{i=1}^{N} x_{0i}^2 = N$$
$$\sum x_{1i}^2 = \sum x_{2i}^2 = 2^2+2\alpha^2$$
$$\sum x_{0i} x_{1i}^2 = \sum x_{0i} x_{2i}^2 = 2^2+2\alpha^2$$
$$\sum x_{1i}^2 x_{2i}^2 = 2^2$$
$$\sum x_{1i}^4 = \sum x_{2i}^4 = 2^2+2\alpha^4$$

The total number of observations in the matrix of a composite design for k factors (Table 5.14) is given by

$$\begin{aligned} N &= 2^k+2k+n_0 \qquad k < 5 \\ N &= 2^{k-1}+2k+n_0 \quad k \geqslant 5 \end{aligned} \tag{5.52}$$

Table 5.14. Second-order Composite Design for k Factors

Design point	x_0	x_1	x_2	x_3		x_k
1	+1	+1	−1	+1	...	−1
2	+1	−1	−1	−1	...	+1
3	+1	+1	−1	+1	...	−1
4	+1	−1	+1	−1	...	+1
5	+1	+1	−1	+1	...	−1
.	.	.	.	.	...	.
.	.	.	.	.	...	.
.	.	.	.	.	...	.
n_k	+1	−1	−1	−1	...	−1
n_k+1	+1	$+\alpha$	0	0	...	0
n_k+2	+1	$-\alpha$	0	0	...	0
.	.	.	.	.	...	.
.	.	.	.	.	...	.
.	.	.	.	.	...	.
n_k+2k	+1	0	0	0	...	$-\alpha$
.	.	.	.	.	...	.
.	.	.	.	.	...	.
.	.	.	.	.	...	.
N	+1	0	0	0	...	0

The matrix X^TX that corresponds to the design of Table 5.14 is given by (5.53)

$$
\begin{array}{c}
 \\ b_0 \\ b_1 \\ \vdots \\ b_k \\ b_{12} \\ b_{13} \\ \vdots \\ b_{k-1,k} \\ b_{11} \\ b_{22} \\ \vdots \\ b_{kk}
\end{array}
\begin{array}{cccc:cccc:cccc}
b_0 & b_1, & \ldots, & b_k & b_{12} & b_{13}, & \ldots, & b_{k-1,k} & b_{11} & b_{22}, & \ldots, & b_{kk} \\
\sum x_{0i}^2 & & & 0 & & & & & \sum x_{0i} x_{1i}^2 & \sum x_{0i} x_{2i}^2 & \ldots & \sum x_{0i} x_{ki}^2 \\
 & \sum x_{1i}^2 & & & & & & & & & & \\
 & & \ddots & & & & 0 & & & & & \\
0 & & & \sum x_{ki}^2 & & & & & & 0 & & \\
\hdashline
 & & & & \sum x_{1i}^2 x_{2i}^2 & & & & & & & \\
 & 0 & & & & \sum x_{1i}^2 x_{3i}^2 & & 0 & & & & \\
 & & & & & & \ddots & & & 0 & & \\
 & & & & 0 & & & \sum x_{k-1}^2 x_{ki}^2 & & & & \\
\hdashline
\sum x_{1i}^2 x_{0i} & & & & & & & & \sum x_{1i}^4 & \sum x_{1i}^2 x_{2i}^2 & \ldots & \sum x_{1i}^2 x_{ki}^2 \\
\sum x_{2i}^2 x_{0i} & & & & & & & & \sum x_{2i}^2 x_{1i}^2 & \sum x_{2i}^4 & \ldots & \sum x_{2i}^2 x_{ki}^2 \\
\vdots & & 0 & & & & 0 & & \cdots & \cdots & \cdots & \cdots \\
\sum x_{ki}^2 x_{0i} & & & & & & & & \sum x_{ki}^2 x_{1i}^2 & \sum x_{ki}^2 x_{2i}^2 & \ldots & \sum x_{ki}^4
\end{array}
\qquad (5.53)
$$

where

$$\sum_{i=1}^{N} x_{0i}^2 = N$$

$$\sum x_{1i}^2 = \sum x_{2i}^2 = \ldots = \sum x_{ki}^2 = \begin{cases} 2^k+2\alpha^2 & k < 5 \\ 2^{k-1}+2\alpha^2 & k \geq 5 \end{cases}$$

$$\sum x_{0i} x_{1i}^2 = \sum x_{0i} x_{2i}^2 = \ldots = \sum x_{0i} x_{ki}^2 = \begin{cases} 2^k+2\alpha^2 & k < 5 \\ 2^{k-1}+2\alpha^2 & k \geq 5 \end{cases}$$

$$\sum x_{1i}^2 x_{2i}^2 = \sum x_{1i}^2 x_{3i}^2 = \ldots = \sum x_{(k-1)i}^2 x_{ki}^2 = \begin{cases} 2^k & k < 5 \\ 2^{k-1} & k \geq 5 \end{cases}$$

$$\sum x_{1i}^4 = \sum x_{2i}^4 = \ldots = \sum x_{ki}^4 = \begin{cases} 2^k+2\alpha^4 & k < 5 \\ 2^{k-1}+2\alpha^4 & k \geq 5 \end{cases}$$

Thus we see that second-order composite designs are not orthogonal

$$\sum_{i=1}^{N} x_{0i} x_{ji}^2 \neq 0 \qquad j = 1, 2, \ldots, k$$
$$\sum_{i=1}^{N} x_{ji}^2 x_{ui}^2 \neq 0 \qquad j = 1, 2, \ldots, k, \ u \neq j$$

The magnitude of the star arm, α, and the number of runs at the center points, n_0, depend on the design selected.

5.5. Orthogonal Second-order Designs

Composite designs can readily be transformed into orthogonal ones by an appropriate selection of the star arm, α. To this end, the matrix (5.53) is inverted [10] in a general form. To do this, only that portion of the matrix ought to be inverted which is connected with the x_0 and x_j^2 columns (Table 5.14), i. e. with coefficients b_0 and b_{jj}; and the value of α is found equating the off-diagonal element of the inverted matrix to zero

$$\begin{aligned} \alpha^4+2^k\alpha^2-2^{k-1}(k+0.5n_0) &= 0 \qquad k < 5 \\ \alpha^4+2^{k-1}\alpha^2-2^{k-2}(k+0.5n_0) &= 0 \qquad k \geq 5 \end{aligned} \tag{5.54}$$

The values of α^2 given by Eqs. (5.54) are listed in Table 5.15.

Table 5.15. Values of α^2

n_0	k = 2	3	4	5*	n_0	k = 2	3	4	5*
1	1.00	1.476	2.00	2.39	6	1.742	2.325	2.950	3.31
2	1.160	1.650	2.164	2.58	7	1.873	2.481	3.140	3.49
3	1.317	1.831	2.390	2.77	8	2.00	2.633	3.310	3.66
4	1.475	2.00	2.580	2.95	9	2.113	2.782	3.490	3.83
5	1.606	2.164	2.770	2.14	10	2.243	2.928	3.66	4.00

* Half-replicate, $x_5 = x_1x_2x_3x_4$.

Given a certain value of α (Table 5.15) and the x_j^2 columns linearly converted by

$$x_j' = x_j^2 - \overline{x_j^2} = x_j^2 - \frac{1}{n}\sum_{i=1}^{N} x_{ji}^2 \tag{5.55}$$

one obtains an orthogonal matrix. For example, a second-order orthogonal design for $k = 2$ and $n_0 = 1$ takes the form shown in Table 5.16.

Table 5.16. Second-order Orthogonal Plan for $k = 2$

N	x_0	x_1	x_2	x_1x_2	x_1'	x_2'
1	+1	+1	+1	+1	+1/3	+1/3
2	+1	+1	−1	−1	+1/3	+1/3
3	+1	−1	−1	+1	+1/3	+1/3
4	+1	−1	+1	−1	+1/3	+1/3
5	+1	+1	0	0	+1/3	−2/3
6	+1	−1	0	0	+1/3	−2/3
7	+1	0	+1	0	−2/3	+1/3
8	+1	0	−1	0	−2/3	+1/3
9	+1	0	0	0	−2/3	−2/3

Since the design matrix is orthogonal, all regression coefficients are determined independently of one another using the following formula

$$b_j = \frac{\sum_{i=1}^{N} x_{ji} y_i}{\sum_{i=1}^{N} x_{ji}^2} \tag{5.56}$$

and $s_{b_j}^2$ is given by

$$s_{b_j}^2 = \frac{s_e^2}{\sum_{i=1}^{N} x_{ji}^2} \tag{5.57}$$

As a result, the following estimated regression equation is derived from the matrix with transformed columns for the quadratic terms

$$\hat{y} = b_0' + b_1x_1 + b_2x_2 + \ldots + b_kx_k + b_{12}x_1x_2 + \ldots$$
$$+ b_{(k-1)k}x_{k-1}x_k + b_{11}(x_1^2 - \overline{x_1^2}) + \ldots + b_{kk}(x_k^2 - \overline{x_k^2}) \tag{5.58}$$

To change back to normal notation, b_0 is found from

$$b_0 = b_0' - b_{11}\overline{x_1^2} - \ldots - b_{kk}\overline{x_k^2} \tag{5.59}$$

and estimated with a variance

$$s^2b_0 = s_{b_0'}^2 + (\overline{x_1^2})^2 s_{b_{11}}^2 + \ldots + (\overline{x_k^2})^2 s_{b_{kk}}^2 \tag{5.60}$$

Knowing the error mean square (replication variance), one can test the equation

$$\hat{y} = b_0+b_1x_1+b_2x_2+\ldots+b_kx_k+b_{12}x_1x_2+\ldots$$
$$+b_{(k-1)k}x_{k-1}x_k+b_{11}x_1^2+\ldots+b_{kk}x_k^2$$

for significance of its coefficients and for adequacy of fit.

In the goodness-of-fit test, use is made of F-ratio, $F = s_g^2/s_e^2$. The equation fits the experimental data adequately, if the variance ratio thus formed is smaller than the tabulated value of F for a significance level α (usually, $\alpha = 0.05$) and the degrees of freedom for the adequacy and replication variances

$$F < F_{1-\alpha}(\nu_1, \nu_2)$$

where $\nu_1 = N-l$ is the number of degrees of freedom for the adequacy variance, ν_2 is the number of degrees of freedom for the error (replication) variance, N is the number of observations in the design matrix, and l is the number of coefficients in the estimated second-order regression equation.

The coefficients are tested for significance by the Student t-test, $t_j = |b_j|/s_{b_j}$. A coefficient is significant, if $t_j > t_\alpha(\nu_2)$, where ν_2 is the number of degrees of freedom of the error variance.

The coefficients of estimated regression equations derived by means of orthogonal second-order designs are estimated to a varying degree of accuracy. In agreement with Eq. (5.33), we have

$$s_{b_0} = s_e/\sqrt{N}$$
$$s_{b_j} = s_e/\sqrt{2^k+2\alpha^2} \qquad j = 1, 2, \ldots, k \qquad k < 5$$
$$s_{b_j} = s_e/\sqrt{2^{k-1}+2\alpha^2} \qquad k \geqslant 5$$
$$s_{b_{uj}} = s_e/\sqrt{2^k} \qquad k < 5$$
$$u, j = 1, 2, \ldots, k$$
$$u \neq j$$
$$s_{b_{uj}} = s_e/\sqrt{2^{k-1}} \qquad k \geqslant 5 \tag{5.61}$$
$$s_{b_{jj}} = \frac{s_e}{\sqrt{2^k(1-\overline{x_j^2})^2+2(\alpha^2-\overline{x_j^2})^2+(2k-2+n_0)(\overline{x_j^2})^2}} \qquad k < 5$$
$$j = 1, 2, \ldots, k$$
$$s_{b_{jj}} = \frac{s_e}{\sqrt{2^{k-1}(1-\overline{x_j^2})^2+2(\alpha^2-\overline{x_j^2})^2+(2k-2+n_0)(\overline{x_j^2})^2}} \qquad k \geqslant 5$$

Note that orthogonal second-order designs are not rotatable.

Example 5.2. [19]. We seek the conditions that maximize the decomposition of borates by a mixture of sulphuric and phosphoric acids. The factors controlling the degree of decomposition, y, are chosen to be: z_1, the reaction temperature in °C; z_2, the reaction time in minutes; z_3, the stoichiometric ratio of phosphoric acid in %; and z_4, the percent concentration of phosphoric acid in terms of P_2O_5.

The principal level and the incremental changes in the factors are listed in a table below.

	z_1	z_2	z_3	z_4
z_j^0	55	37.5	80	32.8
Δz_j	25	22.5	20	18.8

Solution. From the previous experiments it is known that the optimum process conditions lie within the range over which the factors under study vary. Therefore, we shall use a second-order orthogonal design to derive an estimated regression equation (Table 5.17). The number of observations in the design matrix for $k = 4$ is 25; the star arm $\alpha = 1.414$, and $n_0 = 1$.

The error mean square is found from four repeated observations ($y_1 = 61.8\%$, $y_2 = 59.3\%$; $y_3 = 58.7\%$, $y_4 = 69\%$)

$$\bar{y} = \frac{1}{4}\sum^{4} y_u = 60.95$$

$$s_e^2 = \frac{1}{3}\sum^{4} (y_u - \bar{y})^2 = 5.95$$

The number of degrees of freedom for the error mean square is $\nu = 4 - 1 = 3$.

Using Eqs. (5.56) and (5.57), we estimate the coefficients of the second-order regression equation and the standard errors of estimate of these coefficients

$$\begin{array}{llll} b_0 = 61.54 & b_{11} = 4.5 & b_{12} = 2.18 & s_{b_j} = 0.545 \\ b_1 = 17.37 & b_{22} = 1.5 & b_{13} = 0.2 & s_{b_{uj}} = 0.61 \\ b_2 = 6.4 & b_{33} = 4.09 & b_{14} = 1.2 & s_{b_{jj}} = 0.864 \\ b_3 = 4.7 & b_{44} = -5.34 & b_{23} = 0.56 & \\ b_4 = -4.37 & & b_{24} = 0.76 & \\ & & b_{34} = 1.9 & \end{array}$$

The significance of the coefficients is tested with the Student t-test

$$\begin{array}{ll} t_1 = 17.37/0.545 = 31.9 & t_{12} = 2.18/0.61 = 3.57 \\ t_2 = 6.4/0.545 = 11.7 & t_{34} = 1.9/0.61 = 3.8 \\ t_3 = 4.70/0.545 = 8.64 & t_{13} = 0.2/0.61 = 3.18 \\ t_4 = 4.37/0.545 = 8.04 & t_{14} = 1.2/0.61 = 1.97 \\ t_{11} = 4.5/0.864 = 5.2 & t_{23} = 0.56/0.61 = 0.91 \\ t_{22} = 1.5/0.864 = 1.5 & t_{24} = 0.76/0.61 = 1.25 \\ t_{33} = 4.09/0.864 = 4.73 & \\ t_{44} = 5.34/0.864 = 6.22 & \end{array}$$

The tabulated value of the t-ratio for the significance level $\alpha = 0.05$ and $\nu = 3$ degrees of freedom is $t_\alpha(\nu) = 3.18$.

Table 5.17. Second-Order Orthogonal Design, $k = 4, n_0 = 1$

N	x_0	x_1	x_2	x_3	x_4	x_1'	x_2'	x_3'	x_4'	x_1x_2	x_1x_3	x_1x_4	x_2x_3	x_2x_4	x_3x_4	y
1	+1	+1	+1	+1	+1	0.2	0.2	0.2	0.2	+1	+1	+1	+1	+1	+1	86.9
2	+1	−1	−1	+1	+1	0.2	0.2	0.2	0.2	+1	−1	−1	−1	−1	+1	40.0
3	+1	+1	−1	−1	+1	0.2	0.2	0.2	0.2	−1	−1	+1	+1	−1	−1	66.0
4	+1	−1	+1	−1	+1	0.2	0.2	0.2	0.2	−1	+1	−1	−1	+1	−1	34.4
5	+1	+1	−1	+1	−1	0.2	0.2	0.2	0.2	−1	+1	−1	−1	+1	−1	76.6
6	+1	−1	+1	+1	−1	0.2	0.2	0.2	0.2	−1	−1	+1	+1	−1	−1	55.7
7	+1	+1	+1	−1	−1	0.2	0.2	0.2	0.2	+1	−1	−1	−1	−1	+1	91.0
8	+1	−1	−1	−1	−1	0.2	0.2	0.2	0.2	+1	+1	+1	+1	+1	+1	47.6
9	+1	+1	−1	+1	+1	0.2	0.2	0.2	0.2	−1	+1	+1	−1	−1	+1	74.1
10	+1	−1	+1	+1	+1	0.2	0.2	0.2	0.2	−1	−1	−1	+1	+1	+1	52.0
11	+1	+1	+1	−1	+1	0.2	0.2	0.2	0.2	+1	−1	+1	−1	+1	−1	74.5
12	+1	−1	−1	−1	+1	0.2	0.2	0.2	0.2	+1	+1	−1	+1	−1	−1	29.6
13	+1	+1	+1	+1	−1	0.2	0.2	0.2	0.2	+1	+1	−1	+1	−1	−1	94.8
14	+1	−1	−1	+1	−1	0.2	0.2	0.2	0.2	+1	−1	+1	−1	+1	−1	49.6
15	+1	+1	−1	−1	−1	0.2	0.2	0.2	0.2	−1	−1	−1	+1	+1	+1	68.6
16	+1	−1	+1	−1	−1	0.2	0.2	0.2	0.2	−1	+1	+1	−1	−1	+1	51.8
17	+1	0	0	0	0	−0.8	−0.8	−0.8	−0.8	0	0	0	0	0	0	61.8
18	+1	+1.414	0	0	0	1.2	−0.8	−0.8	−0.8	0	0	0	0	0	0	95.4
19	+1	−1.414	0	0	0	1.2	−0.8	−0.8	−0.8	0	0	0	0	0	0	41.7
20	+1	0	+1.414	0	0	−0.8	1.2	−0.8	−0.8	0	0	0	0	0	0	79.0
21	+1	0	−1.414	0	0	−0.8	1.2	−0.8	−0.8	0	0	0	0	0	0	42.4
22	+1	0	0	+1.414	0	−0.8	−0.8	1.2	−0.8	0	0	0	0	0	0	77.6
23	+1	0	0	−1.414	0	−0.8	−0.8	1.2	−0.8	0	0	0	0	0	0	58.0
24	+1	0	0	0	+1.414	−0.8	−0.8	−0.8	1.2	0	0	0	0	0	0	45.6
25	+1	0	0	0	−1.414	−0.8	−0.8	−0.8	1.2	0	0	0	0	0	0	52.3

After the insignificant coefficients for which the t-ratio is less than the tabulated value are dropped the estimated regression equation in dimensionless form becomes:

$$\begin{aligned}\hat{y} &= 61.54+17.37x_1+6.4x_2+4.7x_3-4.37x_4+2.18x_1x_2\\ &\quad +1.9x_3x_4+4.5(x_1^2-0.8)+4.09(x_3^2-0.8)-5.34(x_4^2-0.8)\\ &= 58.9+17.37x_1+6.4x_2+4.7x_3-4.37x_4+2.18x_1x_2\\ &\quad +1.9x_3x_4+4.5x_1^2+4.09x_3^2-5.34x_4^2\end{aligned}$$

To test the equation for the adequacy of fit, we determine the residual mean square

$$s_{res}^2 = \frac{\sum_{i=1}^{N}(y_i-\hat{y}_i)^2}{N-l} = \frac{396.2}{25-10} = 26.4$$

and find the F (variance) ratio

$$F = s_{res}^2/s_e^2 = 26.4 \div 5.95 = 4.4$$

The tabulated value of the Fisher F for a significance level of $\alpha = 0.05$ and $\nu_1 = 15$ and $\nu_2 = 3$ degrees of freedom is 8.6. Thus, $F < F_{1-\alpha}(\nu_1, \nu_2)$ and the estimated regression equation fits the experimental data adequately converted back to the natural scale, the estimated regression equation takes the form

$$\begin{aligned}\hat{y} &= 90.64-0.242z_1-0.07z_2+1.5z_3+0.354z_4\\ &\quad +0.00388z_1z_2+0.00506z_3z_4+0.072z_1^2\\ &\quad +0.0102z_3^2-0.015z_4^2\end{aligned}$$

The conditions answering $y_{max} = 100\%$ as found from the last regression equation by the Gauss-Seidel method on a computer were: $z_{1,\,op} = 90$ °C, $z_{2,\,op} = 50$ min, $z_{3,\,op} = 90\%$, and $z_{4,\,op} = 32.5\%$.

Under the optimum conditions thus obtained, check observations were run. The borates were found to have been 98.5% decomposed by 30.3% dry-process phosphoric acid and 98.8% decomposed by wet-process phosphoric acid with a concentration of 29.6%.

5.6. Box-Hunter Second-order Rotatable Designs

A second-order orthogonal design is not rotatable. The amount of information, defined as the reciprocal of s_y^2, is different for equidistant points. Figure 5.5 shows contours of constant information for 2^2 design of Table 5.16. The surfaces of constant information for a large number of factors are of a very complicated nature. Box and Hunter [20] have suggested that second-order rotatable designs should be considered an optimum. A design is said to be rotatable if its variance-covariance matrix $(X^TX)^{-1}$ is invariant to the orthogonal rotation of the coordinates. A second-order design is assumed to be rotatable, if the moments of its information matrix satisfy the following relations:

$$\sum_{i=1}^{N} x_{ji}^2 = N\lambda_2 \quad j = 1, 2, \ldots, k \tag{5.62}$$

$$\sum_{i=1}^{N} x_{ji}^4 = 3\sum_{i=1}^{N} x_{ji}^2 x_{ui}^2 = 3N\lambda_4 \; u, \quad j = 1, 2 \ldots, k, \; u \neq j \tag{5.63}$$

where λ_2 and λ_4 are arbitrary constants.

The remaining moments of the information matrix are equal to zero.

$$
N^{-1}X^TX = \begin{array}{c} \\ b_0 \\ b_1 \\ \cdot \\ \cdot \\ \cdot \\ b_k \\ b_{12} \\ b_{13} \\ \cdot \\ \cdot \\ \cdot \\ b_{k-1,k} \\ b_{11} \\ b_{22} \\ \cdot \\ \cdot \\ \cdot \\ b_{kk} \end{array}
\begin{array}{c}
b_0 b_1, \ldots, b_k \; b_{12} b_{13}, \ldots, b_{k-1,k} \; b_{11} b_{22}, \ldots, b_{kk} \\
\left[\begin{array}{ccc:ccc:cccc}
1 & & 0 & & & & & & & \\
\lambda_2 & & & & & & \lambda_2 & \lambda_2 & \ldots & \lambda_2 \\
 & \cdot & & & & & & & & \\
 & & \cdot & & 0 & & & 0 & & \\
 & & & \cdot & & & & & & \\
0 & & \lambda_2 & & & & & & & \\
\hdashline
 & & & \lambda_4 & 0 & & & & & \\
 & & & & \lambda_4 & & & & & \\
 & & & 0 & & \cdot & & & & \\
 & 0 & & & & \cdot & & 0 & & \\
 & & & & & \cdot & & & & \\
 & & & & & \lambda_4 & & & & \\
\hdashline
\lambda_2 & & & & & & 3\lambda_4 & \lambda_4 & \ldots & \lambda_4 \\
\lambda_2 & & & & & & \lambda_4 & 3\lambda_4 & \ldots & \lambda_4 \\
\cdot & & & & & & \ldots & \ldots & \ldots & \ldots \\
\cdot & & 0 & & 0 & & \ldots & \ldots & \ldots & \ldots \\
\cdot & & & & & & \ldots & \ldots & \ldots & \ldots \\
\lambda_2 & & & & & & \lambda_4 & \lambda_4 & \ldots & 3\lambda_4
\end{array}\right]
\end{array}
\qquad (5.64)
$$

The information profiles vary with λ_2 and λ_4. Figure 5.6 shows the profiles for $k = 2$, $\lambda_2 = 1$ at different values of λ_4. Similar profiles are produced

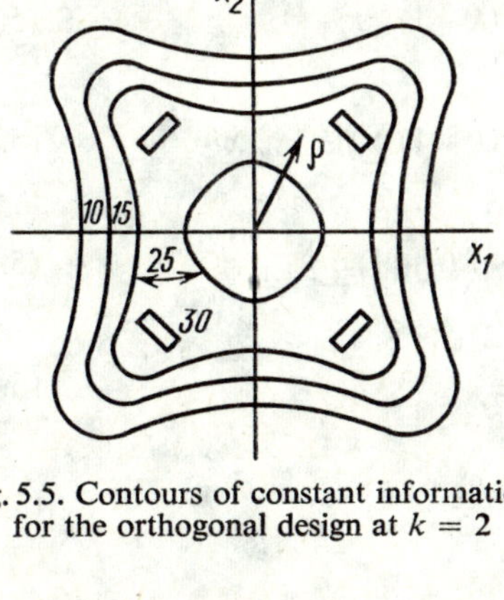

Fig. 5.5. Contours of constant information for the orthogonal design at $k = 2$

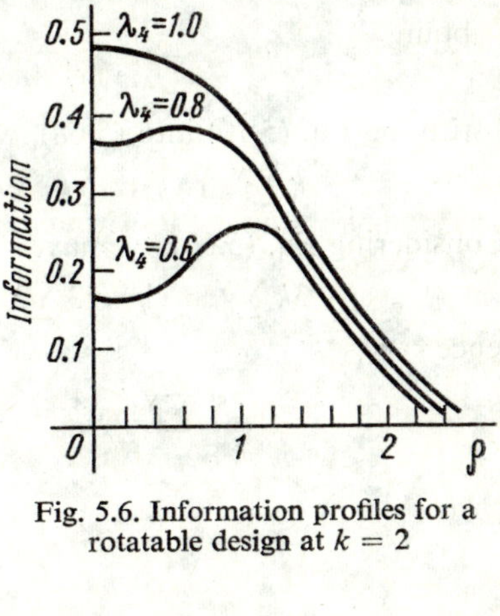

Fig. 5.6. Information profiles for a rotatable design at $k = 2$

at other values of k. The value of λ_4 is selected so that the information is maintained constant within the range $0 \leqslant \varrho \leqslant 1$ [20]. Such designs are known as uniform precision rotatable designs.

There appears a multiplier $(2+k)\lambda_4/\lambda_2^2-k$ in the determinant of matrix (5.64). For the matrix $(\boldsymbol{X}^T\boldsymbol{X})^{-1}$ to exist, the condition

$$(2+k)\lambda_4/\lambda_2^2-k \neq 0$$

should be satisfied, or otherwise

$$\lambda_4/\lambda_2^2 > k/(k+2) \tag{5.65}$$

Consider the cases when the condition given by Eq. (5.65) is not satisfied. To yield a relation of constants λ_2 and λ_4 to the number of factors, k, we multiply Eq. (5.62) by k

$$k\sum_{i=1}^{N} x_{ji}^2 = \sum_{j=1}^{k}\sum_{i=1}^{N} x_{ji}^2 = \sum_{i=1}^{N}\sum_{j=1}^{k} x_{ji}^2 = \sum_{i=1}^{N} \varrho_i^2 = k\lambda_2 N \tag{5.66}$$

whence

$$\lambda_2 = \frac{1}{kN}\sum_{i=1}^{N} \varrho_i^2 \tag{5.67}$$

where ϱ_i is the radius of the ith point in a k-dimensional space.

Multiplying Eq. (5.63) in a similar way, we obtain

$$k\sum_{i=1}^{N} x_{ij}^4 = \sum_{i=1}^{N}\sum_{j=1}^{k} x_{ji}^4 \tag{5.68}$$

Now we express the sum $\sum_{j=1}^{k} x_{ij}^4$ in terms of the point radius and the number of factors, k. Recalling that for any ith point

$$\left(\sum_{j=1}^{k} x_j^2\right)^2 = \varrho^4 = (x_1^2+x_2^2+\ldots+x_k^2)^2 = \sum_{j=1}^{k} x_j^4+2C_k^2 x_u^2 x_j^2$$

$$=\sum_{j=1}^{k} x_j^4+2\frac{k\,!}{2\,!\,(k-2)\,!}x_u^2 x_j^2 = \sum_{j=1}^{k} x_j^4+k(k-1)x_u^2 x_j^2 \tag{5.69}$$

we obtain

$$\sum_{j=1}^{k} x_j^4 = \varrho^4-k(k-1)x_u^2 x_j^2 \tag{5.70}$$

Substituting Eq. (5.70) into (5.68), we get

$$k\sum_{i=1}^{N} x_{ji}^4 = \sum_{i=1}^{N} \varrho_i^4-k(k-1)\sum_{i=1}^{N} x_{ui}^2 x_{ji}^2 \tag{5.71}$$

Considering Eq. (5.63), we have

$$3k\sum_{i=1}^{N} x_{ji}^4 = 3\sum_{i=1}^{N} \varrho_i^4-k(k-1)\sum_{i=1}^{N} x_{ji}^4 \tag{5.72}$$

whence

$$\sum_{i=1}^{N} x_{ji}^4 = \frac{3\sum \varrho_i^4}{k\,(k+2)} = 3N\lambda_4 \tag{5.73}$$

and

$$\lambda_4 = \frac{\sum_{i=1}^{N} \varrho_i^4}{Nk\,(k+2)} \tag{5.74}$$

This leads to a relation

$$\frac{\lambda_4}{\lambda_2^2} = \frac{\sum \varrho_i^4}{(\sum \varrho_i)^2}\frac{kN}{k+2} \tag{5.75}$$

If all N points of a rotatable design are on a single sphere, i. e.

$$\varrho_1^2 = \varrho_2^2 = \ldots = \varrho_N^2 \text{ and } \varrho_1^4 = \varrho_2^4 = \ldots = \varrho_N^4$$

we obtain from (5.75)

$$\lambda_4/\lambda_2^2 = k/(k+2)$$

and the determinant of matrix $X^T X$ is equal to zero. Thus we see that the design points should be positioned on several spheres. For a certain number of factors, the sphere on which the points of the design kernel lie coincides with another sphere embracing the star arm points. To provide for non-singularity of information matrix, a rotatable design is augmented with the points lying on a sphere of zero radius, namely, with n_0 points at the design center. Let N design points be on s spheres, n_u points belonging to each sphere. Then,

$$\begin{aligned} \textstyle\sum_{i=1}^{N} \varrho_i^2 &= \textstyle\sum_{u=1}^{s} n_u \varrho_u^4 \\ \textstyle\sum_{i=1}^{N} \varrho_i^2 &= \textstyle\sum_{u=1}^{s} n_u \varrho_u^2 \end{aligned} \tag{5.76}$$

With $\lambda_2 = 1$, we have

$$\lambda_4 = \frac{\sum_{u=1}^{s} n_u \varrho_u^4}{(\sum_{u=1}^{s} n_u \varrho_u^2)^2}\frac{kN}{k+2} \tag{5.77}$$

If the design kernel points n_k and star points n_α are on a single sphere, $\varrho_k = \varrho_\alpha$, and if there are n_0 points at the center point of the design, then we have

$$N = n_\alpha + n_k + n_0 = n + n_0$$

and Eq. (5.77) gives

$$\lambda_4 = \frac{kN}{k+2}\frac{n\varrho_\alpha^4}{n^2\varrho_\alpha^4} = \frac{k(n+n_0)}{(k+2)\,n} > \frac{k}{k+2} \tag{5.78}$$

Thus the condition given by Eq. (5.65) is satisfied when n_0 points are present at the center point of the design. The value of the star arm in a rotatable design is derived from Eq. (5.63)

$$2^k + 2\alpha^4 = 3\times 2^k \qquad \alpha = 2^{k/4} \qquad k < 5 \tag{5.79}$$

$$2^{k-1} + 2\alpha^4 = 3\times 2^{k-1} \qquad \alpha = 2^{\frac{k-1}{4}} \qquad k \geqslant 5 \tag{5.80}$$

Table 5.18 lists the values of α, n_0 and the radius of the sphere bearing the design kernel points for different values of k in uniform precision rotatable designs.

Table 5.18. Characteristics of Uniform Precision Rotatable Designs

Characteristic	k								
	2	3	4	5	5	6	6	7	7
kernel	2^2	2^3	2^4	2^5	2^{5-1}	2^6	2^{6-1}	2^7	2^{7-1}
ϱ_k	1.41	1.73	2.00	2.24	2.27	2.45	2.45	2.64	2.64
α	1.41	1.68	2.00	2.38	2.00	2.83	2.38	3.36	2.83
n_0	5	6	7	10	6	15	9	21	14

The matrix of a second-order rotatable design is not orthogonal because

$$\sum_{i=1}^{N} x_{0i}\, x_{ji}^2 \neq 0 \qquad j = 1, 2, \ldots, k$$
$$\sum_{i=1}^{N} x_{ji}^2 x_{ui}^2 \neq 0 \qquad j \neq u \quad j, u = 1, 2, \ldots, k$$

The b_{jj} coefficients are cross-correlated and correlated with the intercept, b_0. Therefore, in order to estimate the coefficients of the regression equation, one needs to solve a system of simultaneous normal equations by inverting the matrix $\boldsymbol{X}^T\boldsymbol{X}$

$$\boldsymbol{B} = (\boldsymbol{X}^T\boldsymbol{X})^{-1}\boldsymbol{X}^T\boldsymbol{Y}$$

In view of the specific nature of matrix $(\boldsymbol{X}^T\boldsymbol{X})$ for rotatable designs, it is possible to obtain its inverse in the general form and to derive equations for the estimation of regression coefficients and of their variances

$$b_0 = \frac{A}{N}\left[2\lambda_4^2(k+2)\sum_{i=1}^{N} y_i - 2\lambda_4 C \sum_{i=1}^{N} x_{ji}^2 y_i\right]$$

$$b_j = \frac{C}{N}\sum_{i=1}^{N} x_{ji}\, y_i \qquad j = 1, 2, \ldots, k$$

$$b_{uj} = \frac{C^2}{N\lambda_4}\sum_{i=1}^{N} x_{ui}x_{ji}y_i \qquad u \neq j \quad u, j = 1, 2, \ldots, k$$

$$b_{jj} = \frac{A}{N}\left\{[C^2(k+2)\lambda_4 - k]\sum_{i=1}^{N} x_{ji}^2 y_i + C^2(1-\lambda_4)\sum_{j=1}^{k}\sum_{i=1}^{N} x_{ji}^2 y_i - 2\lambda_4 C\sum_{i=1}^{N} y_i\right\} \tag{5.81}$$

$$s_{b_0}^2 = \frac{2A\lambda_4^2(k+2)}{N}s_e^2$$

$$s_{b_j}^2 = \frac{C}{N}s_e^2$$

$$s_{b_{uj}}^2 = \frac{C^2}{\lambda_4 N}s_e^2$$

$$s_{b_{jj}}^2 = \frac{A[(k+1)\lambda_4 - (k-1)]C^2}{N}s_e^2$$

where k is the number of factors

$$C = N/\sum_{i=1}^{N} x_{ji}^2$$

$$A = \frac{1}{2\lambda_4[(k+2)\lambda_4 - k]}$$

The constant λ_4 is given by Eq. (5.77). Equations (5.81) reduce to a more simple form by combining the constants

$$\begin{aligned} b_0 &= a_1 \sum_{i=1}^{N} y_i - a_2 \sum_{j=1}^{k} \sum_{i=1}^{N} x_{ji}^2 y_i \\ b_j &= a_3 \sum_{i=1}^{N} x_{ji} y_i \qquad j = 1, 2, \ldots, k \\ b_{uj} &= a_4 \sum_{i=1}^{N} x_{ui} x_{ji} y_i \qquad u \neq j \quad j, u = 1, 2, \ldots, k \\ b_{jj} &= a_5 \sum_{i=1}^{N} x_{ji}^2 y_i + a_6 \sum_{j=1}^{k} \sum_{i=1}^{N} x_{ji}^2 y_i - a_7 \sum_{i=1}^{N} y_i \end{aligned} \tag{5.82}$$

$$s_{b_0}^2 = a_1 s_e^2 \quad s_{b_j}^2 = a_3 s_e^2 \quad s_{b_{uj}}^2 = a_4 s_e^2 \quad s_{b_{jj}}^2 = (a_5 + a_6) s_e^2 \tag{5.83}$$

The values of the constants in Eqs. (5.82) are listed in Table 5.19.

Table 5.19. Regression Coefficients for Rotatable Designs with $k \leqslant 7$

k	N	n_0	α	a_1	a_2	a_3	a_4	a_5	a_6	a_7
2	13	5	1.412	0.2	0.1	0.125	0.25	0.1251	0.0187	0.1
3	20	6	1.682	0.1663	0.0568	0.0732	0.125	0.0625	0.0069	0.0568
4	31	7	2.00	0.1428	0.0357	0.0417	0.0625	0.0312	0.0037	0.0357
5*	32	6	2.00	0.1591	0.0341	0.0417	0.0625	0.0312	0.0028	0.0341
5	52	10	2.378	0.0988	0.0191	0.0231	0.0312	0.156	0.0015	0.0191
6*	53	9	2.378	0.1108	0.0187	0.0231	0.0312	0.0156	0.0012	0.0187
6	91	15	2.828	0.0725	0.0098	0.0125	0.0156	0.0078	0.0005	0.0098
7*	92	14	2.828	0.0703	0.0098	0.0125	0.0156	0.0078	0.0005	0.0098
7	163	21	3.333	0.0398	0.0052	0.0066	0.0078	0.0039	0.0002	0.0052

* Half-replicate.

In using second-order rotatable designs no replication is needed in order to find the error mean square. Instead, this can be done by replicating the center point. Accordingly, the estimated regression equation derived from a second-order rotatable design is tested for adequacy of fit as follows.

To begin with, one finds the error mean square

$$s_e^2 = \frac{1}{n_0-1} \sum_{u=1}^{n_0} (y_u^0 - \bar{y}^0)^2$$
$$\bar{y}^0 = \frac{1}{n_0} \sum_{u=1}^{n_0} y_u^0 \tag{5.84}$$

for which the number of degrees of freedom is $\nu_e = (n_0 - 1)$.

The residual mean square is given by

$$s_{\text{res}}^2 = \frac{\sum_{i=1}^{N} (y_i - \hat{y}_i)^2}{N-l}$$

for which the number of degrees of freedom is $\nu_{\text{res}} = N - l$.

To test the estimated regression equation for the goodness of fit, use is made of the Fisher F-test

$$F = s_g^2 / s_e^2 \tag{5.85}$$

where s_g^2 is the adequacy mean square given by

$$s_g^2 \nu_g = s_{\text{res}}^2 \nu_{\text{res}} - s_e^2 \nu_e$$
$$s_g^2 = \frac{s_{\text{res}}^2 \nu_{\text{res}} - s_e^2 \nu_e}{\nu_g} \tag{5.86}$$

and ν_g is the number of degrees of freedom for the adequacy mean square, $\nu_g = \nu_{\text{res}} - \nu_e$.

The equation is considered adequate, if $F < F_{1-\alpha}(\nu_1, \nu_2)$, where ν_1 is the number of degrees of freedom for the adequacy mean square and ν_2 is the number of degrees of freedom for the error mean square.

The estimated coefficients of the regression equation are tested for significance using the Student t-test. If one of the second order effects fails to be significant, it is deleted from the equation, and the remaining coefficients should be recalculated.

Example 5.3 [21]. We seek the functional relationship between the impurities in wet-process phosphoric acid and the degree of decomposition (y%) of the flotation concentrate of the Karatau phosphorite. The factors governing the decomposition are chosen to be: z_1, the process temperature in °C; z_2 through z_5, the MgO, SO_3, Al_2O_3 and F contents (by weight per cent) of phosphoric acid, respectively.

The principal level, incremental changes, and the region of exploration are given in the table below.

	z_1	z_2	z_3	z_4	z_5
z_j^0	50	2.1	2.0	1.33	0.75
Δz_j	20	0.9	1.0	0.37	0.25
+2	90	3.9	4.0	2.07	1.25
−2	10	0.3	0.0	0.59	0.25

The range of changes in factors corresponds to the range of impurity contents of commercial wet-process phosphoric acid.

Solution. A second-order rotatable design will be used to derive an estimated regression equation (Table 5.20).

Table 5.20. Second-order Rotatable Design for $k = 5$

N	x_1	x_2	x_3	x_4	x_5	y_1, %	N	x_1	x_2	x_3	x_4	x_5	y_1, %
1	+1	+1	+1	+1	+1	34.7	17	−2	0	0	0	0	25.0
2	−1	+1	+1	+1	−1	40.0	18	+2	0	0	0	0	33.3
3	+1	−1	+1	+1	−1	39.0	19	0	−2	0	0	0	49.2
4	−1	−1	+1	+1	+1	39.2	20	0	+2	0	0	0	42.0
5	+1	+1	−1	+1	−1	26.6	21	0	0	−2	0	0	17.5
6	−1	+1	−1	+1	+1	29.5	22	0	0	+2	0	0	41.0
7	+1	−1	−1	+1	+1	30.0	23	0	0	0	−2	0	35.6
8	−1	−1	−1	+1	−1	34.5	24	0	0	0	+2	0	27.2
9	+1	+1	+1	−1	−1	32.2	25	0	0	0	0	−2	39.0
10	−1	+1	+1	−1	+1	41.4	26	0	0	0	0	+2	30.0
11	+1	−1	+1	−1	+1	33.7	27	0	0	0	0	0	35.4
12	−1	−1	+1	−1	−1	40.9	28	0	0	0	0	0	36.4
13	+1	+1	−1	−1	+1	23.9	29	0	0	0	0	0	33.2
14	−1	+1	−1	−1	−1	33.3	30	0	0	0	0	0	32.4
15	+1	−1	−1	−1	−1	27.7	31	0	0	0	0	0	37.7
16	−1	−1	−1	−1	+1	35.9	32	0	0	0	0	0	36.9

The number of observations in the design matrix for $k = 5$ is 32. The kernel of the design is a half-replicate of a 2^{5-1} design with a defining relation $x_5 = x_1x_2x_3x_4$. The star arm $\alpha = 2$ is found from Table 5.18. The natural variables are transformed into coded x's by Eq. (5.3).

The error mean square as found by the observations at the center point of the design is $s_e^2 = 4.466$ for which the number of degrees of freedom is $\nu_e = n_0 - 1 = 5$.

Using Table 5.20, we estimate the coefficients of the second-order regression equation and their standard errors of estimate

$$
\begin{array}{llll}
b_0 = 35.41, & b_1 = 1.07794, & b_2 = -0.146, & b_3 = 4.5098 \\
b_4 = -0.542, & b_5 = -1.3, & b_{11} = -1.5, & b_{22} = 2.66 \\
b_{33} = -1.47, & b_{44} = -0.93, & b_{55} = -0.15, & b_{12} = 0.147 \\
b_{13} = 0.256, & b_{14} = 1.61, & b_{15} = 0.0534, & b_{23} = 0.736 \\
b_{24} = -0.198, & b_{25} = 0.403, & b_{34} = 0.401, & b_{35} = 0.256 \\
b_{45} = 0.93, & s_{b_j} = 0.43, & s_{b_{uj}} = 0.53, & s_{b_{jj}} = 0.394
\end{array}
$$

The significance of the coefficients is tested, using the t-ratio

$$
t_1 = \frac{1.07}{0.43} = 2.48 \qquad t_{12} = \frac{0.147}{0.53} = 0.278
$$

$$
t_2 = \frac{0.146}{0.43} = 0.34 \qquad t_{13} = \frac{0.256}{0.53} = 0.483
$$

$$t_3 = \frac{4.51}{0.43} = 10.4 \qquad t_{14} = \frac{1.61}{0.53} = 3.04$$

$$t_4 = \frac{0.542}{0.43} = 1.26 \qquad t_{15} = \frac{0.0534}{0.53} = 0.1$$

$$t_5 = \frac{1.3}{0.43} = 3.02 \qquad t_{23} = \frac{0.736}{0.53} = 0.1375$$

$$t_{11} = \frac{1.5}{0.394} = 3.82 \qquad t_{24} = \frac{0.198}{0.53} = 0.374$$

$$t_{22} = \frac{2.66}{0.394} = 6.75$$

$$t_{33} = \frac{1.47}{0.394} = 3.73 \qquad t_{25} = \frac{0.403}{0.53} = 0.762$$

$$t_{44} = \frac{0.93}{0.394} = 2.36 \qquad t_{34} = \frac{0.401}{0.53} = 0.758$$

$$t_{55} = \frac{0.15}{0.394} = 0.38 \qquad t_{45} = \frac{0.93}{0.53} = 1.75$$

For a significance level of $\alpha = 0.05$ and $\nu = 5$ degrees of freedom, the tabulated value of the t-ratio $t_{0.05}(5) = 3.18$. After the deletion of the coefficients which are not significant and for which the t-ratio is smaller than the tabulated value, the estimated regression equation takes the form

$$\hat{y} = 36.2 + 4.51x_3 - 1.3x_5 - 1.45x_1^2 + 2.82x_2^2 - 1.53x_3^2 + 1.01x_1x_4$$

The Fisher F-test shows that the above equation fits the experimental data adequately

$$s_e^2 = 4.466 \qquad s_g^2 = 15.35 \qquad F = 3.43 \qquad F_{0.95}(20.5) = 4.5$$

Converted back to the natural scale, the equation takes the form

$$\hat{y} = 44.04 + 0.086z_1 - 13.8z_2 + 10.39z_3 - 10.9z_4 - 5.2z_5 - 0.00375z_1^2 + 3.28z_2^2 - 1.4z_3^2 + 0.217z_1z_4$$

The above equation gives an idea about the degrees of decomposition of the flotation concentrate of the Karatau phosphorite at different temperatures, depending on the impurity contents in the acid.

5.7. Optimality Criterions for Experimental Designs

The optimality criterions developed by Box and his coworkers were based on the empirical/intuitional method. First, orthogonal designs were treated, by them as an optimum but later they did so in relation to rotatable designs. A design is said to be orthogonal, if it is described by a diagonal information matrix. The parameter estimates associated with orthogonal designs are independent of one another. A rotatable design has its covariance matrix held invariant in the case of orthogonal rotation of the coordinates. When this condition is satisfied, the response surface can be estimated with the same degree of accuracy in any direction from the center point of the design.

The designs possessing the properties of orthogonality and rotatability are very useful in practical applications. Note that 2^k and 2^{k-p} linear orthogonal designs are rotatable, but Box-Hunter composite rotatable designs are not orthogonal. If one takes the orthogonality as an optimality criterion for discrimination between experimental designs, this always decreases to some extent the accuracy of estimation of the parameters and regression function.

There is another, purely theoretical, trend in the development of the optimality criterions for experimental designs, which is based mainly on the work of Kiefer. His concept of *D*-optimality stems from Fisher's theory of effective estimates which are connected merely with an optimum method of handling the experimental data. In the case of least squares estimation, effective estimates of the coefficients of a linear regression equation are found. Here the estimate scatter ellipsoid is related to the determinant of the information matrix by

$$V_k = \frac{(k+2)^{k/2}\pi^{k/2}}{\Gamma\left(\frac{k}{2}+1\right)\sqrt{|\mathbf{X}^T\mathbf{X}|}} \tag{5.87}$$

where $\Gamma\left(\frac{k}{2}+1\right)$ is the gamma function.

According to Kiefer, the estimate effectiveness depends on an optimum distribution of the experimental points. In the case of a *D*-optimal design, the scatter ellipsoid has a minimal volume for a set of designs in the region of exploration. A maximal determinant of the information matrix should correspond to a *D*-optimal design.

Kiefer has devised a number of design optimality criterions. Like the *D*-optimality criterion, they deal with certain requirements to be met by the form of the variance-covariance and, consequently, the information matrix. The information matrix of an *A*-optimal design has its trace (the sum of the diagonal elements) kept at a minimum. Using an *A*-optimal design, one can minimize the variance of the parameter estimate. In the case of an *E*-optimal design, the maximal eigenvalue of its variance matrix is kept at a minimum. This means that an *E*-optimal design minimizes the maximal axis of the estimate scatter ellipsoid. A *G*-optimal design provides a maximal variance of the predicted values of y for all designs in the region of exploration. Therefore, there are no points within that region at which the error of estimate of the response surface is too low.

Box and Draper [51] used a design optimality criterion for minimization of the systematic and general bias that occurs in approximating the response surface by a polynomial having a lower order than an adequately approximating polynomial.

The *D*- and *G*-optimal designs are based on a most reliable theory [46]. There is no general solution to the problem of construction of *D*-optimal designs. *D*-optimal designs proved useful in the case of the single parameter estimation. Kiefer, Wolfowitz, Hoel, and Kono introduced the concept of a continuous design and constructed continuous *D*-optimal designs for first and second regression polynomials in the case of constraints on a hypercube and a k-dimensional sphere and also for a trigonometric regression function by segment, various weighting functions [46]. Here the experimental region is often depicted by a hypercube. A 2^k full factorial design is an adequate means of constructing first-order *D*-optimal designs with constraints on a cube. Other *D*-optimal designs are some fractional replicates of a full fractional design and Plackett-Burman designs for k factors that satisfy a condition in which $k+1$ is a multiple of 4. These designs are both orthogonal and rotatable.

Kiefer and Wolfowitz [46] developed continuous second-order *D*-optimal designs which are constructed on 2- to 5-dimensional cubes for second-order polynomials. As a rule, such designs involve a very large number of observations; a precise design of the described type requires more than 1 500 observations for $k = 5$. A way out is to find by a computer the second-order asymmetric designs which involve a relatively small number of experimental points and resemble *D*-optimal designs as to their information matrix determinants and variance of the predicted value of the optimization parameter [47]. Some second-order composite designs with constraints on a cube for $k = 4$, 5, and 6 were tested in terms of *D*-optimality. The selection of a specific design depends on the objective and operational capabilities of the experiment.

5.8. Canonical Analysis of the Response Surface. Solution of the Optimization Problem

An estimated second-order regression equation which is adequately fit to the "nearly stationary" region is analyzed to determine the coordinates of the optimum. In addition, it is interesting to investigate the properties of the response surface in the vicinity of the optimum. The usual procedure in this case is to transform the second-order polynomial derived from the experimental data into a standard, canonical equation

$$\hat{y}-y_S = \lambda_{11}X_1^2+\lambda_{22}X_2^2+\ldots+\lambda_{kk}X_k^2, \qquad (5.88)$$

where y_S is the value of the predicted response at the center of the response surface; $X_1, X_2, \ldots, X_k$ are the canonical variables that are linearly related to the factors $x_1, x_2, \ldots, x_k$; $\lambda_{11}, \lambda_{22}, \ldots, \lambda_{kk}$ are the canonical form coefficients (the transformed estimated parameters).

The first transformation deals with the translation of the origin of the old coordinates to the extremum of the response surface, *S*, whose coordinates are given by

$$\frac{\partial y}{\partial x_1} = 0, \ \frac{\partial y}{\partial x_2} = 0, \ \ldots, \ \frac{\partial y}{\partial x_k} = 0 \qquad (5.89)$$

When a second-order polynomial serves as an approximation of the response surface, a system of *k* simultaneous linear equations must be solved. If the determinant of the system is equal to zero, then the response surface has no centre. In this case, there are two possible approaches: the origin of the old coordinates is translated to the point where the best predicted response occurs; the origin is not translated at all. With an off-central surface, the optimum will lie at the boundary of the experimental region. If the response surface has a centre point, then the origin of the old coordinates is translated to that point. Owing to the translation, first-order terms are eliminated and the free term is changed. Note that quadratic and interaction terms are not affected by the translation.

The second transformation of the estimated regression equation is the rotation of the new coordinate system to get rid of the interaction, or cross-product, terms: this procedure does not affect, however, the free term. The two transformations yield a new expression represented by Eq. (5.88).

On the basis of canonical equations (Fig. 5.7), one can evaluate the nature at response surfaces as follows:

(1) All canonical form coefficients have the same sign. The response surface is an elliptical paraboloid. (Fig. 5.7*a*) The centre of the figure is a maximum if $\lambda_{ii} < 0$, and a minimum if $\lambda_{ii} > 0$.

(2) The coefficients have unlike signs. The response surface is a hyperbolic paraboloid (a saddle) shown in Fig. 5.7*b*. There is a minimax point at the centre of the surface.

(3) One or more coefficients (but not all of them) are close to zero. The centre of the surface is far outside the region of exploration. Such a surface is called a rising ridge.

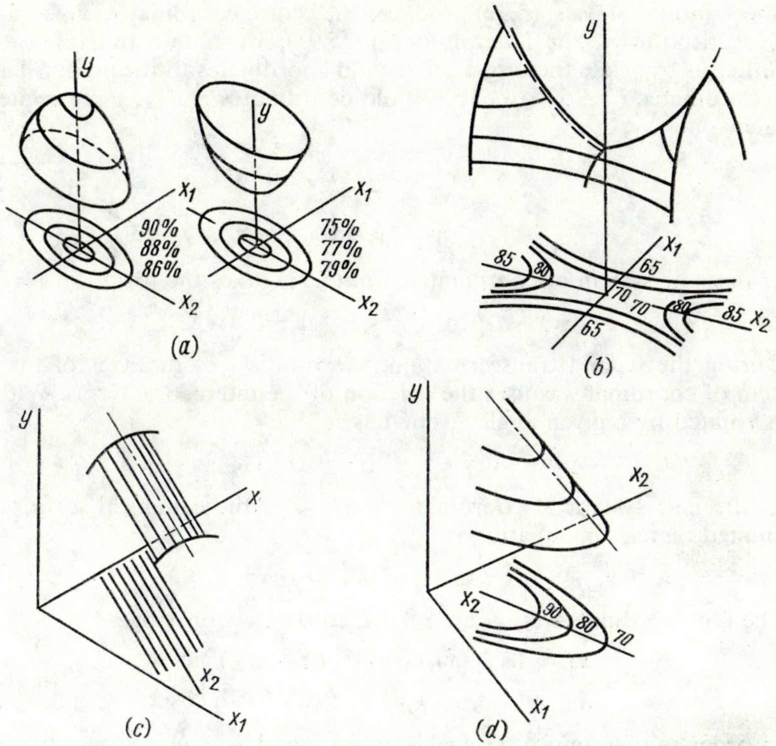

Fig. 5.7. Canonical surfaces and the respective contours at $k = 2$

(4) The response surface is a degenerate one described by parallel straight lines (Fig. 5.7*c*), the case that has no practical value.

If $\lambda_{22} = 0$ (Fig. 5.7*d*), then the translation of the origin of coordinates to the point S (usually near the centre point of the design) produces the equation of a parabola

$$\hat{y}-y_S = \lambda_{11}X_1^2+\lambda_2 X_2 \tag{5.90}$$

Now, we transform the estimated second-order regression equation for $k = 2$ obtained from the experimental data

$$\hat{y} = b_0+b_1x_1+b_2x_2+b_{12}x_1x_2+b_{11}x_1^2+b_{22}x_2^2 \tag{5.91}$$

to its canonical form, Eq. (5.88). In order to determine the coordinates of the point S, the centre or the response surface, one must solve a system of simultaneous equations

$$\begin{aligned} \frac{\partial y}{\partial x_1} &= 0 \qquad 2b_{11}x_1+b_{12}x_2+b_1 = 0 \\ \frac{\partial y}{\partial x_2} &= 0 \qquad b_{12}x_1+2b_{22}x_2+b_2 = 0 \end{aligned} \tag{5.92}$$

The solution of Eqs. (5.92) produces the centre coordinates, x_{1S} and x_{2S}. On substitution of the latter into Eq. (5.91), we obtain the value of y_S at S. Let us translate the origin of the old coordinates to the point S having the coordinates (y_S, x_{1S}, x_{2S}). The old coordinates x_1, x_2, y are related to the new ones x_1', x_2', y' by

$$\begin{aligned} x_1 &= x_{1s}+x_1' \\ x_2 &= x_{2s}+x_2' \\ y &= y_s+y' \end{aligned} \tag{5.93}$$

In the new system of coordinates, Eq. (5.91) takes the form

$$\hat{y}-y_s = b_{11}(x_1')^2+b_{22}(x_2')^2+b_{12}x_1'x_2'.$$

During the second transformation, the rotation of the axes of the new system of coordinates causes the deletion of the interaction terms. With the axes rotated by a given angle α, one has

$$\operatorname{ctg} 2\alpha = (b_{11}-b_{22})/b_{12}. \tag{5.94}$$

In the new system of coordinates X_1, X_2, y, the canonical form of the estimated regression equation is

$$\hat{y}-y_S = \lambda_{11}X_1^2+\lambda_{22}X_2^2 \tag{5.95}$$

The old coordinates x_1, x_2 are related to the new ones by

$$\begin{aligned} x_1 &= (X_1+x_{1s}) \cos \alpha-(X_2+x_{2s}) \sin \alpha \\ x_2 &= (X_1+x_{1s}) \sin \alpha +(X_2+x_{2s}) \cos \alpha \end{aligned} \tag{5.96}$$

In order to determine the coefficients λ_{11} and λ_{12}, one can use two invariants of the equation which are the coefficient functions whose values are held constant regardless of any coordinate transformation

$$\begin{aligned} I_1 &= b_{11}+b_{22} = \text{constant} \\ I_2 &= \begin{vmatrix} b_{11} & \frac{1}{2} b_{12} \\ \frac{1}{2} b_{12} & b_{22} \end{vmatrix} = \text{constant} \end{aligned} \tag{5.97}$$

It follows from (5.95) that

$$\begin{aligned} b_{11}+b_{22} &= \lambda_{11}+\lambda_{22} \\ b_{11}b_{22}-\frac{1}{4} b_{12}^2 &= \lambda_{11}\lambda_{22}-\frac{1}{4} \lambda_{12}^2 \end{aligned} \tag{5.98}$$

Since in this transformation $\lambda_{12} = 0$, one has the following relations that produce λ_{11} and λ_{12}:

$$\begin{aligned} b_{11}+b_{22} &= \lambda_{11}+\lambda_{22} \\ b_{11}b_{22}-\frac{1}{4} b_{12}^2 &= \lambda_{11}\lambda_{22} \end{aligned} \tag{5.99}$$

Using the Vieta theorem, one can find the roots of the following quadratic equation

$$\lambda^2-(b_{11}+b_{22})\lambda+\left(b_{11}b_{22}-\frac{1}{4}b_{12}^2\right) = 0 \tag{5.100}$$

In the case of a multidimensional factor space, canonical analysis uses methods of linear algebra. In the example given below, one can employ the coefficients of the estimated second-order regression equation derived from the experimental data

$$\hat{y} = b_0+b_1x_1+\ldots+b_kx_k+b_{12}x_1x_2+\ldots$$
$$+b_{k-1,k}x_{k-1}x_k+b_{11}x_1^2+\ldots+b_{kk}x_k^2$$

to form the following quadratic matrix

$$B = \begin{bmatrix} b_{11} & \frac{1}{2}b_{12} & \ldots & \frac{1}{2}b_{1k} \\ \frac{1}{2}b_{21} & b_{22} & \ldots & \frac{1}{2}b_{2k} \\ \cdot & \cdot & & \cdot \\ \cdot & \cdot & & \cdot \\ \cdot & \cdot & & \cdot \\ \frac{1}{2}b_{k1} & \frac{1}{2}b_{k2} & \ldots & b_{kk} \end{bmatrix} \tag{5.101}$$

in which $b_{ji} = b_{ij}$.

To find the coefficients $\lambda_{11}, \ldots, \lambda_{kk}$ of the canonical form of the equation, Eq. (5.88), one must determine the roots of the polynomial $P_k(\lambda)$ of the matrix $\boldsymbol{B}$

$$P_k(\lambda) = |\boldsymbol{B}-\lambda\boldsymbol{I}| \tag{5.102}$$

where $\boldsymbol{I}$ is the identity matrix.

Thus, we obtain

$$P_k(\lambda) = \begin{vmatrix} b_{11}-\lambda & \frac{1}{2}b_{12} & \ldots & \frac{1}{2}b_{1k} \\ \frac{1}{2}b_{21} & b_{22}-\lambda & \ldots & \frac{1}{2}b_{2k} \\ \cdot & \cdot & & \cdot \\ \cdot & \cdot & & \cdot \\ \cdot & \cdot & & \cdot \\ \frac{1}{2}b_{k1} & \frac{1}{2}b_{k2} & \ldots & b_{kk}-\lambda \end{vmatrix} \tag{5.103}$$

In the case of X coordinates, the orthogonal linear transformation is given by a set of simultaneous equations as follows

$$\begin{aligned} X_1 &= m_{11}(x_1-x_{1s})+m_{12}(x_2-x_{2s})+\ldots+m_{1k}(x_k-x_{ks}) \\ X_2 &= m_{21}(x_1-x_{1s})+m_{22}(x_2-x_{2s})+\ldots+m_{2k}(x_k-x_{ks}) \\ X_k &= m_{k1}(x_1-x_{1s})+m_{k2}(x_2-x_{2s})+\ldots+m_{kk}(x_k-x_{ks}). \end{aligned} \tag{5.104}$$

The coefficients m_{ik} provide a solution of k sets of simultaneous homogeneous equations. For λ_i, the set becomes

$$\begin{aligned} (b_{11}-\lambda_i)m_{i1}+\frac{1}{2}b_{12}m_{i2}+\ldots+\frac{1}{2}b_{1k}m_{ik} &= 0 \\ \cdot \qquad \cdot \qquad \cdot \\ \cdot \qquad \cdot \qquad \cdot \\ \cdot \qquad \cdot \qquad \cdot \\ \frac{1}{2}b_{k1}m_{i1}+\frac{1}{2}b_{k2}m_{i2}+\ldots+(b_{kk}-\lambda_i)m_{ik} &= 0 \end{aligned} \tag{5.105}$$

Since the solutions of the equations are merely proportional to those quantities m_{ij} which ensure the orthogonalization of the linear transformation, Eq. (5.10), recalculation is necessary which takes into account the property of orthogonality

$$m_{i1}^2+m_{i2}^2+\ldots+m_{ik}^2 = 1 \qquad i = 1, 2, \ldots, k \tag{5.106}$$

After the coefficients of the estimated regression equation in canonical form have been estimated, one can thus ascertain the shape of the response surface, which determines the strategy of search for the extremum. If the response surface is an elliptical paraboloid, and $\lambda_{11} < 0$ and $\lambda_{22} < 0$, the surface centre is a maximum. When $\lambda_{11} > 0$ and $\lambda_{22} > 0$, the optimization parameter is maximal at the boundary of the experimental region.

If the response surface is a hyperbolic paraboloid given by

$$\hat{y}-y_s = \lambda_{11}X_1^2-\lambda_{22}X_2^2 \tag{5.107}$$

and the conditions are sought for which the optimization parameter is a maximum, then, given the relations $\hat{y} > y_S$ at $X_2 = 0$, one can move along the canonical axis X_1 having a positive canonical coefficient. Under the circumstances, one tries to check whether the constraint $x_j = \pm\alpha$ is valid by substituting into Eqs. (5.96) the quantities

$$X_1 = \pm\sqrt{\frac{\hat{y}-y_s}{\lambda_{11}}} \qquad X_2 = 0 \tag{5.108}$$

In the case of a multidimensional space, the optimum conditions of the process are found by a computer on the basis of a suitable method of non-linear programming.

Example 5.4. Given the estimated regression equation that depicts the degree of decomposition of the flotation concentrate of the Karatau phosphorite, influenced by the process temperature and contents of the impurities contained in the phosphoric acid

$$\hat{y} = 35.4 + 4.51x_3 - 1.3x_5 - 1.5x_1^2 + 2.66x_2^2 - 1.47x_3^2 + 1.61x_1x_4$$

we seek the conditions in which the degree of decomposition (y_{max}) is at a maximum in the presence of constraints imposed by a sphere with a radius equal to a given star arm (Table 5.20).

Solution. We can take certain values of the factors explicitly described by the equation: $x_2 = +2$ and $x_5 = -2$. The positive linear term and a negative quadratic term represent the effect of SO_3 content in the phosphoric acid. An optimum value of x_3 equal to 1.533 is found from a point in which y has an extremum on x_3. Given the above values of the factors x_1, x_3, and x_5, we obtain the following form of the estimated regression equation

$$\hat{y} = 52.12 - 1.5x_1^2 + 1.61x_1x_4.$$

Let us now transform the equation to its canonical form. The coordinates of the response surface centre, S, are given by

$$\frac{\partial \hat{y}}{\partial x_1} = -1.5 \times 2x_1 + 1.61x_4 = 0$$

$$\frac{\partial \hat{y}}{\partial x_4} = 1.61x_1 = 0$$

$$x_{1s} = 0, \quad x_{4s} = 0, \quad y_s = 52.12.$$

Thus we see that the surface centre coincides with the centre point of the design. Under these circumstances, the characteristic polynomial is

$$P_k(\lambda) = \begin{vmatrix} -1.5-\lambda & 0.805 \\ 0.805 & 0-\lambda \end{vmatrix}$$

or

$$P_k(\lambda) = \lambda^2 + 1.5\lambda - 0.64 = 0$$

The roots of the polynomial are $\lambda_1 = 0.35$ and $\lambda_2 = -1.85$. The canonical form of the equation is

$$\hat{y} - 52.12 = 0.35X_1^2 - 1.85X_4^2$$

Now the response surface is a hyperbolic paraboloid. On passing planes of constant value, y = constant, through the response surface, we obtain hyperbolas (Fig. 5.8). At S is a minimax point. The linear transformation is given by

$$X_1 = 0.920x_1 + 0.39x_4$$

$$X_4 = -0.39x_1 + 0.92x_4$$

To find the maximum degree of decomposition, we leave the minimax point and move along the axis X_1 (the canonical form coefficient is positive), X_4 being equated to zero

$$X_1 = \pm\sqrt{\frac{\hat{y} - 52.12}{0.35}}, \qquad X_4 = 0.$$

As the quantity $\hat{y}$ increases, one checks to see that the condition $x_1 = x_4 \leqslant 2$ is satisfied. Thus we obtain the maximum degree of decomposition equal to 53.5% ($x_1 = \pm 1.82$ and $x_4 = \pm 0.795$).

Increasing $\hat{y}$ to 54%, we have $x_1 > 2$. For the optimum conditions so obtained ($x_1 = +1.82$, $x_2 = +2$, $x_3 = +1.533$, $x_4 = +0.795$, $x_5 = -2$) and ($x_1 = -1.82$, $x_2 = +2$, $x_3 = +1.533$, $x_4 = -0.795$, $x_5 = -2$), check tests were conducted that gave the maxi-

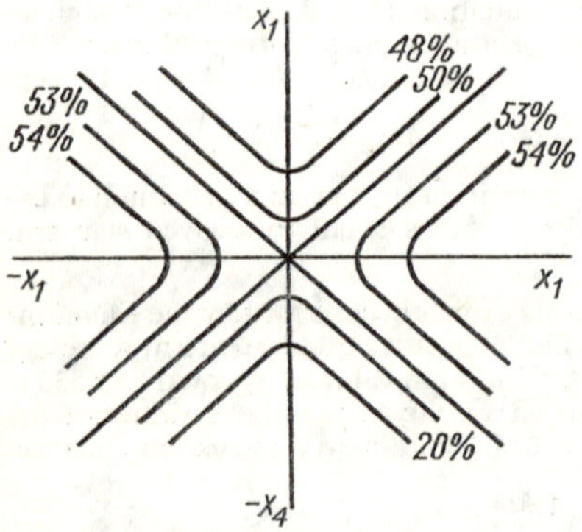

Fig. 5.8. Hyperbolas of equal yield

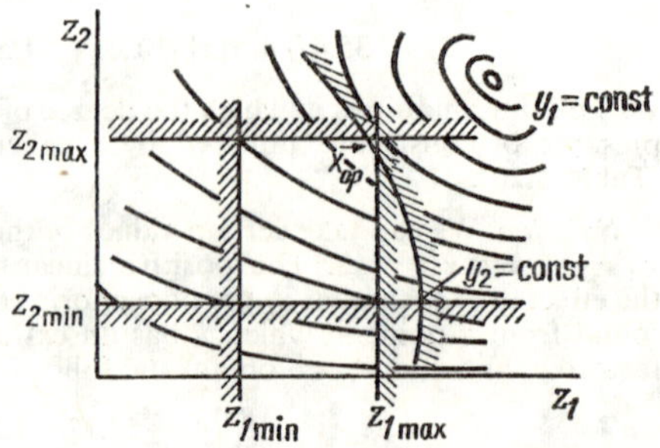

Fig. 5.9. Search for an optimum under conditions of constraints

mum degree of decomposition equal to 55.8% and 53.7%, respectively. Therefore, the deviation from the computed value ($\bar{y} = 54.7\%$) is within the limits of experimental error ($s_e = \sqrt{4.466} = 2.1$).

If the process is described by several regression equations, a trade-off is to find the extremum of one response function in the presence of constraints resulted from the other response functions and the boundaries of the experimental region (Fig. 5.9). Let one be interested in the extremum of the function $y = f(x_1, \ldots, x_k)$, which depends on k variables x_j ($j = 1, \ldots, k$) related to one another by

$$\varphi_u(x_i, \ldots, x_k) = 0, \qquad u = 1, \ldots, m. \tag{5.109}$$

The extremum that the function $f(x_1, \ldots, x_k)$ reaches on the assumption that the relationships of Eq. (5.109) are satisfied is called conditional or relative. Analytically, the search of the conditional extremum is carried out by using Lagrange multipliers. Formally, the problem of finding the conditional extremum of the function f reduces to determination of the unconditional extremum of the following Lagrange function

$$\Phi(x, \lambda) = f(x) + \sum_{u=1}^{m} \lambda_u \varphi_u(x), \tag{5.110}$$

this Lagrange function being a function of $k+m$ variables, where λ_u represents Lagrange multipliers.

To illustrate the application of Lagrange multipliers, below is given an example which deals with the conversion of calcium nitrate and phosphoric acid into a solid monocalcium phosphate and nitric acid in the presence of n-butyl alcohol [22]. We have the following estimated regression equations for a second-order rotatable design:

$$\begin{aligned}\hat{y}_1 = {} & 60.9 + 14.5x_1 + 3.83x_3 - 4.09x_4 + 2.14x_1x_3 \\ & + 2.71x_1x_4 + 2.21x_1x_5 + 1.28x_2x_3 + 2.48x_3x_4 \\ & + 0.68x_2^2 + 0.68x_4^2 + 0.8x_5^2\end{aligned} \tag{5.111}$$

$$\hat{y}_2 = 1.682 - 0.85x_1 + 0.261x_3 + 0.062x_4 - 0.041x_1x_4 - 0.034x_1x_5 + 0.032x_3x_4 - 0.024x_3^2 - 0.015x_4^2 \quad (5.112)$$

where $\hat{y}_1$ is the degree of conversion, $\hat{y}_2$ is the ratio of nutrients in the fertilizer in terms of P_2O_5 and N contents, x_1 is the original concentration of phosphoric acid, x_2 is the time in contact, x_3 is the stoichiometric ratio of phosphoric acid in the solution, x_4 is the acid/alcohol volume ratio, and x_5 is the conversion temperature.

If one takes into account the constraints imposed on the factors during the first stage of the process in which phosphates undergo acid decomposition ($x_1 = -2$ and $x_5 = 0$) and also the necessity for higher production rate ($x_2 = 0$), the estimated regression equations take the form

$$\hat{y}_1 = 31.9 - 0.45x_3 - 9.51x_4 + 2.48x_3x_4 + 0.68x_4^2 \quad (5.113)$$

$$\hat{y}_2 = 2.1 + 0.2722x_3 + 0.144x_4 - 0.0235x_3^2 - 0.015x_4^2 + 0.032x_3x_4 \quad (5.114)$$

Taking into account the requirements of agrobiology, constraints are imposed on the ratios of the nutrients in the fertilizer as follows:

$$P_2O_5 : N = 1:1$$

$$P_2O_5 : N = 1.5:1$$

$$P_2O_5 : N = 2:1$$

The preferable fertilizer would be a balanced one with the 1 : 1 above ratio. We use indefinite Lagrangian multipliers in order to find the values of $x_{3,\mathrm{op}}$ and $x_{4,\mathrm{op}}$ that provide for the maximum degree of conversion with constraints on the fertilizer nutrient ratio. The Lagrange function is

$$\Phi = \hat{y}_1 + \lambda(\hat{y}_2 - 2.1 - 0.2722x_3 - 0.144x_4 + 0.0235x_3^2 + 0.015x_4^2 - 0.032x_3x_4) \quad (5.115)$$

Thus, we have a set of simultaneous equations with which optimum operating conditions can be found

$$\frac{\partial \Phi}{\partial x_3} = -0.45 + 2.48x_4 - \lambda(0.2722 - 0.047x_3 + 0.032x_4) = 0$$

$$\frac{\partial \Phi}{\partial x_4} = -9.51 + 2.48x_3 + 1.36x_4 - \lambda(0.144 - 0.03x_4 + 0.032x_3) = 0 \quad (5.116)$$

$$\frac{\partial \Phi}{\partial \lambda} = 2.1 + 0.2722x_3 - 0.144x_4 - 0.0235x_3^2 - 0.015x_4^2 + 0.032x_3x_4 - \hat{y}_2 = 0$$

Eqs. (5.116) are solved through the use of a computer. The experimental region determines the constraints imposed on x_3 and x_4: $x_3 = \pm\alpha = \pm 2$ and $x_4 = \pm\alpha = \pm 2$. The values of $\hat{y}_2$ are tied as follows: 1 : 1, 1. 5 : 1, and 2 : 1. It is found that within the region of exploration one can obtain fertilizers with nutrient ratios 1.5 : 1 and 2 : 1. The computation results are as follows: $\hat{y}_1^{\max} = 61.3\%$ and $\hat{y}_2 = 2.03$ for $x_{3,\mathrm{op}} = 1.7$ and $x_{4,\mathrm{op}} = -2.0$; $\hat{y}_1^{\max} = 49.58\%$ and $\hat{y}_2 = 1.54$ for $x_{3,\mathrm{op}} = -0.9$ and $x_{4,\mathrm{op}} = -2.0$.

When the optimum conditions of a process are determined, the extrapolation sometimes moves beyond the experimental region. Thus in all cases the computed optimum conditions should be tested by actual experiment.

5.9. The Desirability Function

Process optimization concerned with several responses is usually reduced to the optimization procedure utilizing one criterion with inequality and equality constraints. Depending on the shape of the response surface and the character of the constraints, there exist a number of optimization methods such as Lagrange multipliers, linear and nonlinear programming, ridge analysis [10], etc. These methods are disadvantageous in that the associated computations are very complicated. For example, if one tries to describe the response surface using second-order polynomials, a set of simultaneous nonlinear equations must be solved in order to find the conditional extremum with the help of Lagrange multipliers. A useful approach to the optimization of processes with several responses is the concept of *over-all desirability*, D [23]. This concept deals with a dimensionless scale or scale of d to which the measured responses (physical properties) are transformed so that they may be interpreted in terms of utility and desirability for any specific application. For a commercial product it is suggested that such a scale is the inherent value of these properties to the customer.

Although many such scales are possible, a useful scale ranges between 0.00 and 1.00 such that $d = 0.00$ corresponds to a completely undesirable level of quality of the product and $d = 1.00$ corresponds to a completely acceptable level of quality, i. e. an improvement in quality would serve no useful purpose. Intermediate values are identified in Table 5.21.

Table 5.21. Values of the Scale of *d*

Desirability value	Description of responses (quality equivalents)
0.80-1.00	Acceptable and excellent
0.63-0.80	Acceptable and good
0.37-0.63	Acceptable but poor
0.20-0.37	Borderline
0.00-0.20	Unacceptable

Because of the mathematical treatment chosen, it is rational, convenient and practical to assign a desirability value of 0.37 to any property at its specification value, maximum or minimum, assuming that realistic specification limits exist for this property. The number 0.37 is approximately $1/e$ where e is the base of the natural logarithms. A second such useful landmark is the value of a property corresponding to the best commercial quality (existing or anticipated), for which a desirability value of 0.63

$$d = 0.63 \approx 1 - 1/e \tag{5.117}$$

is appropriate.

The simplest sort of transformation of a given response to the scale of d is that in which there exist lower and/or upper specification limits. The latter are the sole and unalterable criteria of quality. Outside these limits the value of d is 0.00; and within, the value of d is 1.00. In the case of a one-sided specification, the value of d (Fig. 5.10a) is expressed as

$$d = \begin{cases} 0 & y < y_{min} \\ 1 & y \geqslant y_{min} \end{cases} \tag{5.118}$$

With the upper specification limit, the value of d is determined in a like manner. When the given property has a midspecification (Fig. 5.10b), the value of d is given by

$$d = \begin{cases} 0, & y < y_{min}, \quad y > y_{max} \\ 1 & y_{min} \leqslant y \leqslant y_{max} \end{cases} \tag{5.119}$$

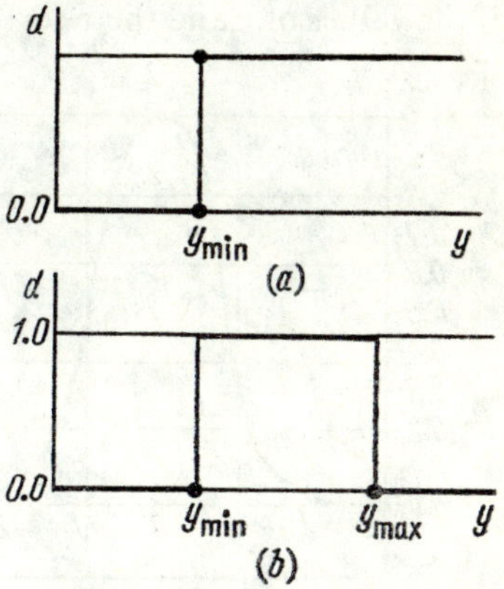

Fig. 5.10. Specifications for desirability values of d

It is nearly always desirable from the manufacturer's standpoint to stay not only within the specification limits but also some distance away from them in order to avoid substandard quality due to the inherent process variability. Furthermore, because of sampling and testing unprecision, it is quite impossible to separate borderline quality into two unequivocal groups, the acceptable and the unacceptable product. Thus, the transformation of y to d obeys more sophisticated laws in the general case. If midspecification is of the form $y_{min} \leqslant y \leqslant y_{max}$, then the measured response y is transformed into the scale of d (Fig. 5.11) by using the following expression

$$d = \exp\left[-(|y'|)^n\right] \tag{5.120}$$

where n is a positive number ($0 < n < \infty$), not necessarily an integer, and

$$y' = \frac{2y - (y_{max} + y_{min})}{y_{max} - y_{min}} \tag{5.121}$$

The exponent, n, may be calculated by selecting a value of $y = d$ (preferably in the range $0.6 < d < 0.9$), finding the corresponding y' and substituting in the equation

$$n = \frac{\ln \ln (1/d)}{\ln |y'|} \tag{5.122}$$

By relating the steepness of the desirability (transformation) curve to a given check point, one can take into account the importance of some specific properties of the product. For such properties, the value of n is of greater importance: a small variation of the property in the vicinity of the specification limits will correspond to an abrupt change in the desirability value. The exponent n determines the slope of the curve, and as n becomes large, the curve approaches the limiting case of $d = 0.00$ outside the specification limits and $d = 1.00$ within the limits (Fig. 5.10*b*). If there is no specification, it is good practice to statistically test the value of n, using the selected range of the values of y and their corresponding d's.

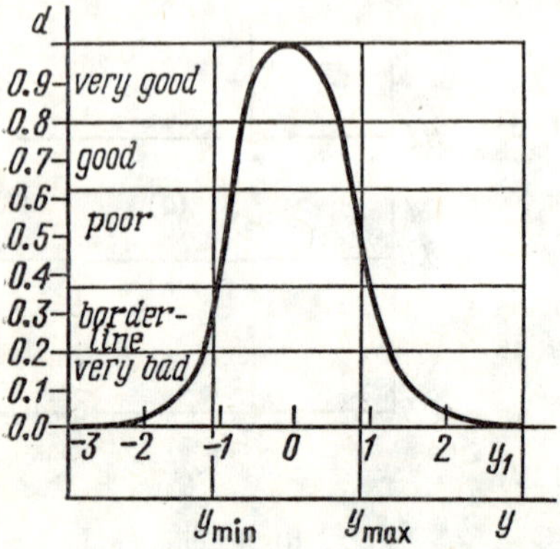

Fig. 5.11. Graphical representations of a midspecification

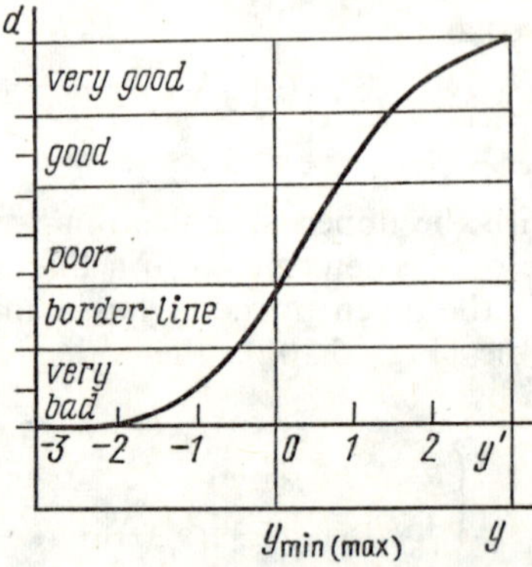

Fig. 5.12. Graphical representation of a one-sided specification

In the case of a one-sided specification such as $y \leqslant y_{max}$ or $y \geqslant y_{min}$, one can utilize another exponential relationship (Fig. 5.12) to transform y into d

$$d = \exp\left[-\exp\left(-y'\right)\right] \tag{5.123}$$

In Eq. (5.123)

$$y' = b_0 + b_1 y \tag{5.124}$$

The two constants, b_0 and b_1, are determined by assigning to two values of y the corresponding values of d (preferably in the range $0.2 < d < 0.8$).

A nonlinear transformation of y into y' is applicable to the case when a given property is critical, when the boundary conditions cannot be described, and when a small change in the property in the vicinity of the specification limit results in a sudden variation of the desirability value. In practice, one often deals with a one-sided specification.

Having transformed the several measures of quality to the dimensionless scale of d, it is now possible to combine these several d's by any of the operations of arithmetic to measure the over-all desirability, D, of the product. A basic premise is this—if any one property is so poor that the product is not suitable to the application, that product will not be acceptable ($D = 0$), regardless of the remaining properties. It is also true that customer reaction to a product is based very largely on the less desirable properties of that product because these are the focus of potential trouble.

The mathematic model analogous to these psychological reactions is the geometric mean of the component d's, or

$$D = \sqrt[k]{d_1 d_2 \ldots d_k} \tag{5.125}$$

It is clear that if any d_i is zero, the associated D will also be zero. Furthermore, D is strongly weighted by the smaller d's. On the other hand, D equals unity only when all component d's are unity, $d_i = 1$ ($i = 1, 2, \ldots, k$). Another important point is that Eq. (5.125) makes it possible to describe d's and D in terms of the same desirability values as given in Table 5.21. For example, with $d_1 = d_2 = \ldots = d_k = 0.37$, one has $D = 0.37$. Like any response in the process, D can be subjected to all computational operations. During investigation and optimization of the process, D can serve as an optimization criterion. One must only keep in mind the restriction that the distribution of D is doubly bounded (at 0.00 and 1.00), requiring mathematical methods appropriate to these restrictions. The concept of D proved useful in developing the compositions of and fabrication processes for new polymer materials.

Example 5.5. A second-order Latin cube is used to develop the composition of a new polymer material based on a high-pressure polyethylene (see Table 3.19). The responses are as follows: y_1, modulus of bending, kgf cm^{-2}; ultimate tensile stress, kgf cm^{-2}; y_3, percent elongation; and D, over-all desirability. Given below are computation steps with which the value of D is determined.

Solution. The comparative estimation of the quality of different compositions is given by

$$D = \sqrt[3]{d_1 d_2 d_3} \tag{5.126}$$

where d_1, d_2, d_3 are component d's.

To obtain the values of d_1, d_2, d_3, it is necessary first to transform the measured response y into an even dimensionless scale of y'. In this case, the constraints take the form $y \geqslant y_{\min}$. The quality of the material being developed is characterized by three properties which provide a measure of the utility of the material for processing and use. Thus we select the responses y_1, y_2, and y_3 that correspond each to two desirability values of the scale of d (Table 5.21).

The value of y is transformed into the corresponding d in accordance with Eq. (5.123). The constants, b_0 and b_1, are found from the data identified in the table below.

	y_1, kgf cm^{-2}		y_2, kgf cm^{-2}		y_3, %	
Property value	430	320	110	60	200	100
Desirability value	0.63	0.2	0.63	0.2	0.63	0.2

On substitution of the values of d into Eq. (5.123), we obtain

$$0.63 = \exp\left[-\exp\left(-y'\right)\right], \quad 0.2 = \exp\left[-\exp\left(-y'\right)\right] \qquad (5.127)$$

$$\frac{1}{0.63} = 1.587 = \exp\left[\exp\left(-y'\right)\right], \quad \frac{1}{0.2} = 5 = \exp\left[\exp\left(-y'\right)\right]$$

Taking twice the logarithms of the Eqs. (5.127), we get

$$-y' = \ln\left(\ln 1.587\right) \qquad -y' = \ln\left(\ln 5\right)$$

or

$$-y' = -0.755 \qquad -y' = 0.326. \qquad (5.128)$$

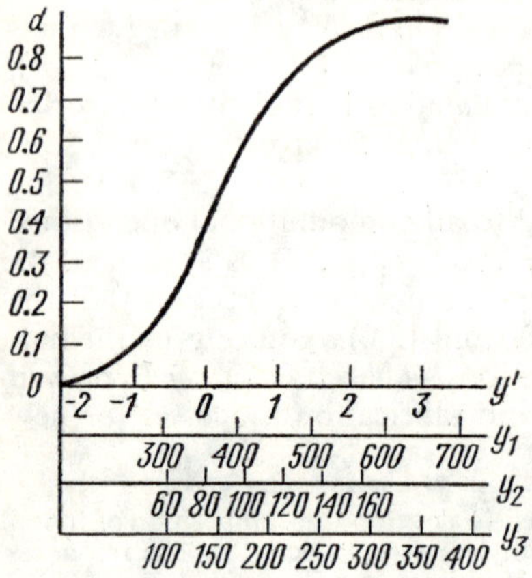

Fig. 5.13. d-versus-y' relationship

Thus, the constants b_0 and b_1 are related by

$$b_0 + 430b_1 = 0.755,$$
$$b_0 + 320b_1 = -0.326, \qquad (5.129)$$

Solving Eqs. (5.129), we get $b_1 = 0.0098$ and $b_0 = -3.445$.

The desirability value of d_1 is

$$d_1 = \exp\left[-\exp\left(-3.445 + 0.0098\, y_1\right)\right] \quad (5.130)$$

The desirability values of d_1 and d_2 are found in a similar manner

$$d_2 = \exp\left[-\exp\left(-1.45 + 0.02y_2\right)\right]$$
$$d_3 = \exp\left[-\exp\left(-1.25 + 0.01y_3\right)\right] \qquad (5.131)$$

For all compositions listed in the table below, the component d's can be found from Eqs. (5.130) and (5.131) or from the data in Fig. 5.13.

N	d_1	d_2	d_3	D	N	d_1	d_2	d_3	D
1	0.410	0.67	0.97	0.645	15	0.260	0.49	0.95	0.491
2	0.420	0.67	0.98	0.647	16	0.720	0.71	0.91	0.773
3	0.423	0.55	0.96	0.610	17	0.850	0.62	0.29	0.535
4	0.730	0.746	0.96	0.810	18	0.630	0.78	0.93	0.768
5	0.419	0.68	0.97	0.650	19	0.930	0.57	0.17	0.210
6	0.270	0.63	0.97	0.550	20	0.930	0.72	0.07	0.350
7	0.640	0.53	0.97	0.686	21	0.890	0.52	0.64	0.668
8	0.370	0.71	0.98	0.638	22	0.917	0.64	0.06	0.430
9	0.371	0.71	0.97	0.638	23	0.790	0.45	0.08	0.304
10	0.740	0.63	0.92	0.759	24	0.760	0.64	0.30	0.530
11	0.720	0.53	0.73	0.650	25	0.930	0.53	0.08	0.340
12	0.760	0.31	0.24	0.381	26	0.920	0.57	0.17	0.445
13	0.780	0.55	0.93	0.732	27	0.870	0.68	0.71	0.749
14	0.860	0.58	0.17	0.440					

The over-all desirability, D, in the table above is given by Eq. (5.126) as follows:

$$D = \exp\left\{-\frac{1}{3}\left[\exp(-3.445+0.0098y_1)+\exp(-1.45+0.02y_2)\right.\right.$$
$$\left.\left.+\exp(-1.25+0.01y_3)\right]\right\} \quad (5.132)$$

The trial 4 gives a maximum value of D equal to 0.810. Good compositions are prepared in trials 10, 13, 16, 18, and 27. After the analysis of factors and variance has been carried out, the optimum composition is determined using the over-all desirability.

5.10. Composite Designs. 2^{2k} Factorial Design Superposed on Latin Square

To determine the optimum combination of qualitative factors, use is made of hyper-Graeco-Latin squares and cubes (see Ch. 3). When an l^2 factorial design is superposed on an $l \times l$ orthogonal Latin square, the experimenter uses all factors at four levels, with the result that the effects obtained from $l+1$ factors can be investigated.

In many cases, qualitative factors are combined with quantitative ones, the latter may be sufficiently large in number. If the same number of levels, $l > 2$, is assigned to all factors, then a large number of observations is required or the number of factors to be included in the design must be limited to the value of $l+1$. Moreover, some qualitative factors cannot be given more than two levels. A useful approach to the situation is a combined design, i. e. a 2^{2k} factorial design superposed on a $2^k \times 2^k$ Latin square [11]. As a result, several factors can be included in such a design at $l = 2^k$ levels, while a relatively large number of quantitative and qualitative factors, at two levels, can be used by the experimenter. Such techniques are applicable only to a 2^{2k} factorial design with the number of observations equal to the full square of the number 2^k, $k = 2, 3, \ldots$.

In order to superpose a 2^{2k} factorial design on a Latin square, it is good practice to represent that design in the form of a table having 2^{k+1} inputs, on which a $2^k \times 2^k$ Latin square is superposed (see, for example, Table 5.22).

We see that the factor introduced in a Latin square is orthogonal to $2k$ factors that determine a full factorial design. Indeed, the rate of occurrence in the design of each of all $L = 2^k$ levels for that factor is the same; also, each such level meets any one of the levels of the original 2^k factors the same number of times.

One can superpose the original design on a $2^k \times 2^k$ Graeco-Latin square or even on a hyper-Graeco-Latin square, which in turn is constructed by superposing one 2^k-1 orthogonal Latin square on another, provided that there exist a complete series of orthogonal Latin squares for $l = 2^k$. Under the circumstances, the introduced 2^k-1 factors are orthogonal to the original $2k$ factors and to all interactions that determine the square columns. One can treat the design as a saturated one, if all the interactions are assumed to be insignificant and are used to introduce extra factors in the design at two levels.

Table 5.22. 2^4 Factorial Design Superposed on 4×4 Latin Square

		$x_2(-1)$		$x_2(+1)$	
		$x_1(-1)$	$x_1(+1)$	$x_1(-1)$	$x_1(+1)$
$x_4(-1)$	$x_3(-1)$	A	B	C	D
	$x_3(+1)$	B	A	D	C
$x_4(+1)$	$x_3(-1)$	D	C	B	A
	$x_3(+1)$	C	D	A	B

Important to the experimenter are very diverse saturated orthogonal designs that are constructed by superimposing a 2^{2k} factorial design on one Latin square, on two orthogonal Latin squares and so on, including 2^k-1 orthogonal Latin squares. Every factor introduced in the design at $l = 2^k$ levels has 2^k-1 degrees of freedom and is confounded with different 2^k-1 interactions resulted from $2k$ factors in a full factorial design. If m factors, with $m < 2^k-1$, are introduced in the design at 2^k levels, they are confounded with $m\,(2^k-1)$ interactions produced by the original factors. For a 2^{2k} full factorial design, the total number of interactions is $2^{2k}-2k-1$. Thus, $(2^{2k}-2k-1)-m(2^k-1)$ interactions are free from confounding with the main effects produced by $2k+m$ factors. One can use these interactions to introduce in the design extra factors at two levels. In this case, a saturated design comprises $n = 2^{2k}-m2^k+2m-1$ factors with m factors thereof being included at $l = 2^{2k}$ levels, and $n-m$ factors at two levels. Most important to the experimenter are the designs for $k = 2$ in which case $N = 16$ and $l = 4$. One can also construct useful designs for $k = 3 (N = 64$ and $l = 8)$. With $k = 4$, the designs require a very large number of observations ($N = 256$).

In the case of a composite design having qualitative factors fixed at two levels, it is possible to calculate the main effects which coincide due to orthogonality of the design with the effects computed by the least squares method; after that, the steepest ascent can be carried out. During the ascent, the quantitative factors are fixed at those levels which produce the best effects.

The effects of the factors included in the design at two levels can be computed according to the following procedure. Assume that $N = 2^{2k}$ observations are run in a composite design. Further assume that n factors are included in the design and m factors thereof are fixed at $l = 2^k$ levels, while $n-m$ factors of these n factors are fixed at two levels. The responses obtained are $y_1, y_2, \ldots, y_N$. It is seen that the main effect of factor x_i

($i = 1, 2, \ldots, n-m$) is the difference between the sum of the responses from those observations in which x_i is fixed at the upper level, x_i', and the sum of the responses from the observations in which x_i is fixed at the lower level x_i^0, divided by the number N of observations in the design

$$m_i = \frac{1}{N}\Big[\sum y(x_1, \ldots, x_i', \ldots, x_{n-m}) - \sum y(x_1, \ldots, x_i^0, \ldots, x_{n-m})\Big] \quad (5.133)$$

The ratio of m_i to $\sigma(m_i) = \sigma_e/\sqrt{N}$, where σ_e is the standard error of the response measurement, has the t-distribution. This ratio provides a means to test the significance of the computed effects. In the case of an unsaturated design, the value of $\sigma(m_i)$ can be estimated with the help of the interaction effects that are not confounded with the main factors.

The effects of the factors included in the design at $l = 2^k$ levels are computed for each respective level. The effect of the factor x_j ($j = n-m+1, \ldots, n$) at the qth level ($q = 0, 1, 2, \ldots, l-1$) is equal to the sum of the responses from the observations, in which x_j is fixed at the qth level, divided by the number $l = 2^k$ of inclusions in the design of x_j at the qth level

$$x_j(q) = \frac{\sum y(x_1, \ldots, x_j^l, \ldots, x_n)}{l} \quad (5.134)$$

If one can rest on the assumption that s_e^2 in the measured responses from all observations are homogeneous, the t-test may be used in order to estimate the significance of the difference between the responses produced by the factors at different levels. Unfortunately, in this test one can use only some portion of information available; this occurs, for example, when the estimation procedure is applied to x_j fixed at two levels, l and $l+1$. The situation is improved when Duncan's multiple range test is employed. Here greater reliability is attained during the estimation of the difference between the effects of the factors included in the design at $l > 2$ levels. This is due to the fact that whole information available from the experiment is used simultaneously.

Significance of the main effects of the factors included in the design at both two and $l > 2$ levels can be tested by a multiway ANOVA and the factor analysis. The results of the effect analysis form a base for the steepest ascent.

In the case of a linear model, we have

$$y_{i \ldots jq \ldots f} = \mu + x_{1i} + \ldots + x_{(n-m)j} + x_{(n-m+1)q} + x_{nf} + \varepsilon_{i \ldots jq \ldots f} \quad (5.135)$$

where $y_{i \ldots jq \ldots f}$ is the response from a certain trial; μ is the total effect from all observations; x_{1i} is the effect from x_1 at the ith level ($i = 0, 1$); $x_{(n-m)j}$ is the effect from x_{n-m} at the jth level ($j = 0, 1$); $x_{(n-m+1)q}$ is the effect of x_{n-m+1} at the qth level ($q = 0, 1, \ldots, 2^k-1$); x_{nf} is the effect of x_n at the fth level ($f = 0, 1, 2, \ldots, 2^k-1$); $\varepsilon_{i \ldots jq \ldots f}$ is the error of response measurement.

Table 5.23. Analysis of Variance of a Combined Design

Source of variation	Degrees of freedom	Sum of squares	Mean square	Expected mean square
x_1	1	$SS_{x_1} = \sum_{i=0}^{1} s_i^2 - SS_y$	SS_{x_1}	$\sigma_e^2 + 2^{2k-1}\sigma_{x_1}^2$
⋮	⋮	⋮	⋮	⋮
x_{n-m}	1	$SS_{x_{n-m}} = \sum_{j=0}^{1} s_j^2 - SS_y$	$SS_{x_{n-m}}$	$\sigma_e^2 + 2^{2k-1}\sigma_{x_{n-m}}^2$
x_{n-m+1}	$2^k - 1$	$SS_{x_{n-m+1}} = \sum_{q=0}^{2k-1} s_q^2 - SS_y$	$SS_{x_{n-m}}/(2^k - 1)$	$\sigma_e^2 + 2^k\sigma_{x_{n-m+1}}^2$
⋮	⋮	⋮	⋮	⋮
x_n	$2^k - 1$	$SS_{x_n} = \sum_{f=0}^{2k-1} s_f^2 - SS_y$	$SS_{x_n}/(2^k - 1)$	$\sigma_e^2 + 2^k\sigma_{x_n}^2$
Error	$2^{2k} - m2^k + 2m - n - 1$	$SS_e = SS_{tot} - SS_{x_1} - \ldots - SS_{x_n}$	$SS_e/(2^{2k} - m2^k + 2m - n - 1)$	σ_e^2
Total	$2^{2k} - 1$	$SS_{tot} = \sum\sum y_{i\ldots je\ldots f}^2 - SS_y$		

Table 5.23 shows the analysis of variance with the designations used as described below.

s_i^2 is the square of the sum of the responses from the trials in which factor x_1 is fixed at the ith level, divided by the number 2^{2k-1} of inclusions of x_1 at the ith level:

$$s_i^2 = \frac{(\sum y_{i \ldots jq \ldots f})_i^2}{2^{2k-1}} \qquad (i = 0, 1) \tag{5.136}$$

s_q^2 is the square of the sum of the responses from the observations in which factor x_{n-m-1} is fixed at the qth level, divided by the number 2^k of inclusions of x_{n-m+1} at the qth level:

$$s_q^2 = \frac{(\sum y_{i \ldots jq \ldots f})_q^2}{2^k} \qquad (q = 0, 1, 2, \ldots, 2^k - 1) \tag{5.137}$$

s_f^2 is the square of the sum of the responses from the trials in which factor x_n is fixed at the fth level, divided by the number 2^k of inclusions of x_n in the design at the fth level:

$$s_f^2 = \frac{(\sum y_{i \ldots jq \ldots f})_f^2}{2^k} \qquad (f = 0, 1, 2, \ldots, 2^k - 1) \tag{5.138}$$

$\sum \sum y_{i \ldots jq \ldots f}^2$ represents the sum of the squares of the responses from all $N = 2^{2k}$ observations;

SS_y represents the mean square of the sum of the responses from all $N = 2^{2k}$ observations:

$$SS_y = \frac{(\sum \sum y_{i \ldots jq \ldots f})^2}{2^{2k}} \tag{5.139}$$

Example 5.6 [24]. A single-stage process is investigated in which water-soluble polyelectrolytes are produced by radical polymerization of vinylpyridine salts without their intermediate precipitation. The process is governed by a large number of quantitative and qualitative factors. We seek to find the optimum conditions of the process.

Solution. We shall use a composite design, which is a 2^{6-2} fractional replicate superposed on two Latin squares (Table 5.24). We seek the optimum conditions of a process that uses 2-methyl-5-vinylpyridine (MVP) in producing water-soluble polyelectrolytes. There are 8 factors that influence the yield of the polymer, y, as follows: two quantitative factors—alkyl halides, RX, and solvents—are changed at four levels, and one qualitative factor—initiator—at two levels. The levels of the factors are shown in Table 5.25. Owing to the useful combination properties of the design and the appropriately selected factors, already during the construction of the design matrix the conditions are established in which the polymer yields meet the process requirements (trials 4, 5, 8 and 13).

In order to determine how different model parameters influence the product yield, the experimental results are analyzed by means of the factor effects method and subjected to the analysis of variance. In the case of the factor effects method, the value of $s_e = 1.94$ obtained from the previous experimentation is used. The number of degrees of freedom is chosen to be $\nu_e = 6$.

The effects of the factors included in the design at two levels are calculated from Eq. (5.133). They are listed in Table 5.25. The effects are tested for significance by virtue of the Student t. The tabulated value of the Student t is found to be $t_{0.05}(6) = 2.45$. The effect of factor x_2 (the ratio of the components taking part in the reaction) is proved to be insignificant. This means that an excess of a given alkyl halide does not influence the polymer yield. In Table 5.25, the insignificant effect is replaced by zero. The composite analysis

Table 5.24. Design Matrix

N	x_1	x_2	x_3	x_4	x_5	x_6	x_7	x_8	y
1	−	−	−	−	−	+	A	0	30.155
2	+	−	−	−	+	−	B	1	31.761
3	−	+	−	−	+	−	C	2	12.899
4	+	+	−	−	−	+	D	3	80.715
5	−	−	+	−	+	+	B	3	88.876
6	+	−	+	−	−	−	A	2	16.806
7	−	+	+	−	−	−	D	1	12.463
8	+	+	+	−	+	+	C	0	81.443
9	−	−	−	+	−	+	C	1	24.959
10	+	−	−	+	+	−	D	0	19.602
11	−	+	−	+	+	−	A	3	21.761
12	+	+	−	+	−	+	B	2	52.953
13	−	−	+	+	+	+	D	2	78.220
14	+	−	+	+	−	−	C	3	20.953
15	−	+	+	+	−	−	B	0	1.684
16	+	+	+	+	+	+	A	1	48.910

Table 5.25. Factor Analysis

Factors		Levels		Effect
		+1	−1	
Temperature, °C	x_1	70	60	+4.633
MVP/RX ratio, mol/mol	x_2	1/1.3	1/1.1	0
Concentration of initiator, %	x_3	1.2	0.8	+4.159
MVP/solvent mass fractions	x_4	1/2	1/1	−5.88
Reaction time, hours	x_5	12	6	+8.425
Initiator	x_6	ABD	BP	17.24
Solvent	x_7	Butanol	A	27.41
		Ethanol	B	43.82
		Propanol	C	35.06
		Isopropanol	D	47.75
Alkyl halide	x_8	Ethyl bromide	0	33.22
		Ethyl iodide	1	27.52
		Propyl iodide	2	40.22
		Methyl iodide	3	53.08

of variance is used to test the main effects of the factors included in the design at both two and four levels. To estimate the significance of the effects the ratio of the mean squares associated with the respective factors to the error mean square is formed. This ratio has an F-distribution. The sum of squares for the effect of factor x_2, proved to be not significant, is added with the respective degrees of freedom to the sum of squares for the experimental error. The analysis of variance is shown in Table 5.26. We see that the results of the factor effects analysis are in good agreement with the ANOVA.

Table 5.26. Analysis of Variance

Source of variance	Degrees of freedom	Sum of squares	Mean square	Significance test
x_1	1	342.6461	342.6461	$\sigma^2_{x_1} \neq 0$
x_2	1	2.6518	2.6518	$\sigma^2_{x_2} \neq 0$
x_3	1	276.0360	276.0360	$\sigma^2_{x_3} \neq 0$
x_4	1	552.3729	552.3729	$\sigma^2_{x_4} \neq 0$
x_5	1	1136.1904	1136.1904	$\sigma^2_{x_5} \neq 0$
x_6	1	7238.9555	7238.9555	$\sigma^2_{x_6} \neq 0$
x_7	3	2980	933.991	$\sigma^2_{x_7} \neq 0$
x_8	3	4420	1454.35	$\sigma^2_{x_8} \neq 0$
Error (1)	3	361.3800	120.4597	
Total	15	185378.610	12358.571	
Error (2)	4	364.03	91.007	

Table 5.25 shows the effects of the factors at two and four levels. The significance of the difference between the effects of these factors at different levels was tested by means of the Duncan's multiple range test, a confidence coefficient was chosen to be $1-\alpha = 0.95$. The standard deviation is

$$s_{\bar{y}} = \frac{9.55}{\sqrt{4}} = 4.775$$

The effects of the factors at different levels are given in ascending order.

Factor x_7

I $\bar{y}^{(A)} = 27.41$		$\bar{y}^{(C)} = 35.06$	$\bar{y}^{(B)} = 43.82$	$\bar{y}^{(D)} = 47.75$
II		2	3	4
III ranges r		3.93	4.01	4.02
IV ranges multiplied by standard deviation, $r \cdot s_{\bar{y}}$		18.83	19.15	19.25

$\bar{y}^{(D)} - \bar{y}^{(A)} = 20.34 > 19.25$ – a significant difference

$\bar{y}^{(D)} - \bar{y}^{(C)} = 12.69 < 19.15$ – an insignificant difference

$\bar{y}^{(D)} - \bar{y}^{(B)} = 3.93 < 18.8$ – an insignificant difference

$\bar{y}^{(B)} - \bar{y}^{(A)} = 6.41 < 19.15$ – an insignificant difference

$\bar{y}^{(B)} - \bar{y}^{(C)} = 8.46 < 18.8$ – an insignificant difference

$\bar{y}^{(C)} - \bar{y}^{(A)} = 7.65 < 18.8$ – an insignificant difference

We see that the synthesis of strong-base polyelectrolytes has different effectivenesses for butanol and isopropanol. In the case of the remaining solvents, the difference is insignificant.

Factor x_8

$$\bar{y}^{(1)} = 27.52 \quad \bar{y}^{(0)} = 33.22 \quad \bar{y}^{(2)} = 40.22 \quad \bar{y}^{(3)} = 53.08$$

$\bar{y}^{(3)} - \bar{y}^{(1)} = 25.44 > 19.25$ – a significant difference

$\bar{y}^{(3)} - \bar{y}^{(0)} = 19.74 > 19.15$ – a significant difference

$\bar{y}^{(3)} - \bar{y}^{(2)} = 12.74 < 18.8$ – an insignificant difference

$\bar{y}^{(2)} - \bar{y}^{(1)} = 12.7 < 19.15$ – an insignificant difference

$\bar{y}^{(2)} - \bar{y}^{(0)} = 7.0 < 18.8$ – an insignificant difference

$\bar{y}^{(0)} - \bar{y}^{(1)} = 5.7 < 18.8$ – an insignificant difference

The polymer yield considerably decreases when the most active alkyl halide, methyl iodide, is replaced with either ethyl iodide or ethyl bromide. The difference between the remaining alkyl halides is insignificant.

The analysis shows that the polymer yield increases with increasing process temperature and reaction time; on the other hand, the reaction rate decreases with increasing amount of the solvent. Asoisobutyrodinitrile (ABD) is found to be the most effective initiator. Benzoyl peroxide (BP) used as initiator causes a decrease in the rate of radical polymerization. The best solvents are ethanol and isopropanol. The most active alkyl halide is methyl iodide. However, the effects of the remaining alkyl halides are also substantial.

The optimum conditions of the synthesis of halogen-containing water-soluble polyelectrolytes are as follows: the process temperature, 70 °C; the MVP/RX ratio, 1 : 1.1; the MVP/solvent ratio, 1 : 1; the reaction time, 12 hours; the concentration of asoisobutyrodinitrile, 1.2%; the solvent, propanol or ethanol. The optimum conditions were used to produce water-soluble polyelectrolytes based on 2-methyl-5-vinylpyridine and various alkyl halides. The polymer yield thus obtained is

Polymer	MVP-CH_3	MVP-C_2H_5	MVP-C_3H_7	MVP-C_2H_5
Yield, %	99.0	97.0	86.0	98.0

Since the effect of the process temperature is found to be favourable (Table 5.25), the experimenter decides to increase it further so as to cut down the reaction time. For the MVP-C_2H_5 reaction with ethanol as a solvent, the kinetic curves are obtained for the process temperature of 60, 70, 75, and 80 °C (Fig. 5.14). Increasing the temperature up to 80 °C gives the reaction time of 7-8 hours. Higher temperatures are bounded by the boiling temperature of the solvent and would destroy partially the polymer being produced.

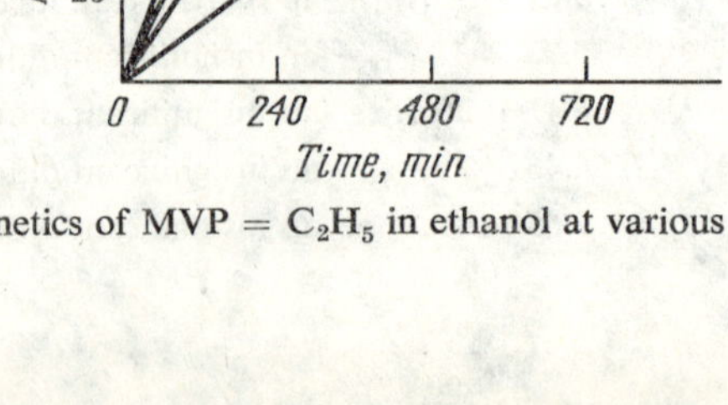

Fig. 5.14. Kinetics of MVP = C_2H_5 in ethanol at various temperatures

5.11. Sequential Simplex Design

In the previous sections we have discussed 2^k and 2^{k-p} designs whose experimental points are located at the corners of a multidimensional cube. Now we consider a regular simplex that may also be used for experimental design [25]. A simplex is defined as a convex polyhedron that has $k+1$ vertices produced by k intersecting hyperplanes in k-dimensional space. In two-dimensional space, a simplex is a triangle. In three-dimensional space, a simplex is a regular pyramid having its four vertices produced by intersecting the planes of the respective three faces.

A regular k-simplex is defined as a set of $k+1$ equidistant vertices. A two-dimensional regular simplex is an equilateral triangle while a three-dimensional regular simplex is a regular tetrahedron. In planning the experiments, regular simplex designs usually are employed. It should be noted, however, that the property of regularity of a simplex, like the direction of the gradient in the steepest ascent method and the rotatability of designs, is not invariant to the scaling of the coordinates of factor space. A change in scaling would result in a non-regularity of a regular simplex. On the other hand, one can by an appropriate transformation of the coordinates make a non-regular simplex become regular.

In practice, simplex designs are applicable to the solution of the optimization problems when the experimenter moves towards a "nearly stationary" region. Using a linear transformation described by the relation below, one can obtain, in this case, a regular simplex

$$x_j = \frac{z_j - z_j^0}{\Delta z_j}$$

where z_j^0 is the jth coordinate of the centre point of the design and Δz_j is the interval of variation referred to the jth factor.

Experimental optimization proceeds by using the following important property of simplex: opposite to any vertex A_j is only one face on which a new simplex can be constructed which differs from the original by a new vertex $\tilde{A}_j$ of a different location. But the remaining vertices of the both simplices coincide with one another. In the sequential simplex search, movement factor space is made via reflection and contraction discarding one vertex on each move.

The sequential simplex design includes the steps as follows. In the search for an optimum, trial values of the experimental factors can be selected at points in multidimensional factor space located at the vertices of the regular simplex. After the first cycle of observations, the vertex associating with the conditions that yield the poorest response is identified. The point is now deleted, and a new point which is a mirror reflection from the opposite face of the simplex is added. The new point forms together with those remained in the old simplex a new regular simplex termed a reflection, in which the centroid is displaced in the direction from the poorest-response point towards the centroid of the remaining points. In the general case, this direction is not the path of steepest ascent; yet, it points towards a better performance of the process.

Continuation of this procedure, always deleting the point that yields the poorest response and replacing it with a mirror image, leads to a region close to the extremum.

Figure 5.15 shows how the experimenter reaches the extremum on the response surface using the conventional method of steepest ascent and the sequential simplex search. Let us consider the path towards the extremum for the case of an objective function in two factors.

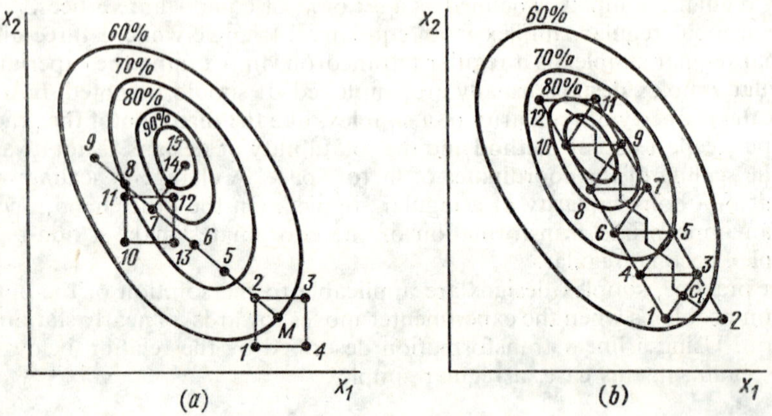

Fig. 5.15. Patterns to advance to the optimum area by: (*a*) the steepest ascent, and (*b*) the regular simplex method

When the steepest ascent is used (Fig. 5.15*a*), a 2^2 full factorial design if constructed in the vicinity of point *M* (points *1-4*) in which the value os the objective function is known and trials *5* through *9* are conducted so as to move in the direction of the gradient until the value of the objective function worsens. Then the experimenter constructs a 2^2 design (points *10-13*) whose centre is at the point *7* that yields the best response, and reaches the extremum of the objective function moving in the direction of the gradient (points *14, 15*).

In the case of the regular simplex design (Fig. 5.15*b*), the poorest-response point is point *2* of the original simplex (points *1-3*). Point *4* is a mirror image of point *2* as viewed with respect to c_1, the centre of the face *1-3*. In the new simplex that uses points *1, 3* and *4*, point *1* gives the poorest-response. The search for the optimum is completed when a simplex is built around points *9, 10* and *11*. Note that the both methods require approximately the same number of trials.

In Fig. 5.15 the reader can see that cycling around the optimum might occur when using the regular simplex design. As the optimum is approached, the technique will either reach a point where no reflections are possible or the design will pivot on one of the points and cycle around the optimum. If the new simplex is so located with respect to the response surface that the new point (like the old one) again yields the poorest response, the experimenter then must return to the previous simplex and try to move in another favorable direction. The errors that occur in determining the proper response hinder the movement towards the extremum.

There are a number of ways in which the original simplex can be oriented in factor space. It can be shown that with the dimensionless variables arranged in matrix X

$$\mathbf{X} = \begin{bmatrix} x_1 & x_2 & \dots & x_j & \dots & x_{k-1} & x_k \\ -x_1 & x_2 & \dots & x_j & \dots & x_{k-1} & x_k \\ 0 & -2x_2 & \dots & x_j & \dots & x_{k-1} & x_k \\ \cdot & \cdot & & \cdot & & \cdot & \cdot \\ \cdot & \cdot & & \cdot & & \cdot & \cdot \\ \cdot & \cdot & & \cdot & & \cdot & \cdot \\ 0 & 0 & \dots & -jx_j & \dots & x_{k-1} & x_k \\ \cdot & \cdot & & \cdot & & \cdot & \cdot \\ \cdot & \cdot & & \cdot & & \cdot & \cdot \\ \cdot & \cdot & & \cdot & & \cdot & \cdot \\ 0 & 0 & \dots & 0 & \dots & -(k-1)x_{k-1} & x_k \\ 0 & 0 & \dots & 0 & \dots & 0 & -kx_k \end{bmatrix} \quad (5.140)$$

where

$$x_j = \sqrt{\frac{1}{2j(j+1)}} \quad (5.141)$$

the points of the matrix will form a regular k-dimensional simplex with the centre at the origin of coordinates and with an edge of unity.

The height of the simplex (that is, the distance from a vertex to the opposite face) is

$$h_k = \frac{k+1}{\sqrt{2k(k+1)}} \quad (5.142)$$

The number of observations in a simplex matrix for k factors is $N = k+1$ which is equal to the number of coefficients in the regression equation. This is why simplex designs are called saturated.

The matrix satisfies the condition

$$\sum_{i=1}^{N} x_{ji} x_{li} = 0, \; j \neq l; j, l = 1, 2, \dots, k \quad \text{and} \quad \sum_{i=1}^{N} x_{ji} = 0 \quad (5.143)$$

but fails to satisfy the condition $\sum_{i=1}^{N} x_{ji}^2 = N$.

Only for column x_0, which has all its elements equal to unity, we have

$$\sum_{i=1}^{N} x_{0i}^2 = N$$

For any jth column

$$\sum_{i=1}^{N} x_{ji}^2 = j\frac{1}{2j(j+1)} + j^2\frac{1}{2j(j+1)} = 0.5 \qquad (5.144)$$

Therefore, for a simplex design the variance-covariance matrix $(X^T X)^{-1}$ takes the form

$$(X^T X)^{-1} = \begin{bmatrix} -1/N & & & & & 0 \\ & 2 & & & & \\ & & 2 & & & \\ & & & \cdot & & \\ & & & & \cdot & \\ & & & & & \cdot \\ 0 & & & & & 2 \end{bmatrix} \qquad (5.145)$$

and the regression coefficients are estimated by the equations

$$b_0 = \frac{1}{N}\sum_{i=1}^{N} y_i, \qquad b_j = 2\sum_{i=1}^{N} x_{ji} y_i \qquad j = 1, 2, \ldots, k \qquad (5.146)$$

Simplex designs are rotatable designs. A major drawback of simplex designs is the lack of D-optimality.

With orthogonal designs, the estimated variance of coefficients is given by

$$s_{b_j}^2 = s_e^2 \Big/ \sum_{i=1}^{N} x_{ji}^2 \qquad (5.147)$$

With simplex designs, we have, in agreement with Eq. (5.147),

$$s_{b_j}^2 = 2s_e^2 \qquad (5.148)$$

For 2^k and 2^{k-p} designs,

$$s_{b_j}^2 = s_e^2/N$$

Thus, the coefficients of the estimated regression equation corresponding to a simplex design are estimated with a lesser degree of accuracy. The experimenter can construct saturated designs with elements ± 1 only for cases where the number of factors is $4a-1$, where a is a positive integer. For example, this can be done for 3, 7, 11, 15, etc. factors.

To facilitate the practical use of the simplex matrix (5.140), the numerical values have been calculated for its elements in advance by using Eq. (5.141).

$$X = \left[\begin{array}{ccccc|c} 0.5 & 0.289 & 0.204 & 0.158 & 0.129 & 0.109 \ldots \\ -0.5 & 0.289 & 0.204 & 0.158 & 0.129 & 0.109 \ldots \\ 0 & -0.578 & 0.204 & 0.158 & 0.129 & 0.109 \ldots \\ 0 & 0 & -0.612 & 0.158 & 0.129 & 0.109 \ldots \\ 0 & 0 & 0 & -0.632 & 0.129 & 0.109 \ldots \\ 0 & 0 & 0 & 0 & -0.645 & 0.109 \ldots \\ \hline 0 & 0 & 0 & 0 & 0 & -0.655 \ldots \\ \cdot & \cdot & \cdot & \cdot & \cdot & \cdot \ \ldots \\ \cdot & \cdot & \cdot & \cdot & \cdot & \cdot \ \ldots \\ \cdot & \cdot & \cdot & \cdot & \cdot & \cdot \ \ldots \end{array}\right] \tag{5.149}$$

In dimensionless form, an experimental design for k factors consists of k columns and $k+1$ rows in the matrix.

As already explained, experimental optimization proceeds by constructing a new simplex on the face of the old one, opposite the worst point. The new point added to the simplex is the location of the new experiment (new experimental conditions): hence, the design incorporates the old experiments and a new one on each move. To find the coordinates of the added vertex take twice the average of all the coordinates of the common points (those retained in the simplex) and subtract the coordinates of the deleted vertex. In mathematical form:

$$x_j^{(k+2)} = 2x_j^{(c)} - x_j^{(l)}, \qquad j = 1, 2, \ldots, k \tag{5.150}$$

where $x_j^{(l)}$ is the jth coordinate of the poorest-response vertex l, $x_j^{(k+2)}$ is the jth coordinate of the added vertex, and $x_j^{(c)}$ is the jth coordinate of the centre on the opposite face, defined as

$$x_j^{(c)} = \frac{1}{k} \sum_{i=1}^{k+1} x_j^{(i)}, \qquad i \neq l. \tag{5.151}$$

The original k-dimensional simplex may be completed to a $(k+1)$-dimensional simplex by adding one more point. The need for this arises when at its first stage a process was examined as one dependent on k factors, while actually it depends on $k+1$ factors. For some reason, the value of factor $k+1$ was not varied in the experiment. Thus, all points of the k-dimensional simplex are in effect the vertices of the $(k+1)$-dimensional simplex, located in hyperplane $x_{k+1} = d$, where d is a fixed value of the $(k+1)$st factor in dimensionless form. From geometric considerations it follows that to construct a $(k+1)$-dimensional simplex from a k-dimensional simplex, it is necessary to locate the centroid of the k-dimensional simplex in $k+1$-dimensional space and to erect through this point a perpendicular to the hyperplane in which the points of the k-dimensional simplex are positioned. Laying off a segment h_{k+1} long (which is the height of the $(k+1)$-dimensional simplex) will locate a point on the perpendicular: this point together with those retained in the old simplex will form a $k+1$-dimensional simplex. The coordinates of the new point are

$$\{x_1^{(0)}, x_2^{(0)}, \ldots, x_j^{(0)}, \ldots, x_k^{(0)}, d+h_{k+1}\} \tag{5.152}$$

where x_j^0 is the jth coordinate of the centre point of the original simplex.

With conventional factorial designs, adding one more factor doubles the number of observations required, a drawback that one can combat by virtue of the regular simplex design. Another advantage of the latter is that the optimization parameter y can be estimated roughly and ranking will suffice in this case. Under these circumstances, several optimizing factors such as product yield, costs, purity of the product and others may be considered simultaneously. Furthermore, one is allowed not to take aquantitative estimate of the optimization parameter. Besides, the constraints on the approximation of the response surface with a hyperplane are relatively loose. A regular simplex design can serve as an algorithm in process optimization utilizing a control device.

Example 5.7. We seek to compare the effectiveness of the sequential simplex search with the steepest ascent method by using the results obtained from 8 observations as described in Table 5.11.

Solution. A one-sixteenth replicate of a 2^7 full factorial design (see Example 5.1) is a D-optimal simplex in a 7-dimensional space. This design was used as the original simplex for runs 1-8 listed in the table below.

Design point	z_1	z_2	z_3	z_4	z_5	z_6	z_7	y
1	.022	.028	.035	1350	1.5	.152	.333	0
2	.063	.028	.035	1300	2.0	.127	.333	.129
3	.022	.0094	.035	1300	2.0	.152	.5	0
4	.063	.0094	.035	1350	1.5	.127	.5	.177
5	.022	.028	.10	1300	1.5	.127	.5	.295
6	.063	.028	.10	1350	2.0	.152	.5	.404
7	.022	.0094	.10	1350	2.0	.127	.333	.2665
8	.063	.0094	.10	1300	1.5	.152	.333	.4305
9	.069	.031	.109	1360	1.42	.124	.310	.42
10	.082	.013	.1304	1310	1.91	.115	.469	.336
11	.0316	.0129	.1278	1370	1.16	.147	.47	.510
12	.023	.033	.154	1330	1.09	.153	.437	.489
13	.079	.035	.1346	1312	1.02	.149	.520	.2630

The analysis of the results in the above table shows that the lowest response corresponds to observations 1 and 3. Now we replace point 3 with its mirror image, i. e. point 9, and compute the coordinates of the new point from Eqs. (5.150) and (5.151). The coordinates of the centre of the face (point c) formed by points *1, 2, 4, 5, 6, 7* and *8* are

$$z_1^{(c)} = \frac{0.022\times 3 + 0.063\times 4}{7} = 0.0454$$

$$z_2^{(c)} = \frac{0.028\times 4 + 3\times 0.0094}{7} = 0.020$$

$$z_3^{(c)} = \frac{3\times 0.35 + 4\times 0.1}{7} = 0.0721$$

$$z_4^{(c)} = \frac{4\times 1350 + 3\times 1300}{7} = 1330$$

$$z_5^{(c)} = \frac{4\times 1.5 + 3\times 2.0}{7} = 1.71$$

$$z_6^{(c)} = \frac{3\times 0.152 + 4\times 0.127}{7} = 0.138$$

$$z_7^{(c)} = \frac{4\times 0.333 + 0.5\times 3}{7} = 0.405$$

Thus, the coordinates of point 9 are

$$z_1^{(9)} = 2\times 0.0454 - 0.022 = 0.0688$$

$$z_2^{(9)} = 2\times 0.2 - 0.0094 = 0.0306$$

$$z_3^{(9)} = 2\times 0.0721 - 0.035 = 0.109$$

$$z_4^{(9)} = 2\times 1330 - 1300 = 1360$$

$$z_5^{(9)} = 2\times 1.71 - 2 = 1.42$$

$$z_6^{(9)} = 2\times 0.138 - 0.152 = 0.124$$

$$z_7^{(9)} = 2\times 0.405 - 0.5 = 0.310$$

The coordinates of point 10 are obtained in analogous manner when mirror reflection procedure is applied to point 1. The simplex built around points 2, 4, 5, 6, 7, 8, 9 and 10 has the lowest response at point 2. The mirror image of the latter gives the coordinates of point 11, while the mirror image of point 4 the coordinates of point 12. In the simplex formed by points 5, 6, 7, 8, 9, 10, 11 and 12, the lowest response is at point 7 that, when reflected, gives the coordinates of point 13. The latter has lesser yield than point 7. After the mirror image of point 13 has been constructed, the experimenter returns to point 7, which signifies that cycling around the optimum occurs. Let us now find the yield at the centre point of the simplex built on points 5, 6, 7, 8, 9, 10, 11 and 12. The coordinates of the centre of the simplex (point S) are

$$z_1^{(S)} = (2\times 0.022 + 2\times 0.063 + 0.069 + 0.082 + 0.0316 + 0.023) \div 8 = 0.047$$

$$z_2^{(S)} = (0.028\times 2 + 0.0094\times 2 + 0.031 + 0.013 + 0.0129 + 0.033) \div 8 = 0.0206$$

$$z_3^{(S)} = (0.10\times 4 + 0.109 + 0.1304 + 0.1278 + 0.154) \div 8 = 0.115$$

$$z_4^{(S)} = (1300\times 2 + 1350\times 2 + 1360 + 1310 + 1370 + 1330) \div 8 = 1334$$

$$z_5^{(S)} = (1.5\times 2 + 2\times 2.0 + 1.42 + 1.91 + 1.16 + 1.09) \div 8 = 1.57$$

$$z_6^{(S)} = (0.127\times 2 + 0.152\times 2 + 0.124 + 0.115 + 0.147 + 0.153) \div 8 = 0.137$$

$$z_7^{(S)} = (0.5\times 2 + 0.333\times 2 + 0.310 + 0.469 + 0.47 + 0.437) \div 8 = 0.420$$

An observation is conducted at S which gives the value of the optical density $y^{(S)} = 0.570$. Thus, one requires 14 observations in order to obtain the best value of the optimization criterion at the centre point of the simplex, as compared with 15 runs in the steepest ascent method.

Example 5.8 [26]. A reaction proceeding according to the scheme $A+B+C \rightarrow D$ in a water-alcohol solution was investigated. The quality and quantity of the reaction product, D(y), were governed by the reaction time in hours, z_1; the alcohol content of the solution in mole fractions, z_2; the content of the substance C in mole fractions, z_3; the content of the substance B in mole fractions, z_4; and the molar ratio B/A, z_5. The principal level and the incremental changes are

Factor	z_1	z_2	z_3	z_4	z_5
z_j^0	2.0	0.65	0.10	0.25	1.20
Δz_j	0.20	0.15	0.025	0.05	0.20

We seek to find the set of operating conditions which maximize the yield of the product D, (y_{max}).

Solution. We shall use a regular simplex design. For $k = 5$, we select from matrix (5.149) a submatrix containing five columns and six rows. Using the coding convention of Eq. (5.3), we get

$$x_1 = \frac{z_1 - 2.0}{0.20} \qquad x_2 = \frac{z_2 - 0.65}{0.15} \qquad x_3 = \frac{z_3 - 0.10}{0.025}$$

$$x_4 = \frac{z_4 - 0.25}{0.05} \qquad x_5 = \frac{z_5 - 1.20}{0.20}$$

Then the matrix of the original simplex on the natural scale will have the form given in the table below.

N	z_1	z_2	z_3	z_4	z_5	y
1	2.10	0.693	0.105	0.258	1.225	0.760
2	1.90	0.693	0.105	0.258	1.225	0.491
3	2.00	0.564	0.105	0.258	1.225	0.513
4	2.00	0.650	0.085	0.258	1.225	0.675
5	2.00	0.650	0.100	0.218	1.225	0.693
6	2.00	0.650	0.100	0.250	1.075	0.666

As is seen from the table, observation 2 gives the lowest response. We replace point 2 with its mirror reflection, point 7. The coordinates of the new point are given by Eqs. (5.150) and (5.151). Now we find the coordinates of point c, the centre of the face formed by points 1, 3, 4, 5 and 6

$$z_1^{(c)} = \frac{4 \times 2.00 + 2.1}{5} = 2.02$$

$$z_2^{(c)} = \frac{3 \times 0.65 + 0.504 + 0.693}{5} = 0.641$$

$$z_3^{(c)} = \frac{2 \times 0.105 + 0.085 + 2 \times 0.100}{5} = 0.099$$

$$z_4^{(c)} = \frac{3\times0.258+0.218+0.250}{5} = 0.248$$

$$z_5^{(c)} = \frac{4\times1.225+1.075}{5} = 1.195$$

Then the coordinates of point 7 are

$$z_1^{(7)} = 2\times2.02-1.90 = 2.14$$
$$z_2^{(7)} = 2\times0.641-0.693 = 0.589$$
$$z_3^{(7)} = 2\times0.099-0.105 = 0.093$$
$$z_4^{(7)} = 2\times0.248-0.258 = 0.238$$
$$z_5^{(7)} = 2\times1.195-1.225 = 1.165$$

Point 7 forms together with the remaining points 1, 3, 4, 5 and 6 a new regular simplex (see the table below).

N	z_1	z_2	z_3	z_4	z_5	y
1	2.10	0.693	0.105	0.258	1.225	0.760
3	2.00	0.569	0.105	0.258	1.225	0.513
4	2.00	0.650	0.085	0.258	1.225	0.675
5	2.00	0.650	0.100	0.218	1.225	0.693
6	2.00	0.650	0.100	0.250	1.075	0.666
7	2.14	0.589	0.093	0.238	1.165	0.810

An observation with its centre at point 7 gives the poorest response at point 3. Its reflection from the face formed by points 1, 4, 5, 6 and 7 defines the centre for the next design, etc. After observation 7, one more factor, z_6, the speed of the agitator, is added. Until then, z_6 has been held constant, $z_6^0 = 800$ min^{-1}. The coordinates of point 8 in dimensionless form are

$$\{x_1^{(0)}, x_2^{(0)}, \ldots, x_{k+1}^{(0)}+h_{k+1}\}$$

The unit of incremental change is $\Delta z_6 = 100$ min^{-1}, and the principal level is $z_6^0 = 800$ min^{-1}. Then the coding equation for z_6 is

$$x_6 = \frac{z_6-800}{100} \qquad x_6^0 = 0$$

The height of the six-dimensional simplex is given by Eq. (5.142) to be $h = 0.764$. Now we shall find the set of factors for run 8. The first five factors are the coordinates of the centroid of five-dimensional simplex formed by points 1, 3, 4, 5, 6 and 7.

$$z_1^{(s)} = z_1^{(8)} = \frac{2.10+4\times2.0+2.14}{6} = 2.04$$
$$z_2^{(s)} = z_2^{(8)} = 0.633$$
$$z_3^{(s)} = z_3^{(8)} = 0.098$$
$$z_4^{(s)} = z_4^{(8)} = 0.247$$
$$z_5^{(8)} = 1.19$$
$$z_6^{(8)} = 800+100x_6^{(8)} = 800+100(x_6^{(0)}+h_6) = 877 \text{ min}^{-1}$$

Point 8 together with points 1, 3, 4, 5, 6 and 7 forms a six-dimensional simplex (points 1, 3, 4, 5, 6, 7 and 8) identified in the table below.

N	z_1	z_2	z_3	z_4	z_5	z_6	y
1	2.10	0.693	0.105	0.258	1.225	800	0.760
3	2.00	0.564	0.105	0.258	1.225	800	0.513
4	2.00	0.650	0.085	0.258	1.225	800	0.675
5	2.00	0.650	0.100	0.218	1.225	800	0.693
6	2.00	0.650	0.100	0.250	1.075	800	0.666
7	2.14	0.589	0.093	0.238	1.165	800	0.810
8	2.04	0.633	0.098	0.147	1.190	877	

After observation 8, the results should be analyzed, and the procedure is restored for six factors.

5.12. Plackett-Burman Saturated Orthogonal Designs

Two-level D-optimal saturated orthogonal designs can be constructed on the basis of fractional replicates of a full factorial design for $k = 3$ ($N = 4$), $k = 7$ ($N = 8$), $k = 15$ ($N = 16$), $k = 31$ ($N = 32$), etc. The family of saturated orthogonal designs can be extended considerably by using the rigorous rules of construction and analysis of orthogonal designs developed by Plackett and Burman [27]. They found, in particular, that in the case of a saturated design the estimates of the effects computed by the least squares method have a maximum accuracy for a given number N of observations and that the accuracy is the same for all effects provided that the design matrix has orthogonal columns. For a design matrix to be orthogonal, the following conditions are necessary and sufficient: (1) the number of times each factor is fixed to each of its levels must be the same; (2) the number of times every two factors, each at any one of its levels, are encountered must be the same; and (3) the number of observations must be divisible by the square of the number of levels, defined as

$$N = nl^2 \tag{5.153}$$

where n is an integer.

When the conditions stated above are available, to construct an orthogonal matrix (experimental design) requires combinatorial operations only.

If $N = nl^2$, then the number k of factors for which one can compute their effects, given N, is

$$k = (nl^2-1)/(l-1) \tag{5.154}$$

or an integer part of the value given by Eq. (5.154).

If the number of levels for all factors is 2, then an optimum design is obtained by constructing $N \times N$ orthogonal matrix comprising $+1$'s and -1's, where N is a multiple of 4, i. e. $N = 4n$. The maximum number of factors which can be included in the design is $k = N-1$.

To construct saturated designs for $k = 11$, 19, 23 and 35, one may use the rows listed in Table 5.27. Constructing a design for $k = 11$ (see Example

Table 5.27. Sign Combinations for Saturated Designs, k = 11, 19, 23 and 35

k	N	Sign combinations
11	12	+ + − + + + − − − + −
19	20	+ + − − + + + + − + − + − − − − + + −
23	24	+ + + + + − + − + + − − + + − − + − + − − − −
35	36	− + − + + + − − − + + + + + − + + + − − + − −
		− − + − + + − − + −

5.9), the respective row taken from Table 5.27 produces a first column. A second column is then generated from the first by placing its last element in the first position and moving the remaining elements down one position. A third column is produced from the second one by placing the last two elements in the first two positions and shifting the remaining elements down, and the process is continued until column k is generated. The elements of the last row are -1's.

Designs for $k = 19$, 23 and 35 are constructed similarly. In the case of $k = 27$, the design matrix is generated by using blocks A, B and C shown in Table 5.28. The three blocks are written down in the cyclic permutation order

$$\begin{matrix} A & B & C \\ C & A & B \\ B & C & A \end{matrix}$$

and a row of -1's again is added to them.

Plackett and Burman have proved that saturated designs for N reaching 100 can be constructed (except $N = 92$) provided that N is a multiple of 4. The Plackett-Burman designs allow one to obtain separate (unbiased) estimates of the linear effects of all factors with an accuracy that is a maximum for a given number of trials and equal for all effects. Any coefficient of the estimated linear regression equation is given by

$$b_j = \frac{1}{N} \sum_{i=1}^{N} x_{ji} y_i \qquad j = 1, 2, \ldots, k$$

Table 5.28. Blocks for Saturated Orthogonal Design, $k = 27$

A									B									C								
+	−	+	+	+	+	−	−	−	−	+	−	−	−	+	−	−	+	+	+	−	+	−	+	+	−	+
+	+	−	+	+	+	−	−	−	−	−	+	+	−	−	+	−	−	−	+	+	+	+	−	+	+	−
−	+	+	+	+	+	−	−	−	+	−	−	−	+	−	−	+	−	+	−	+	−	+	+	−	+	+
+	−	−	+	−	+	+	+	+	−	−	+	−	+	−	−	−	+	+	−	+	+	+	−	+	−	+
−	−	−	+	+	−	+	+	+	+	−	−	−	−	+	+	−	−	+	+	−	−	+	+	+	+	−
−	−	−	−	+	+	+	+	+	−	+	−	+	−	−	−	+	−	−	+	+	+	−	+	−	+	+
−	+	+	−	−	−	+	−	+	−	−	+	−	−	+	−	+	−	+	−	+	+	−	+	+	+	−
+	+	+	−	−	−	+	+	−	+	−	−	+	−	−	−	−	+	+	+	−	+	+	−	−	+	+
+	+	+	−	−	−	−	+	+	−	+	−	−	+	−	+	−	−	−	+	+	−	+	+	+	−	+

The value of s_{b_j} is

$$s_{b_j} = s_e / \sqrt{N}$$

where s_e is the error mean square.

Since the design matrix is orthogonal, this estimate of the linear effects agrees with the estimate produced by the least squares. In addition, the linear effects so obtained are unconfounded with one another and their estimates are unbiased.

The ratio of b_j to $s_e/\sqrt{N}$ has a Student distribution under the null hypothesis, i. e. the true value of β_j equals zero. This ratio can be used to test the effects for significance. When one wants to test the difference between the effects for significance, the following ratio is appropriate

$$t = \frac{b_u - b_j}{s_e/\sqrt{N}} \tag{5.155}$$

which is also distributed as the Student t.

Example 5.9. The possibility of producing a nitrogen phosphate/potash fertilizer by partial replacement of potassium carbonate with ammonia during the neutralization of a nitrous-acid extract is investigated. The neutralization process is given by

$$6\,Ca(NO_3)_2 + 3\,H_3PO_4 + 6\,NH_3 + 3\,K_2CO_3 = 3\,CaHPO_4 + 6\,NH_4NO_3 + 6\,KNO_3 + 3\,CaCO_3$$

There are several sequential steps in which the extract is subjected to neutralization by ammonia and potassium carbonate. Important to the experimenter is the degree of retrogradation of the assimilable forms of phosphorus pentoxide. As the process criterion therefore, we choose the degree of assimilability of the phosphoric compounds formed, y, which is the percent ratio of the amount of water- and citrate-soluble phosphoric compounds to the amount of phosphorus in the reaction products. The factors are chosen to be: z_1, the ammoniation temperature, 25-70 °C; z_2, the ammoniation time, 15-30 min; z_3, the ammonia ratio, 100-150% of the stoichiometric ratio; z_4 through z_7, the $Mg(NO_3)_2$ (0-3.16%), $Fe(NO_3)_2$ (0-0.89%), $Al(NO_3)_3$ (0-0.56%), H_2SiF_6 (0-0.88%) contents of the original extract; z_8, the temperature at which the components of the ammoniated extract react with the potassium carbonate solution, 25-70 °C; z_9, the time during which the potassium carbonate is involved in the reaction, 30-60 min; and z_{10}, the potassium carbonate ratio (100-120% of stoichiometric ratio).

The P_2O_5 (6.9%) and CaO (11%) contents of the extract are held constant during the process.

Solution. The first 10 columns of the Plackett-Burman design (Table 5.29) are used for the experimental design.

Table 5.29. Design Matrix and Experimental Data

N	x_0	x_1	x_2	x_3	x_4	x_5	x_6	x_7	x_8	x_9	x_{10}	$\bar{y}$
1	+1	+1	−1	+1	−1	−1	−1	+1	+1	+1	−1	19.15
2	+1	+1	+1	−1	+1	−1	−1	−1	+1	+1	+1	34.44
3	+1	−1	+1	+1	−1	+1	−1	−1	−1	+1	+1	85.08
4	+1	+1	−1	+1	+1	−1	+1	−1	−1	−1	+1	92.88
5	+1	+1	+1	−1	+1	+1	−1	+1	−1	−1	−1	90.91
6	+1	+1	+1	+1	−1	+1	+1	−1	+1	−1	−1	51.76
7	+1	−1	+1	+1	+1	−1	+1	+1	−1	+1	−1	101.33
8	+1	−1	−1	+1	+1	+1	−1	+1	+1	−1	+1	101.34
9	+1	−1	−1	−1	+1	+1	+1	−1	+1	+1	−1	98.62
10	+1	+1	−1	−1	−1	+1	+1	+1	−1	+1	+1	87.85
11	+1	−1	+1	−1	−1	−1	+1	+1	+1	−1	+1	84.49
12	+1	−1	−1	−1	−1	−1	−1	−1	−1	−1	−1	89.89
b_j	78.19	15.36	−3.43	−2.83	8.40	7.84	7.96	2.64	−13.23	−7.01	2.91	
t_j	312.76	61.44	13.72	11.32	33.60	31.36	31.86	10.58	52.92	28.04	11.64	

The sample means of the degree of assimilability, $\bar{y}$, are determined by using the results of two replicate observations. The error mean square s_e^2 is 1.48. The number of degrees of freedom ν_e is 12. The tabulated value of the Student t, $t_{0.05}$ (12) is 2.18. Thus, all coefficients of the estimated regression equation are significant and the equation has the form

$$\hat{y} = 78.19 - 15.36x_1 - 3.43x_2 - 2.83x_3 + 8.40x_4 + 7.84x_5 + 7.96x_6 + 2.64x_7 - 13.23x_8 - 7.01x_9 + 2.91x_{10} \quad (5.156)$$

Since the experimental design is unsaturated ($N = 12$ and $k = 10$), one degree of freedom is available which allows us to test the estimated regression equation for the adequacy of fit. The adequacy mean square is

$$s_g^2 = \frac{2\sum^{12}(\bar{y}_i - \hat{y}_i)}{12 - 11} = 4.26$$

The value of the Fisher variance ratio, $F = s_a^2/s_e^2$, is 2.88.

The tabulated value of F, $F_{0.95}$ (1,12) is 4.8.

We see that the equation fits the experimental data adequately. Using this equation, one can specify the conditions in which a fertilizer can be produced by using a nitrous acid extract from phosphates, the whole phosphorus contained in this fertilizer being in an assimilable form.

If there are no replicate observations that could provide an estimate for σ_e, one can utilize the effects of dummy factors. The latter are included in the design provided it is unsaturated, which means that $k < N-1$. One may assume that some of the dummy factors included correspond to the vacant columns of the design matrix and compute the effects of these dummy factors by using the rules underlying the computational procedure for the real factors. The effects so computed differ from zero as there is an error in measuring the value of y and the interaction effects are not taken into account. If b_{k+1}, b_{k+2}, b_{N-1} are the effects of the dummy factors, then one may take

$$s_{b_j}^2 = (b_{k+1}^2+b_{k+2}^2+\ldots+b_{N-1}^2)\frac{1}{N-k-1} \tag{5.157}$$

as an estimate of the squared error that occurs in determining the effects of the main factors, and

$$s_e^2 = N(b_{k+1}^2+b_{k+2}^2+\ldots+b_{N-1}^2)\frac{1}{N-k-1} \tag{5.158}$$

as an estimate of the squared error that occurs in measuring the response, assuming that there are no interaction effects.

In some cases, the Plackett-Burman designs are less laborious, as compared with fractional replicates constructed in factor space of the same dimension.

5.13. Screening Experiments. Method of Random Balance

Though given no good grounds, the experimenter often tries to fix the values of some of the factors in order to cut down the number of observations. As a result of optimization, only local extremes can be found. For multifactor problems, it is feasible to conduct factor screening at the first stage of experiment. Since the contribution from a given factor is connected with its range of variation, many factors which are treated as significant ones on the basis of a priori information, may be found to be insignificant during experimentation. Thus, screening experiments can be used effectively not only to investigate new processes but also to serve as a first stage of exploration of a multifactor process having a sufficient a priori information. This is the case with processes whose number of factors is so large that the experimenter cannot plan at once the experiment to seek for the optimum conditions of the process. To screen out qualitative and quantitative factors with the number of levels equal to 2, one may use fractional replicates of a factorial design with a relatively higher order of divisibility and also Plackett-Burman saturated orthogonal designs. Such means allow one to obtain unbiased estimates of the linear effects of all factors with a maximum permissible accuracy for a given number of observations, that accuracy being the same for all effects. The last property is very useful for the stage of screening, since no one knows which effects will be found to be insignificant. The above designs are disadvantageous in that they cannot deal with the interaction effects which are significant.

To assign the main effects produced by linear effects and pair interactions, a *method of random balance* was developed [28] which uses supersaturated designs. In such a design, the number N of observations in the design matrix is less than the number of effects under study, which means that at the beginning of investigation the number of degrees of freedom, ν, is less than zero. The random balance method has no theoretical foundations but has heuristic features. Here a basic premise is this: among a great number of effects under study there are only several ones that actually influence the process, while the remaining factors may be treated as insignificant ones and thus assigned to the noise field. If one arranges the effects in the order of their descending contribution to the variance of the yield, then a ranking diagram is produced (Fig. 5.16). The effects that occupy the right portion of

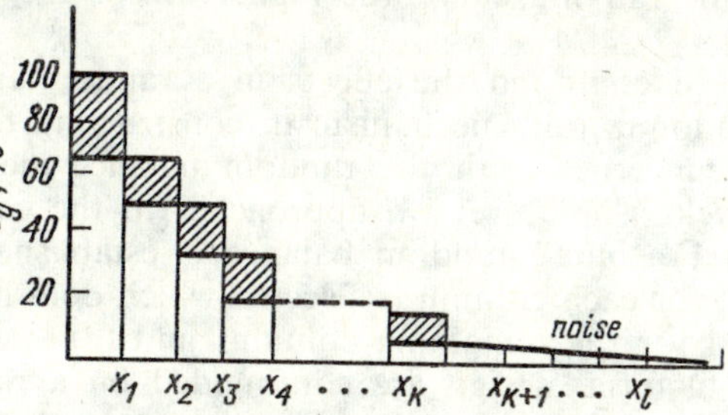

Fig. 5.16. Effects ranging diagram

the diagram belong to the noise field. The noise field concept allows us to split the original model

$$\hat{y} = b_0+b_1x_1+b_2x_2+\ldots+b_kx_k+b_{12}x_1x_2+b_{13}x_1x_3+\ldots$$
$$+b_{k-1,k}x_{k-1}x_k+s_e$$

where s_e is the experimental error.

With certain x_j's standing for pair interactions, the splitted model takes the following linear form:

$$\hat{y} = b_0+\sum_{j=1}^{l-m} b_jx_j+\sum_{j=1}^{m} c_jx_j'+u = b_0+\sum_{j=1}^{l-m} b_jx_j+nf$$

where $l-m$ is the number of significant effects, m is the number of effects included in the noise field, and

$$nf = \sum_{j=1}^{m} c_jx_j'+u$$

Thus we have

$$s_{nf}^2 = \sum_{j=1}^{m} c_j^2s_{xj}^2+s_e^2$$

The estimated significant coefficients have a large error due to the noise field, s_{nf}. Thus we may say that the random balance method has a lesser sensitivity than either full factorial or fractional factorial design. Here by the *sensitivity* of a method is meant its ability to select the regression coefficients that significantly differ from zero. On the other hand, the random balance method has a greater resolution, with the result that one can distinguish between the prevailing effects encountered among a large number of effects available.

The experimenter mainly uses the random balance method to find out the true ranking diagram and then split the model. This allows a supersaturated design to be transformed into an unsaturated one, as to significant effects, and the latter can be estimated quantitatively in the usual fashion. If the number of the effects that are actually significant is large, the property of supersaturation does not necessitate a lesser number of observations. Increasing the number of significant (especially interaction) effects decreases the effectiveness of the random balance method. However, the steeper the descent of the ranking curve, the higher will be the degree of screening.

If one wants to confound the effects in a random fashion, one must utilize some randomization mechanism in constructing the design matrix. In most cases concerned with the random balance method, factors are fixed to two levels. There are two approaches to the construction of the design matrix: (1) "pure" random balance is established, in which case a specific design for each column is selected which does not depend on the designs intended for the remaining columns and (2) systematic fractional replicates of a factorial design are confounded on a random basis. The pure random balance has a smaller effectiveness and is applicable to the situations where the nearest exponent of 2 considerably increases the number of observations. The design matrix so produced should be tested for applicability. One cannot use the matrix that has its columns completely cross-correlated with each other. The stronger the cross-correlation between the columns, the higher the probability that spurious effects will be selected. The following optimality criteria are used in constructing a design matrix: (1) the number of orthogonal columns; (2) the minimal value of the modulus of the correlation coefficient for all possible types of interactions (linear to pair and pair to pair); and (3) for all factors the condition $\sum_{i=1}^{N} x_i = 0$ must be satisfied.

The design matrices shown in Table 5.30 and also in the Appendix Tables 11 and 12 were constructed through the use of a computer on the basis of the above enumerated optimality criteria. In this procedure, 2^{5-1} and 2^{6-1} replicates were confounded randomly and their columns were selected in a random fashion too. These matrices can also be used for a lesser number of factors.

Given below is an example that deals with the extraction of significant effects from 14 linear effects and 91 interaction effects, with $k = 14$. The design matrix and experimental data are arrayed in Table 5.30. In the

Table 5.30. Design Matrix for Random Balance Method

Design point	x_1	x_2	x_3	x_4	x_5	x_6	x_7	x_8	x_9	x_{10}	x_{11}	x_{12}	x_{13}	x_{14}	y	y^{I}	y^{II}	y^{III}	y^{IV}	y^{V}	y^{VI}
1	−	−	+	+	+	−	+	−	+	−	+	+	−	+	38.03	38.03	38.03	38.03	36.03	36.03	34.93
2	+	−	−	+	−	−	−	−	−	+	−	−	−	+	58.0	38.0	33.0	37.6	73.6	36.5	36.5
3	−	+	−	+	+	−	−	+	+	+	+	−	+	−	46.59	46.59	34.79	34.79	34.79	34.65	34.65
4	+	+	+	+	+	+	−	−	+	+	−	+	−	+	78.63	40.63	33.83	38.43	36.43	35.33	34.23
5	−	−	−	−	−	+	+	+	−	−	+	−	−	−	65.41	47.41	35.61	35.61	35.61	34.51	34.51
6	+	−	+	−	+	+	+	−	+	+	+	−	+	+	81.75	43.75	31.95	36.55	34.55	35.51	34.41
7	−	+	+	−	−	+	+	−	+	+	−	+	+	−	59.81	41.81	35.81	36.81	34.81	35.77	34.07
8	+	+	−	−	−	−	−	+	−	−	+	−	−	−	58.39	38.39	33.39	37.99	35.99	35.99	35.99
9	−	−	+	−	−	+	−	+	−	+	+	+	+	+	58.99	40.99	40.99	40.99	38.99	38.85	37.75
10	+	−	−	−	−	+	−	−	−	−	−	−	+	+	72.15	34.15	34.15	34.15	34.15	34.01	34.01
11	−	+	−	−	+	−	+	−	+	−	−	−	−	−	43.0	43.0	36.2	36.2	34.2	33.1	33.1
12	+	+	+	−	+	−	−	−	−	−	−	−	+	−	70.37	50.37	38.57	38.57	36.57	36.43	35.33
13	−	−	−	+	+	+	−	+	−	+	+	+	+	−	54.0	36.0	29.2	33.8	31.8	32.76	32.76
14	+	−	+	+	−	−	+	+	+	−	+	+	−	−	59.85	39.85	33.05	37.65	35.65	34.55	33.45
15	−	+	+	+	−	−	+	+	+	+	−	+	+	+	29.51	29.51	29.51	34.11	34.11	35.07	33.97
16	+	+	−	+	+	+	+	+	−	−	−	+	−	+	74.43	36.43	31.43	36.03	36.03	34.93	34.93

design matrix, the correlation coefficient r^* is equal to zero for columns 1-4; for columns 1-10, $r^* \leqslant 0.517$; and for columns 1-14, $r^* \leqslant 0.7$.

A *scatter diagram* (Fig. 5.17) is a graphical representation of the significant factors obtained from the experimental data. At the first stage of processing these data, a scatter diagram is constructed for linear effects only. The effect is considered significant if it varies markedly within the median interval ΔM. However, this criterion is insufficient since it is not univalued. To consider the effect significant, another condition must be satisfied: there must be a large number of the separated points that are

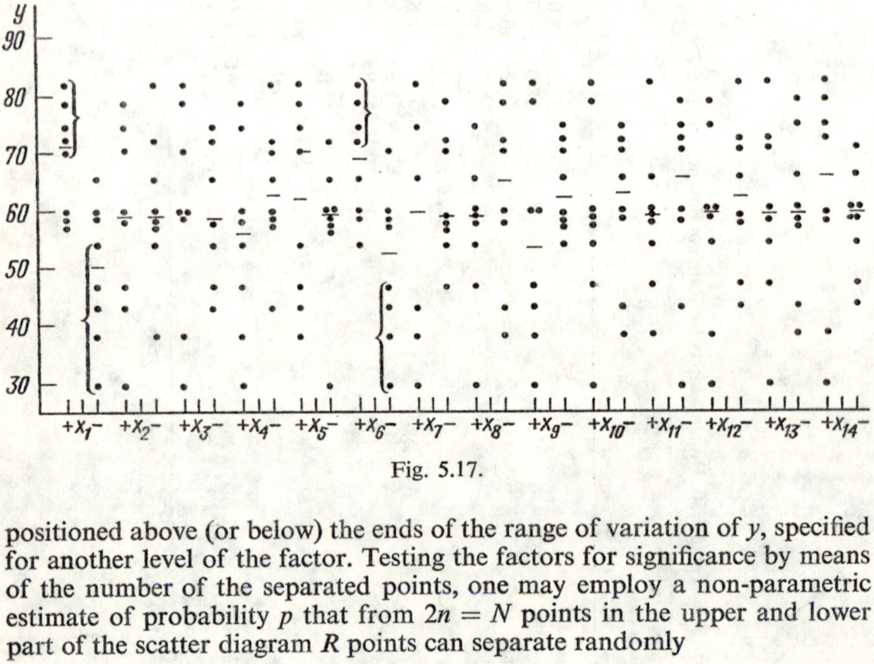

Fig. 5.17.

positioned above (or below) the ends of the range of variation of y, specified for another level of the factor. Testing the factors for significance by means of the number of the separated points, one may employ a non-parametric estimate of probability p that from $2n = N$ points in the upper and lower part of the scatter diagram R points can separate randomly

$$p = \frac{C_{2n-R}^{n} + \sum_{j=1}^{R} C_{2n-1-R}^{n-1-j}}{C_{2n}^{n} - C_{2n-2}^{n-1}} \tag{5.159}$$

To simplify computations, one may take as the significance criterion for a factor the product of the difference between the medians by the number of the deviated points, $g = |\Delta M R|$. The factors x_1, x_6 and x_9, to be subjected to quantitative estimation, are taken from the scatter diagram (Fig. 5.17) on the basis of this criterion. When computations in the quantitative estimation procedure are carried out by hand, use is made of tables with several inputs. The greater the number of inputs to a table, the higher will be the accuracy with which the given effect is estimated. Note, however, that a supersaturated design does not allow one to fill all cells of the table. Such a situation occurs in constructing a table with three inputs for factors x_1, x_6 and x_9 (Table 5.31). This causes one to select a table with two inputs (Table 5.32).

Table 5.31. Table with Three Inputs for Estimation of Linear Effects of Factors x_1, x_6, and x_9

	x_6^+		x_6^-	
	x_1^+	x_1^-	x_1^+	x_1^-
x_9^+	78.63 81.75	59.81	59.85	38.03 46.59 43.0 29.51
x_9^-	72.15 74.43	65.41 58.99 54.0	58.0 58.39 70.37	

Table 5.32. Table with Two Inputs for Estimation of Linear Effects of Factors x_1 and x_6

	x_1^+	x_1^-		x_1^+	x_1^-
x_6^+	78.63 81.75 72.15 74.43	65.41 59.81 58.99 54.0	x_6^-	58.0 58.39 70.37 59.85	38.03 46.59 43.0 29.51
	$\bar{y}_1 = 76.74$	$\bar{y}_2 = 59.65$		$\bar{y}_3 = 61.65$	$\bar{y}_4 = 39.28$

Using the data of Table 5.32, we get the regression coefficients b_1 and b_6

$$b_1 = \left(\frac{\bar{y}_1+\bar{y}_3}{2} - \frac{\bar{y}_2-\bar{y}_4}{2}\right) \div 2 = 9.86$$

$$b_6 = \left(\frac{\bar{y}_1+\bar{y}_2}{2} - \frac{\bar{y}_3+\bar{y}_4}{2}\right) \div 2 = 8.86$$

To test the jth effect for significance, use is made of the Student t-test

$$t_{x_j} = \frac{b_j \cdot 2}{s_p}$$

where

$$s_p = s_w \sqrt{\sum_i 1/n_i}$$

n_i is the number of observations for the ith cell, and s_w^2 is the residual mean square computed on the basis of the data on the dispersion about the means for all cells. The number of degrees of freedom $\nu_w = \sum n_i - a$, where a is the number of means for the table.

Referring to the data in Table 5.32, we get

$$s_p = 5.6\sqrt{4\times\frac{1}{4}} = 5.6$$

The tabulated value of the Student t is $t_{0.05}$ (12) 2.18. Thus one has

$$t_1 = \frac{2\times 9.86}{5.6} = 3.5 > 2.18$$

$$t_6 = \frac{2\times 8.86}{5.6} = 3.1 > 2.18$$

We see that the both coefficients are significant.

Then the effects of the significant factors are deleted, to reveal weaker effects. To this end, the values of $2b_j$ are subtracted from the values of y for which factor x_j is fixed at level $+1$. For example, deleting the effect of x_1, 2×9.86 is subtracted from $y_2, y_4, y_6, y_8, y_{10}, y_{12}, y_{14}$ and y_{16}, and deleting the effect of x_6, 2×8.86 is subtracted from $y_4, y_5, y_6, y_7, y_9, y_{10}, y_{13}, y_{16}$. The values of y^{I} so obtained are listed in Table 5.30. Then the processing of the experimental data is repeated, beginning with the construction of the scatter diagram for the corrected values of y. The Fisher variance ratio is used to detect the end of the factor screening procedure

$$F = s_{\bar{y}}^2 / s_e^2$$

where $s_{\bar{y}}^2$ is the variance of the results about their mean.

In the example under consideration, the over-all procedure is repeated seven times, Table 5.33 lists the defined effects and their estimates.

The error mean square, s_e^2, is 0.48 and ν_e is 5. The calculated value of F-ratio, $F = 1.73/0.48 = 3.62$, is less than the tabulated value $F_{0.95}$ (15.5) $= 4.6$.

Therefore, we see that the difference between $s_{\bar{x}}^2$ and s_e^2 is insignificant.

To process the results produced by the random balance method by hand is very tedious. The literature shows an algorithm for processing such results by a computer, which is based on the so-called branching strategy [29]. Using this algorithm, one can extract the greatest effects from the respective scatter diagrams. To estimate the defined effects quantitatively, use is made of regression analysis. The method allows one to estimate concurrently 20 coefficients as a maximum. The estimation step is very useful. If 20 coefficients (instead of three) are estimated concurrently, then the residual mean square drastically decreases, with the result that the sensitivity of the

Table 5.33. Results of Screening Experiment

Step number	Deviated effects	Estimated regression coefficients	Variance $s^2_{\bar{x}}$
I	x_1	9.86	214.6
	x_6	8.86	
II	x_1x_3	3.14	27.36
	x_6x_7	2.53	
III	x_6x_5	−2.31	10.45
IV	x_6x_{10}	1.0	3.89
V	x_{13}	−0.48	2.84
	x_2x_5	+0.55	
VI	x_3	0.55	2.08
VII	x_7	−0.25	1.73
	x_9	−0.4	

method increases. The significant effects are defined in two steps as follows. First, the effects are screened out which differ from zero by a value less than $3s$, otherwise further screening by the branching method would be a cumbersome task. The remaining effects are then handled by the branching method; in this case, the residual mean square is used as the end-of-screening criterion. During the branching procedure, the experimenter screens the effects in all possible combinations (of one, of two, etc.). This process is continued until a group of effects is selected which give a minimum residual mean square. Owing to the correlatedness of the regression coefficients (resulted from the non-orthogonality of the design matrix), the branching strategy has a marked advantage over the method of independent estimation of the regression coefficients by virtue of the t-test, since the latter often gives spurious effects. The over-all screening process is continued until the residual mean square reaches the value desirable.

5.14. Experimental Design for Formal Kinetics Equations

The literature shows that many researchers use statistical methods to study the mechanisms of complex chemical reactions and determine their rate constants [30, 31, 32]. Described below are very simple techniques that utilize the experimental design in determining the constants of the formal kinetics equations. In common use is a method of processing kinetics data whereby kinetic relationships are linearized with the aid of special transformations. For example, the rate of the reaction

$$\sum_{i=1}^{m} a_i \mathrm{A}_i \rightarrow \sum_{j=1}^{n} b_j \mathrm{B}_j \tag{5.160}$$

where A_i and B_j are the initial substances and the reaction products, respectively, and a_i and b_j are stoichiometric coefficients, can be described by means of the formal kinetics equation

$$W = K\prod_{i=1}^{m} [A_i]^{p_i} \tag{5.161}$$

where K is the rate constant of the reaction, p_i is the order of the reaction with respect to the ith component, and $[A_i]$ is the concentration of the ith component.

If the rate constant is related to temperature by the Arrhenius law, then one has

$$K = K_0 \exp(-E/RT) \tag{5.162}$$

where K_0 is the preexponential factor, E is the activation energy, R is the gas constant, and T is the absolute temperature.

Substituting (5.162) into (5.161) and taking the logarithms of the latter, we get

$$\ln W = \ln K_0 - E/RT + \sum_{i=1}^{m} p_i \ln [A_i] \tag{5.163}$$

With the variables defined as

$$\begin{array}{ll} \ln W = \hat{y} & \ln K_0 = a_0 \\ 1/T = Z_1 & -E/R = a_1 \\ \ln [A]_1 = Z_2 & p_1 = a_2 \\ \dots\dots\dots & \dots\dots\dots \\ \ln [A_m] = Z_{m+1} & p_m = a_{m+1} \end{array} \tag{5.164}$$

one can write a linear equation

$$\hat{y} = a_0 + \sum_{i=1}^{m+1} a_i z_i \tag{5.165}$$

The coefficients of Eq. (5.165) can be determined by using linear orthogonal designs rated for the number of observations $N \geqslant m+2$.

Example 5.10 [33]. The kinetics of copolymerization of α, β, β-trifluorostyrene and methacrylic acid, characterized by a low degree of conversion of the initial monomers into a polymer, is investigated. The formal kinetics equation to describe the process is

$$W = K[C_1 + C_2]^{n_1} [I]^{n_2} \tag{5.166}$$

where $[C_1]$ and $[C_2]$ are the concentrations of the initial monomers, $[I]$ is the concentration of the initiator (benzoyl peroxide), mol/l; n_1, n_2 is the order of the reaction with respect to the total concentration of the monomers and the concentration of the initiator, respectively, E is the activation energy, kJ/mol, and K is the rate constant of the reaction.

Using Eq. (5.166), one can determine the order of the reaction with respect to the total concentration of the initial monomers and the concentration of the initiator; the activation energy; and the rate constant of the reaction. The rate constant K is related to temperature by Eq. (5.162). The ranges of variations are as follows: the concentration of the initial monomers, 8.08-11.18 mol/l; the concentration of the initiator, 0.0504-0.1512 mol/l; and the reaction temperature, 60-80 °C.

Solution. Substituting (5.162) into (5.166) and taking the logarithms of the latter, we obtain

$$\log_{10} W = \log_{10} K_0 + n_1 \log_{10} [C_1 + C_2] + n_2 \log_{10} [I] - \frac{E}{R} \log_{10} e \frac{1}{T} \tag{5.167}$$

or, in the general form,

$$\hat{y} = a_0 + a_1 z_1 + a_2 z_2 - a_3 z_3 \qquad (5.168)$$

where $\hat{y} = \log_{10} W$ $\quad a_0 = \log_{10} K_0 \quad z_1 = \log_{10} [C_1 + C_2]$

$a_1 = n_1 \quad z_2 = \log_{10} [I]$

$a_2 = n_2 \quad z_3 = 1/T$

$a_3 = +(E/R) \log_{10} e$

We use a 2^4 full experimental design. The coding equation, Eq (5.3), is used to linearly relate z_j to x_j. A table below lists the incremental changes and the coordinates of the centre point of the design.

	$z_1 = \log_{10} [C_1 + C_2]$	$z_2 = \log_{10} [I]$	$z_3 = T^{-1}$, K^{-1}
Incremental changes, Δz_j	0.0703	0.2380	0.00008
Centre point of design, z_j^0	0.9777	-1.0591	0.00292

Table 5.34. Copolymerization of α, β, β-trifluorostyrene and Methacrylic Acid. Design Matrix

N	Natural scale			$W \times 10^3$, mol/l s	Coded factors				$y = \lg W_0$
	z_1, mol/l	z_2, mol/l	T, °C		x_0	x_1	x_2	x_3	
1	11.18	0.0504	60	1.11	+	+	−	−	$\bar{3}.0453$
2	8.08	0.0504	60	0.76	+	−	−	−	$\bar{4}.8808$
3	11.18	0.0504	80	6.60	+	+	−	+	$\bar{3}.8195$
4	8.08	0.0504	80	4.46	+	−	−	+	$\bar{3}.6776$
5	11.18	0.1512	60	1.85	+	+	+	−	$\bar{3}.2672$
6	8.08	0.1512	60	1.29	+	−	+	−	$\bar{3}.1106$
7	11.18	0.1512	80	11.30	+	+	+	+	$\bar{2}.0531$
8	8.08	0.1512	80	8.17	+	−	+	+	$\bar{3}.9122$

z_1 = total concentration of initial monomers
z_2 = concentration of initiator (benzoil peroxide)
W = copolimerization rate
W_0 = initial copolymerizaton rate (mean of two replications)

Using the design matrix (Table 5.34), we obtain the kinetic curves that depict how polymerization depends on time. A linear model to relate the degree of conversion, q, to time, t, is

$$q = d_0 + d_1 t$$

The coefficients of this equation are found by the least squares method. Using the equation, we calculate the initial rate of the reaction, W_0, as the product of the coefficient d_1 and the total concentration of the initial monomers. From replicate trials we have s_e equal to 5.43×10^{-2}, while v_e is 8. The coefficients of the estimated linear regression equation

$$\hat{y} = b_0 + b_1 x_1 + b_2 x_2 + b_3 x_3 \qquad (5.169)$$

are given by Eq. (5.9). The experimental data (Table 5.34) give

$$b_0 = -2.5292, \quad b_1 = 0.075, \quad b_2 = 0.115, \quad b_3 = 0.3948$$

The value of s_{b_j} calculated from Eq. (5.14) is

$$s_{b_j} = \frac{s_e}{\sqrt{Nm}} = \frac{5.43\times10^{-2}}{\sqrt{2\times8}} = 1.35\times10^{-2}$$

The coefficients are tested for significance by using the Student t. The test results show that all coefficients significantly differ from zero, since the t-ratio, exceeds the tabulated value equal to 2.31. Thus we may say that Eq. (5.169) adequately fits the experimental data

$$s_g^2 = \frac{2\sum^{8}(\log_{10} W_i - \log_{10}\hat{W}_i)^2}{8-4} = 9\times10^{-3}$$

The Fisher variance ratio is

$$F = s_g^2/s_e^2 = (9\times10^{-3}) \div (2.96\times10^{-3}) = 3.1$$

and is less than the tabulated value, $F_{0.95}$ (4.8) = 3.8. Using Eq. (5.169) and Eq. (5.168), we obtain

$$n_1 = \frac{b_1}{\Delta z_1} = \frac{0.075}{0.0703} = 1.05 \approx 1$$

$$n_2 = \frac{b_2}{\Delta z_2} = \frac{0.115}{0.238} = 0.49 \approx 0.5$$

$$E = -\frac{b_3 R}{\Delta z_3 \log_{10} e} = \frac{0.3948}{8\times10^{-5}} 1.987\times10^{-3}\times2.3\times4.186 = 89.67 \text{ kJ mol}^{-1}$$

$$a_0 = \log_{10} K_0$$

$$a_0 = b_0 + b_1\frac{z_1^0}{\Delta z_1} + b_2\frac{z_2^0}{\Delta z_2} + b_3\frac{z_3^0}{\Delta z_3} = 10.5132 \quad \text{and} \quad K_0 = 0.3253\times10^{11}$$

The formal kinetics equation to describe the copolymerization of α, β, β-trifluorostyrene and methacrylic acid takes the form

$$W = 0.3253\times10^{11}[C_1+C_2][I]^{0.5} \exp(-89.67/RT) \tag{5.170}$$

Since the design is orthogonal, we can find the standard errors of the constants determined

$$s_{n_1} = \frac{s_{b_1}}{\Delta z_1} = \frac{1.35\times10^{-2}}{0.0703} = 0.193$$

$$s_{n_2} = \frac{s_{b_2}}{\Delta z_2} = \frac{1.35\times10^{-2}}{0.238} = 0.057 \tag{5.171}$$

$$s_E = \frac{s_{b_3}}{\Delta z_3}\frac{R}{\log_{10} e} = \frac{1.35\times10^{-2}\times0.00198\times2.3\times4.186}{0.00008} = 0.323$$

$$s_{\log_{10} K_0} = \sqrt{s_{b_0}^2 + \frac{(z_1^0)^2 s_{b_1}^2}{\Delta z_1^2} + \frac{(z_2^0)^2 s_{b_2}^2}{\Delta z_2^2} + \frac{(z_3^0)^2 s_{b_3}^2}{\Delta z_3^2}}$$

$$= 1.35\sqrt{1 + \frac{(0.977)^2}{(0.0703)^2} + \frac{(-1.0591)^2}{(0.238)^2} + \frac{(0.00292)^2}{(0.00008)^2}} = 0.53$$

It follows from Eqs. (5.171) that the errors of the constants depend on the incremental changes of the factors and tend to decrease with increasing incremental changes.

The described method of linearization of kinetic equations leads to a disturbance of the condition of maximum likelihood and gives biased estimates of the constants since the latter are determined on the assumption that the quadratic form is a minimum

$$\Phi_1 = \sum_{i=1}^{n} (y_i - \hat{y}_i)^2 = \sum_{i=1}^{n} (\log_{10} W_i - \log_{10} \hat{W}_i)^2$$

rather than from the relation

$$\Phi_2 = \sum_{i=1}^{n} (W_i - \hat{W}_i)^2$$

In addition, one can see that Eq. (5.161) has a relatively formal character. Assuming a stepwise mechanism of the reactions, the rate constants for each such stage have been suggested to determine the condition of the minimum squared deviations [34]

$$\Phi_3 = \sum_{j=1}^{m} \sum_{i=1}^{n} \frac{1}{s_{c_j}^2} (c_{ji} - \hat{c}_{ji})^2 \tag{5.172}$$

where n is the number of observations, m is the number of components, and s_{c_j} is the error of measurement of the concentration of the jth component.

If the concentrations of the components are determined without a systematic error, then the estimates of the kinetic constants, which minimize the quadratic form Φ_3, will be unbiased. To obtain the value of concentration $\hat{c}_{ij}$, one can use the integral form of the kinetic equation or apply numerical integration to a system of simultaneous kinetic equations.

To find the minimum of the quadratic form Φ_3, one may use a suitable method of linear programming. The convergence of the solution depends on the initial approximation.

There are reactions whose kinetics is described by a system of simultaneous differential equations, the right sides of the latter include polynomials corresponding to the concentrations of the reactants. In such cases, differential equations, rather than their integral forms, are preferable to estimate the reaction constants [35]. In the example given below, a series/parallel reaction of the type

$$\begin{gathered} \mathrm{B} \xleftarrow{k_1} 2\mathrm{A} \xrightarrow{k_2} \mathrm{C} \\ \mathrm{B+A} \xrightarrow{k_3} \mathrm{D} \xrightarrow{k_4} \mathrm{E} \end{gathered} \tag{5.173}$$

is described by a system of simultaneous kinetic equations

$$\begin{aligned} W_1 &= \frac{d[\mathrm{A}]}{dt} = (k_1+k_2)[\mathrm{A}]^2 + k_3[\mathrm{A}][\mathrm{B}] \\ W_2 &= \frac{d[\mathrm{B}]}{dt} = k_1[\mathrm{A}]^2 - k_3[\mathrm{A}][\mathrm{B}] \\ W_3 &= \frac{d[\mathrm{C}]}{dt} = k_2[\mathrm{A}]^2 \\ W_4 &= \frac{d[\mathrm{D}]}{dt} = k_3[\mathrm{A}][\mathrm{B}] - k_4[\mathrm{D}] \\ W_5 &= \frac{d[\mathrm{E}]}{dt} = k_4[\mathrm{D}] \end{aligned} \tag{5.174}$$

It follows from (5.174) that in the general form the kinetics equations for the described reactions can be represented by a quadratic polynomial which is usually applicable to the description of a nearly stationary region

$$\hat{y} = b_0 + \sum_j b_j x_j + \sum_{u<j} b_{uj} x_u x_j + \sum_j b_{jj} x_j^2 \tag{5.175}$$

Here factors x are the concentrations of the reactants and response y is the rate of formation of a component. The rate constants of the first-order reactions are treated as linear effects b_j, the rate constants of the mixed second-order reactions, as interaction effects b_{uj}, and the rate constants of the second-order reactions, as quadratic effects b_{jj}. If the reaction sequence is unknown but the experimenter knows which substances will serve as reactants, then he constructs a design which could provide such experimental data that all unbiased estimates for coefficients of the second-order polynomials be found. After the estimated coefficients are tested for significance, the reaction sequence is determined.

As the reaction proceeds, the concentrations of the reactants vary continuously. Thus the experimenter must use for factors the initial concentrations of the reactants, and for the response the initial reaction rates. To find the initial reaction rates, use is made of graphical or numerical differentiation along the starting portions of the kinetic curves; there is still another method in which the data obtained from continuous flow installations are handled.

Example 5.11 [35]. The kinetics of the reactions in which butenes are subject to catalytic oxidative dehydrogenation to form divinyl is investigated. According to the previous investigations, we assume that there is no reversion of divinyl.

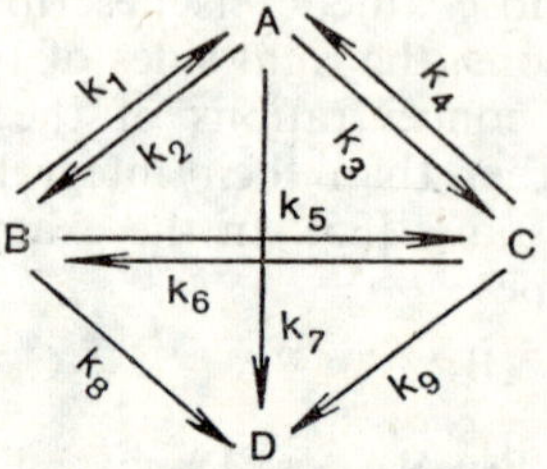

In the above scheme A, B, C, D are butene-1, trans-butene-2, cis-butene-2, and divinyl, respectively.

The order of the reactions with respect to oxygen is zero, while the order of the reactions with respect to butylenes is one. Special experiments are carried out to show that the reaction products do not display inhibitive action. The kinetics of the reactions is described by a set of simultaneous differential equations

$$\begin{aligned}
\frac{d[\mathrm{A}]}{dt} &= -(k_1+k_3+k_7)\,[\mathrm{A}]+k_2\,[\mathrm{B}]+k_4\,[\mathrm{C}] \\
\frac{d[\mathrm{B}]}{dt} &= -(k_1+k_5+k_8)\,[\mathrm{B}]+k_1\,[\mathrm{A}]+k_6\,[\mathrm{C}] \\
\frac{d[\mathrm{C}]}{dt} &= -(k_4+k_6+k_9)\,[\mathrm{C}]+k_3\,[\mathrm{A}]+k_5\,[\mathrm{B}] \\
\frac{d[\mathrm{D}]}{dt} &= k_7\,[\mathrm{A}]+k_8\,[\mathrm{B}]+k_9\,[\mathrm{C}]
\end{aligned} \tag{5.176}$$

where k_i ($i = 1, 2, \ldots, 9$) represents the rate constants for pseudomonomolecular reactions; and [A], [B], [C] and [D] are the concentrations of the reactants in the gas phase.

Using the experimental data, determine the rate constants for nine reactions and show that Eqs. (5.176) adequately describe the kinetics of the reactions.

Solution. The initial concentrations of substances A, B and C are defined as z_1, z_2 and z_3, respectively. If we consider the rates of accumulation of the substances as responses y_u, and the initial concentrations as independent factors, then Eqs. (5.176) give

$$y_u = b_{0_u} + b_{1_u}x_1 + b_{2_u}x_2 + b_{3_u}x_3, \quad u = 1, 2, 3, 4 \tag{5.177}$$

The variables x_1, x_2, x_3 are related to the initial concentrations by

$$x_j = (z_j - z_j^0)/\Delta z_j$$

The free term (intercept) in Eq. (5.177) is related to the linear terms by

$$b_{0_u} = b_{1_u}\frac{z_1^0}{\Delta z_1} + b_{2_u}\frac{z_2^0}{\Delta z_2} + b_{3_u}\frac{z_3^0}{\Delta z_3} \tag{5.178}$$

The kinetics of butene dehydrogenation is examined at different temperatures, a design matrix corresponds to a specific temperature. The description below deals with the case when $T = 669$ K. The experimental region is identified in the table below.

	z_1	z_2	z_3
z_j^0	0.01500	0.01158	0.00993
Δz_j	0.00500	0.00386	0.00331

In this table, z stands for butene concentration in mole fractions.

Any response function in Eqs. (5.176) requires three coefficients only. This means that a half-replicate of a 2^3 full factorial design is appropriate. In our case, however, we use a 2^3 full factorial design (Table 5.35) in order to determine the rate constants with greater accuracy and provide for a more detailed analysis of the deviation errors occurred in the predicted from the experimental values.

To find the error mean square, some observations are replicated.

Table 5.35. Kinetics of Dehydrogenation of Butenes. Design Matrix

N	x_1	x_2	x_3	y_1	y_2	y_3	y_4
1	−1	−1	−1	−0.288	0.046	0.003	0.272
2	−1	−1	+1	−0.263	0.077	−0.102	0.255
3	−1	+1	−1	−0.286	−0.124	0.041	0.257
4	+1	−1	−1	−0.737	0.068	0.033	0.370
5	+1	+1	+1	−0.678	−0.063	−0.060	0.426
6	+1	+1	−1	−0.725	−0.043	0.058	0.342
7	+1	−1	+1	−0.705	0.124	−0.130	0.356
8	−1	+1	+1	−0.276	−0.139	−0.180	0.262

The rate of accumulation of the component, y, in terms of mole fractions per second is given by

$$y_1 = \frac{[A]_\tau - [A]_0}{\tau} \qquad y_2 = \frac{[B]_\tau - [B]_0}{\tau}$$
$$y_3 = \frac{[C]_\tau - [C]_0}{\tau} \qquad y_4 = \frac{[D]_\tau - [D]_0}{\tau} \qquad (5.179)$$

In Eqs. (1.179), the subscript 0 stands for the initial concentrations of butenes, while the subscript τ stands for the concentrations measured at a time τ. Thus a straight line is used as an approximation of the starting portion of the kinetic curve. The higher the chemical activity of the component, the greater will be the shift of the respective rate constants. A table below shows the values of the rate constants and their errors found through the use of the designed experiment and through the nonlinear estimate (NLE) method.

	Rate constant, s^{-1}								
	k_1	k_2	k_3	k_4	k_5	k_6	k_7	k_8	k_9
Designed experiment	7.128	0.74	6.62	5.60	3.47	2.57	10.80	1.08	2.90
NLE method	9.07	0.75	9.38	6.51	3.64	2.58	29.59	1.11	3.10
$s(k)$	1.8	2.6	1.8	3.5	2.6	3.5	1.8	2.6	3.5

The comparison of the rate constants with their errors shows that some of the constants cannot be extracted from the noise background. To decrease the errors, it is necessary to increase the range of variation. The nonlinear estimate method is used to assess the estimated constants more accurately. This method requires that the rate constants be so selected that the sum of squared deviations, Eq. (5.172) be a minimum. The values of concentrations $\hat{c}_{ji}$ are obtained by integrating the set of simultaneous equations, Eqs. (5.176), over the time interval from $t = 0$ to $t = \tau$, subject to the initial conditions (see previous tables). The results of all observations are summed up and the summands are taken with equal weights since it has been found that the errors occurred in the measurement of the concentrations of all components are homogeneous. Taken as the initial approximation are the constants found by the design method. Then Fisher's test is used to determine whether the model described by Eqs. (5.176) is fit to the experimental data

$$s^2_{\text{res}} = \frac{\Phi(k)_{\min}}{8 \times 4 - 9}$$

In our case, the value of F-ratio its close to unity, which means that with Eqs. (5.176) the adequate fit is attained.

Chapter Six
MIXTURE DESIGN

6.1. Simplex Lattice Method

When studying the properties of a q-component mixture, which are dependent on the component ratio only, the factor space is a regular, $(q-1)$-simplex, and for the mixture the relationship holds

$$\sum_{i=1}^{q} x_i = 1 \qquad (6.1)$$

where $x_i \geqslant 0$ is the component concentration, q is the number of components.

For binary systems, ($q = 2$) the simplex of dimension 1 is a straight line segment; and at $q = 3$, the regular 2-simplex is an equilateral triangle with its interior. Each point in the triangle corresponds to a certain composition of the ternary system, and conversely each composition is represented by one distinct point. The composition may be expressed as molar, weight, or volume fraction or percentage. Vertices of the triangle represent straight substances; and sides binary systems. If we draw an altitude from each vertex, then dissect each altitude into ten equal segments and draw through the division points straight lines parallel to the triangle sides, we shall have a triangular network. Approaching from a side to the opposite vertex corresponds to the proportional increasing in content of the "vertex" component, therefore the sequential transfer from one parallel line of two components to another signifies the growth of a third component by 10 per cent. The altitudes of the triangle in Fig. 6.1 have the content of each of pertinent components marked off along them (*e. g.* at point *d*: 30 per cent of *A*, and 70 per cent of *C*). In actual practice though, no altitudes are drawn—instead the component contents are marked off directly on the

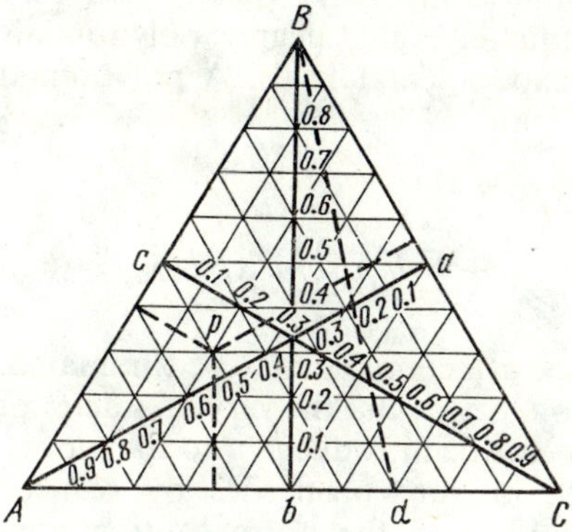

Fig. 6.1. Gibbs's concentration triangle

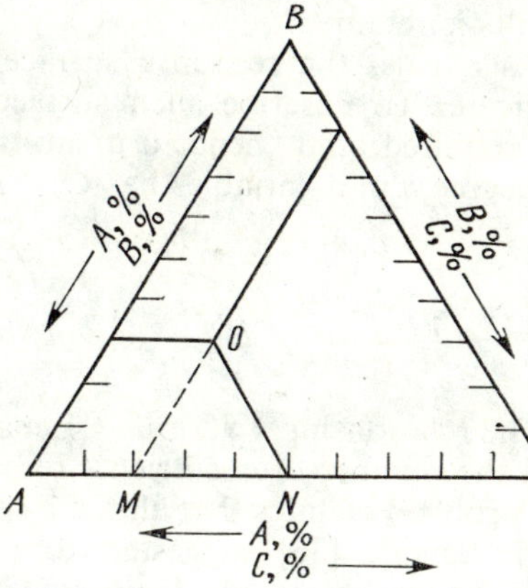

Fig. 6.2. Rozebum's concentration triangle

triangle sides. Such a method of counting is adopted in Gibbs's triangle. In the Rozebum triangle, the composition of a ternary system is read from three segments of one side (Fig. 6.2). In the concentration triangle, points lying on a straight line originating from a vertex correspond to mixtures with a constant ratio of components represented by the other two vertices. The property (y) is normally thought of as projections of lines of constant value on the plane of the concentration triangle.

At $q = 4$, the regular simplex is a tetrahedron where each vertex represents a straight component, an edge represents a binary system, and a face a ternary one. Points inside the tetrahedron correspond to quaternary systems. So, the component x_1 is absent in the face x_2, x_3, x_4, but as tetrahedron sections parallel to the face approach vertex x_1, component x_1 in them grows in concentration.

The representation of property curves of this system in a plane is not possible. Therefore such a system is to be depicted as sections of a three-dimensional simplex with planes perpendicular to one of its axes. The composition of quaternary mixtures belonging to a section plane is then described by a two-dimensional simplex, which enables any changes in system properties to be represented by contour lines. In doing so, only three components of the section are varied. A transition from one section to another signifies a change in the fourth component concentration.

In designing the experiment to attack mixture problems involving composition-property diagrams, the property studied is assumed to be a continuous function of certain arguments and with a sufficient accuracy it can be approximated by a polynomial. When investigating multicomponent systems, the use of experimental design methodologies substantially reduces the volume of an experimental effort. Further, this obviates the need for a spatial representation of complex surfaces, as the wanted properties can be derived from equations, while the possibility to graphically interpret the results is retained.

As a rule, the response surfaces in multicomponent systems are very intricate. To describe such surfaces adequately, high-degree polynomials are required, and hence a great many experimental trials. A polynomial of degree n in q variables has C^n_{q+n} coefficients,

$$\hat{y} = b_0 + \sum_{1 \leqslant i \leqslant q} b_i x_i + \sum_{1 \leqslant i \leqslant j \leqslant q} b_{ij} x_i x_j + \sum_{1 \leqslant i \leqslant j \leqslant k \leqslant q} b_{ijk} x_i x_j x_k + \ldots + \sum b_{i_1 i_2, \ldots, i_n} x_{i_1} x_{i_2} x_{i_n} \quad (6.2)$$

The relationship $\sum_{i=1}^{q} x_i = 1$ enables the qth component to be eliminated, and the number of coefficients reduced to C^n_{q+n-1}. But the very character of the problem dictates that all the q components be introduced into the model.

H. Scheffe [36] suggested to describe mixture properties by *reduced polynomials* obtainable from Eq. (6.2) subject to the normalization condition of Eq. (6.1) for a sum of independent variables. We shall demonstrate below how, for instance, such a reduced second-degree polynomial is derived for a ternary system. The polynomial has the general form

$$\hat{y} = b_0 + b_1 x_1 + b_2 x_2 + b_3 x_3 + b_{12} x_1 x_2 + b_{13} x_1 x_3 + b_{23} x_2 x_3 + b_{11} x_1^2 + b_{22} x_2^2 + b_{33} x_3^2 \quad (6.3)$$

and as

$$x_1 + x_2 + x_3 = 1 \quad (6.4)$$

then

$$b_0 x_1 + b_0 x_2 + b_0 x_3 = b_0 \quad (6.5)$$

Multiplying Eq. (6.4) by x_1, x_2, and x_3 in succession gives

$$\begin{aligned} x_1^2 &= x_1 - x_1x_2 - x_1x_3 \\ x_2^2 &= x_2 - x_1x_2 - x_2x_3 \\ x_3^2 &= x_3 - x_1x_3 - x_2x_3 \end{aligned} \tag{6.6}$$

Substituting Eqs. (6.5) and (6.6) into Eq. (6.3), we obtain after necessary transformations

$$\begin{aligned} \hat{y} = {} & (b_0+b_1+b_{11})x_1 + (b_0+b_2+b_{22})x_2 \\ & + (b_0+b_3+b_{33})x_3 + (b_{12}-b_{11}-b_{22})x_1x_2 \\ & + (b_{13}-b_{11}-b_{33})x_1x_3 + (b_{23}-b_{22}-b_{33})x_2x_3 \end{aligned} \tag{6.7}$$

We denote

$$\beta_i = b_0+b_i+b_{ii}, \quad \beta_{ij} = b_{ij}-b_{ii}-b_{jj} \tag{6.8}$$

Then we arrive at the reduced second-degree polynomial in three variables

$$\hat{y} = \beta_1x_1+\beta_2x_2+\beta_3x_3+\beta_{12}x_1x_2+\beta_{13}x_1x_3+\beta_{23}x_2x_3 \tag{6.9}$$

Thus the number of coefficients has reduced from ten to six. The reduced second-degree polynomial in q variables

$$\hat{y} = \sum_{1 \leq i \leq q} \beta_i x_i + \sum_{1 \leq i < j \leq q} \beta_{ij} x_i x_j \tag{6.10}$$

includes $q + C_q^2 = C_{q+1}^2$ coefficients. The reduced incomplete third-degree polynomial for a ternary mixture has the form

$$\hat{y} = \beta_1x_1+\beta_2x_2+\beta_3x_3+\beta_{12}x_1x_2+\beta_{13}x_1x_2+\beta_{23}x_2x_3+\beta_{123}x_1x_2x_3 \tag{6.11}$$

and for a q-component mixture we have

$$\hat{y} = \sum_{1 \leq i \leq q} \beta_i x_i + \sum_{1 \leq i < j \leq q} \beta_{ij} x_i x_j + \sum_{1 \leq i < j < k \leq q} \beta_{ijk} x_i x_j x_k \tag{6.12}$$

Then the reduced third-degree polynomial for a ternary mixture will be

$$\begin{aligned} \hat{y} = {} & \beta_1x_1+\beta_2x_2+\beta_3x_3+\beta_{12}x_1x_2 \\ & + \beta_{13}x_1x_3+\beta_{23}x_2x_3+\gamma_{12}x_1x_2(x_1-x_2) \\ & + \gamma_{13}x_1x_3(x_1-x_3)+\gamma_{23}x_2x_3(x_2-x_3)+\beta_{123}x_1x_2x_3 \end{aligned} \tag{6.13}$$

and for a q-component mixture

$$\begin{aligned} \hat{y} = {} & \sum_{1 \leq i \leq q} \beta_i x_i + \sum_{1 \leq i < j \leq q} \beta_{ij} x_i x_j \\ & + \sum_{1 \leq i < j \leq q} \gamma_{ij} x_i x_j (x_i - x_j) + \sum_{1 \leq i < j < k \leq q} \beta_{ijk} x_i x_j x_k \end{aligned} \tag{6.14}$$

Further, the reduced fourth-degree polynomial for a ternary mixture is

$$\begin{aligned}\hat{y} = {} & \beta_1 x_1+\beta_2 x_2+\beta_3 x_3+\beta_{12}x_1x_2+\beta_{13}x_1x_3 \\ & +\beta_{23}x_2x_3+\gamma_{12}x_1x_2(x_1-x_2)+\gamma_{13}x_1x_3(x_1-x_3) \\ & +\gamma_{23}x_2x_3(x_2-x_3)+\delta_{12}x_1x_2(x_1-x_2)^2 \\ & +\delta_{13}x_1x_3(x_1-x_3)^2+\delta_{23}x_2x_3(x_2-x_3)^2 \\ & +\beta_{1123}x_1^2x_2x_3+\beta_{1223}x_1x_2^2x_3+\beta_{1233}x_1x_2x_3^2 \end{aligned} \tag{6.15}$$

and, again, for q-components

$$\begin{aligned}\hat{y} = {} & \sum_{1\leq i\leq q}\beta_i x_i+\sum_{1\leq i<j\leq q}\beta_{ij}x_ix_j+\sum_{1\leq i<j\leq q}\gamma_{ij}x_ix_j(x_i-x_j) \\ & +\sum_{1\leq i<j\leq q}\delta_{ij}x_ix_j(x_i-x_j)^2+\sum_{1\leq i<j<k\leq q}\beta_{iijk}x_i^2x_jx_k \\ & +\sum_{1\leq i<j<k\leq q}\beta_{ijjk}x_ix_j^2x_k+\sum_{1\leq i<j<k\leq q}\beta_{ijkk}x_ix_jx_k^2 \\ & +\sum_{1\leq i<j<k<l\leq q}\beta_{ijkl}x_ix_jx_kx_l \end{aligned} \tag{6.16}$$

The non-linear part of these polynomials is called the *synergism* if it gives a higher response as compared with that of the linear part of the equation, and the *antagonism* if gives a lesser response. So, the term β_{ij} in the second-degree polynomial is termed the quadratic coefficient of binary synergism of the components i and j. In the third-degree polynomial the synergism for a ternary system is equal to

$$\begin{aligned} & [\beta_{ij}x_ix_j+\gamma_{ij}x_ix_j(x_i-x_j)]+[\beta_{ik}x_ix_k+\gamma_{ik}x_ix_k(x_i-x_k)] \\ & +[\beta_{jk}x_jx_k+\gamma_{jk}x_jx_k(x_j-x_k)]+\beta_{ijk}x_ix_jx_k \end{aligned} \tag{6.17}$$

where the expressions in brackets are binary synergisms in the ternary system; β_{ijk} is the cubic coefficient of the ternary synergism of the components i, j, and k.

Also, another transformation of the initial polynomial of degree n in q variables is possible. It can be reduced to a so-called *homogeneous* polynomial of degree n multiplying the terms of the degree $s < n$ by $\left(\sum_{1\leq i\leq q} x_i\right)^{n-s} = 1$.

Thus, we reduce the polynomial of Eq. (6.3) to the homogeneous one

$$\begin{aligned}\hat{y} = {} & b_0(x_1+x_2+x_3)^2+b_1x_1(x_1+x_2+x_3) \\ & +b_2x_2(x_1+x_2+x_3)+b_3x_3(x_1+x_2+x_3) \\ & +b_{12}x_1x_2+b_{13}x_1x_3+b_{23}x_2x_3+b_{11}x_1^2+b_{22}x_2^2 \\ & +b_{33}x_3^2 = (2b_0+b_1+b_2+b_{12})x_1x_2 \\ & +(2b_0+b_1+b_3+b_{13})x_1x_3+(2b_0+b_2 \\ & +b_3+b_{23})x_2x_3+(b_0+b_1+b_{11})x_1^2 \\ & +(b_0+b_2+b_{22})x_2^2+(b_0+b_3+b_{33})x_3^2 \end{aligned} \tag{6.18}$$

In Eq. (6.18) we denote

$$\beta'_{ij} = 2b_0+b_i+b_j+b_{ij}, \qquad \beta'_{ii} = b_0+b_i+b_{ii}$$

Thus, we obtain the homogeneous polynomial

$$\hat{y} = \beta'_{12}x_1x_2+\beta'_{13}x_1x_3+\beta'_{23}x_2x_3+\beta'_{11}x_1^2+\beta'_{22}x_2^2+\beta'_{33}x_3^2 \qquad (6.19)$$

The number of unknown coefficients in Eq. (6.19) is the same as in Eq. (6.9).

The reduced polynomials have gained wider acceptance and in what follows we will work with them only. Thus, the minimum number of points required to determine the coefficients of a polynomial of degree n in q variables is equal to C^n_{q+n-1} (Table 6.1).

Table 6.1. Number of Experiments Required to Obtain Polynomials of Various Degrees

Number of components	Degree 2	Degree 3 (incomplete)	Degree 3	Degree 4	Number of components	Degree 2	Degree 3 (incomplete)	Degree 3	Degree 4
3	6	7	10	15	6	21	41	56	126
4	10	14	20	35	8	36	92	120	330
5	15	25	35	70	10	55	175	220	715

6.2. Scheffe's Simplex-Lattice Designs

At present the most common are the *simplex-lattice* designs proposed by H. Scheffe [37, 38]. These designs provide a uniform scatter of points over the $(q-1)$-simplex. The points form a $\{q, n\}$-lattice on the simplex, where

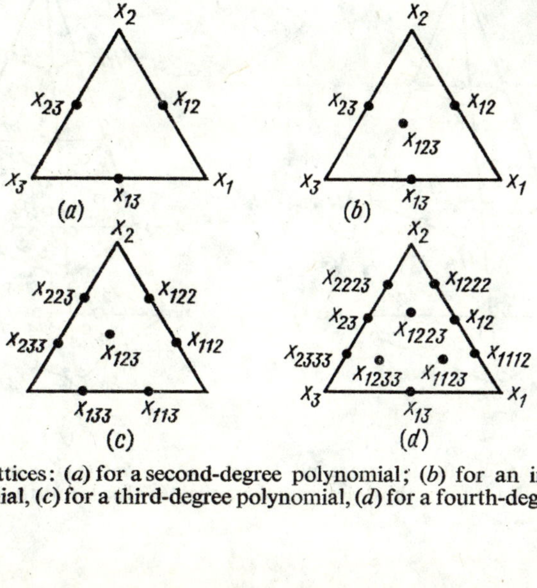

Fig. 6.3. $\{3, n\}$-lattices: (*a*) for a second-degree polynomial; (*b*) for an incomplete-third-degree polynomial, (*c*) for a third-degree polynomial, (*d*) for a fourth-degree polynomial

q is the number of mixture components, n is the degree of polynomial. Simplex-lattice designs are saturated. For each component there exist $(n+1)$ similar levels $x_i = 0, 1/n, 2/n, \ldots, 1$, and all possible combinations are derived with such values of component concentrations. So, for instance, for the quadratic lattice $\{q, 2\}$ approximating the response surface with second-degree polynomials ($n = 2$) the following levels of every factor must be used: 0, 1/2, and 1; for the cubic ($n = 3$): 0, 1/3, 2/3, and 1, *etc.* Some of $\{3, n\}$-lattices are depicted in Fig. 6.3, and $\{4, n\}$-lattices

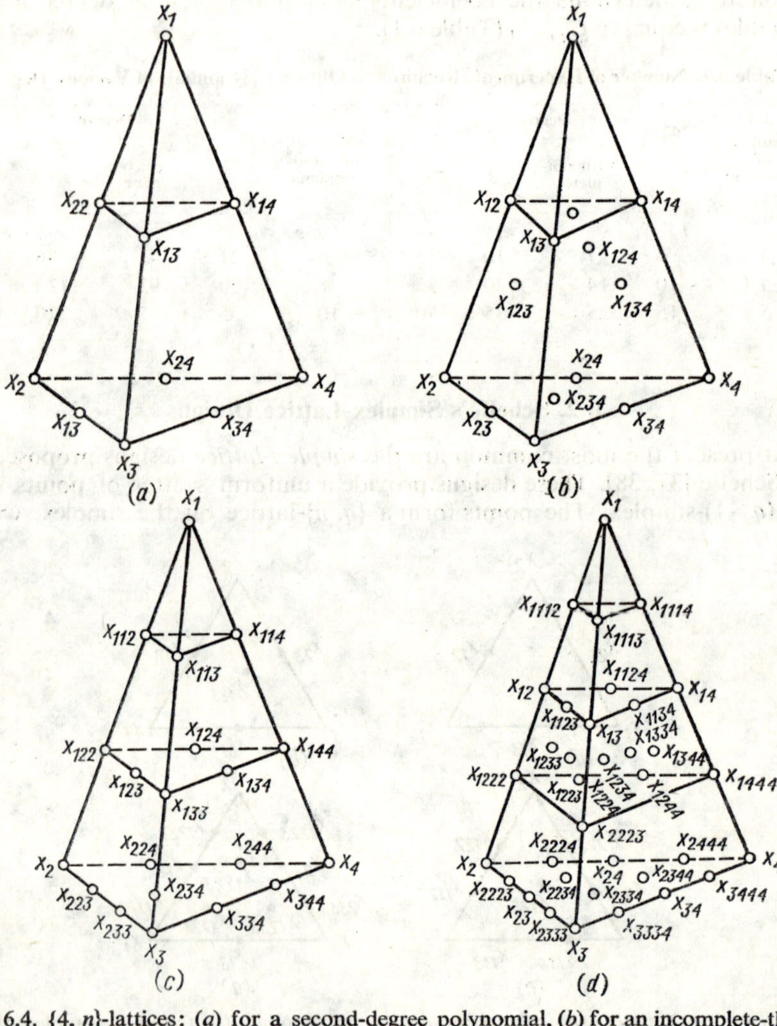

Fig. 6.4. $\{4, n\}$-lattices: (*a*) for a second-degree polynomial, (*b*) for an incomplete-third-degree polynomial, (*c*) for a third-degree polynomial, (*d*) for a fourth-degree polynomial

in Fig. 6.4. These designs are in part compositional. The incomplete cubic lattice {3, 3*}, for example, can be had from the {3, 2} lattice through the addition of only one point at the center of simplex, and {3, 4} lattice through the addition of points to the {3, 2} lattice. Having written the coordinates of points of the simplex lattice, we obtain the design matrix. We build the design matrix for the lattices {3, 2}, {3, 3}, and {3, 4}.

Subscripts of the mixture property symbols indicate the relative content of each component in the mixture. For example, the mixture 1 (Table 6.2) contains the component x_1 alone, the property of this mixture is denoted by y_1, mixture 4 includes $(1/2)x_1$ and $(1/2)x_2$, the property being designated as y_{12}. The design matrix for the simplex lattice {3, 3} is shown in Table 6.3,

Table 6.2. Design Matrix for {3, 2} Lattice

N	x_1	x_2	x_3	y_{exp}	N	x_1	x_2	x_3	y_{exp}
1	1	0	0	y_1	4	1/2	1/2	0	y_{12}
2	0	1	0	y_2	5	1/2	0	1/2	y_{13}
3	0	0	1	y_3	6	0	1/2	1/2	y_{23}

Table 6.3. Third-order Simplex-lattice Design for Ternary Mixture

N	x_1	x_2	x_3	y	N	x_1	x_2	x_3	y
1	1	0	0	y_1	6	0	2/3	1/3	y_{223}
2	0	1	0	y_2	7	0	1/3	2/3	y_{233}
3	0	0	1	y_3	8	2/3	0	1/3	y_{113}
4	2/3	1/3	0	y_{112}	9	1/3	0	2/3	y_{133}
5	1/3	2/3	0	y_{122}	10	1/3	1/3	1/3	y_{123}

and Table 6.4 is the array of the design matrix for building the fourth-degree polynomial in ternary system.

Coefficients of these polynomials are derived using the design saturation property. To obtain the coefficients of the polynomial

$$\hat{y} = \beta_1 x_1 + \beta_2 x_2 + \beta_3 x_3 + \beta_{12} x_1 x_2 + \beta_{13} x_1 x_3 + \beta_{23} x_2 x_3$$

we substitute in succession into the equation the coordinates of all the six points of the design matrix (Table 6.2). Then, substituting the coordinates of the first point ($x_1 = 1, x_2 = 0, x_3 = 0$) gives

$$y_1 = \beta_1 \tag{6.20}$$

Table 6.4. Design Matrix for {3, 4} Lattice

N	x_1	x_2	x_3	y	N	x_1	x_2	x_3	y
1	1	0	0	y_1	9	3/4	0	1/4	y_{1113}
2	0	1	0	y_2	10	1/4	0	3/4	y_{1333}
3	0	0	1	y_3	11	0	3/4	1/4	y_{2223}
4	1/2	1/2	0	y_{12}	12	0	1/4	3/4	y_{2333}
5	1/2	0	1/2	y_{13}	13	1/2	1/4	1/4	y_{1123}
6	0	1/2	1/2	y_{23}	14	1/4	1/2	1/4	y_{1223}
7	3/4	1/4	0	y_{1112}	15	1/4	1/4	1/2	y_{1233}
8	1/4	3/4	0	y_{1222}					

Hence

$$\beta_2 = y_2 \quad \text{and} \quad \beta_3 = y_3 \tag{6.21}$$

And the substitution of the coordinates of the fourth point yields

$$y_{12} = \beta_1 1/2 + \beta_2 1/2 + \beta_{12} 1/4 \tag{6.22}$$

But as $\beta_i = y_i$, then

$$y_{12} = y_1 1/2 + y_2 1/2 + \beta_{12} 1/4 \tag{6.23}$$

Thus

$$\beta_{12} = 4y_{12} - 2y_1 - 2y_2 \tag{6.24}$$

and

$$\beta_{13} = 4y_{13} - 2y_1 - 2y_3$$
$$\beta_{23} = 4y_{23} - 2y_2 - 2y_3 \tag{6.25}$$

The three points defining the coefficient β_{ij} lie on one edge. The coefficients of the reduced second-degree polynomial for q-component mixture

$$\hat{y} = \sum_{1 \leq i \leq q} \beta_i x_i + \sum_{1 \leq i < j \leq q} \beta_{ij} x_i y_i$$

are determined in a similar manner:

$$\beta_i = y_i \qquad \beta_{ij} = 4y_{ij} - 2y_i - 2y_j \tag{6.26}$$

Substituting in succession into the third-order polynomial for the ternary system

$$\hat{y} = \beta_1 x_1 + \beta_2 x_2 + \beta_3 x_3 + \beta_{12} x_1 x_2 + \beta_{13} x_1 x_3$$
$$+ \beta_{23} x_2 x_3 + \gamma_{12} x_1 x_2 (x_1 - x_2)$$
$$+ \gamma_{13} x_1 x_3 (x_1 - x_3) + \gamma_{23} x_2 x_3 (x_2 - x_3) + \beta_{123} x_1 x_2 x_3$$

the coordinates of design matrix points (see Table 6.3) gives

$$\beta_1 = y_1,\quad \beta_2 = y_2,\quad \beta_3 = y_3 \tag{6.27}$$

$$y_{112} = y_1 2/3 + y_2 1/3 + \beta_{12} 2/9 + \gamma_{12} 2/27 \tag{6.28}$$

$$y_{122} = y_1 1/3 + y_2 2/3 + \beta_{12} 2/9 - \gamma_{12} 2/27 \tag{6.29}$$

Having summed up Eqs. (6.28) and (6.29) we obtain

$$\beta_{12} = 9/4(y_{112} + y_{122} - y_1 - y_2) \tag{6.30}$$

and subtracting Eq. (6.28) from Eq. (6.29), we obtain

$$\gamma_{12} = 9/4(3y_{112} - 3y_{122} - y_1 + y_2) \tag{6.31}$$

Substituting into the reduced polynomial the coordinates of points 6 and 7 gives

$$\begin{aligned} \beta_{23} &= 9/4(y_{223} + y_{233} - y_2 - y_3) \\ \gamma_{23} &= 9/4(3y_{223} - 3y_{233} - y_2 + y_3) \end{aligned} \tag{6.32}$$

In a similar manner, for points 8 and 9, we obtain

$$\begin{aligned} \beta_{13} &= 9/4(y_{113} + y_{133} - y_1 - y_3) \\ \gamma_{13} &= 9/4(3y_{113} - 3y_{133} - y_1 + y_3) \end{aligned} \tag{6.33}$$

We have for the last point, considering Eqs. (6.27)-(6.33),

$$\begin{aligned} y_{123} = {} & y_1 1/3 + y_2 1/3 + y_3 1/3 + 1/4(y_{112} + y_{122} - y_1 - y_2) \\ & + 1/4(y_{113} + y_{133} - y_1 - y_3) + 1/4(y_{223} + y_{233} - y_2 - y_3) \\ & + 1/27\beta_{123} \end{aligned}$$

$$\begin{aligned} \beta_{123} = {} & 27y_{123} - 27/4(y_{112} + y_{122} + y_{113} + y_{133} + y_{223} + y_{233}) \\ & + 9/2(y_1 + y_2 + y_3) \end{aligned} \tag{6.34}$$

The coefficients of the third-degree polynomial for the q-component mixture are

$$\begin{aligned} \hat{y} = {} & \sum_{1 \le i \le q} \beta_i x_i + \sum_{1 \le i < j \le q} \beta_{ij} x_i x_j + \sum_{1 \le i < j \le q} \gamma_{ij} x_i x_j (x_i - x_j) \\ & + \sum_{1 \le i < j \le q} \beta_{ijk} x_i x_j x_k \end{aligned}$$

$$\beta_i = y_i \tag{6.35}$$

$$\beta_{ij} = 9/4(y_{iij} + y_{ijj} - y_i - y_j) \tag{6.36}$$

$$\gamma_{ij} = 9/4(3y_{iij} - 3y_{ijj} - y_i + y_j) \tag{6.37}$$

$$\begin{aligned} \beta_{ijk} = {} & 27y_{ijk} - 27/4(y_{iij} + y_{ijj} + y_{iik} + y_{ikk} \\ & + y_{jjk} + y_{jkk}) + 9/2(y_i + y_j + y_k) \end{aligned} \tag{6.38}$$

The relationships for coefficients of any reduced polynomial with any number of components are derived in a similar manner. So, for the incomplete-third-degree polynomial and ternary system we obtain

$$\hat{y} = \beta_1 x_1 + b_2 x_2 + b_3 x_3 + \beta_{12}' x_1 x_2 + \beta_{13} x_1 x_3 + \beta_{23} x_2 x_3 + \beta_{123} x_1 x_2 x_3$$

$$\beta_i = y_1, \text{ etc.} \tag{6.39}$$

$$\beta_{12} = 4y_{12} - 2y_1 - 2y_2, \text{ etc.} \tag{6.40}$$

$$\beta_{123} = 27y_{123} - 12(y_{12} + y_{13} + y_{23}) + 3(y_1 + y_2 + y_3) \tag{6.41}$$

For q components

$$\hat{y} = \sum_{1 \leqslant i \leqslant q} \beta_i x_i + \sum_{1 \leqslant i < j \leqslant q} \beta_{ij} x_i x_j + \sum_{1 \leqslant i < j < k \leqslant q} \beta_{ijk} x_i x_j x_k$$

$$\beta_i = y_i \tag{6.42}$$

$$\beta_{ij} = 4y_{ij} - 2y_i - 2y_j \tag{6.43}$$

$$\beta_{ijk} = 27y_{ijk} - 12(y_{ij} + y_{ik} + y_{jk}) + 3(y_i + y_j + y_k) \tag{6.44}$$

The fourth-degree polynomial for the ternary mixture is

$$\begin{aligned} \hat{y} = {} & \beta_i x_1 + \beta_2 x_2 + \beta_3 x_3 + \beta_{12} x_1 x_2 + \beta_{13} x_1 x_3 + \beta_{23} x_2 x_3 \\ & + \gamma_{12} x_1 x_2 (x_1 - x_2) + \gamma_{13} x_1 x_3 (x_1 - x_3) + \gamma_{23} x_2 x_3 (x_2 - x_3) \\ & + \delta_{12} x_1 x_2 (x_1 - x_2)^2 + \delta_{13} x_1 x_3 (x_1 - x_3)^2 + \delta_{23} x_2 x_3 (x_2 - x_3)^2 \\ & + \beta_{1123} x_1^2 x_2 x_3 + \beta_{1223} x_1 x_2^2 x_3 + \beta_{1233} x_1 x_2 x_3^2 \end{aligned} \tag{6.45}$$

$$\beta_1 = y_1, \text{ etc.} \tag{6.46}$$

$$\beta_{12} = 4y_{12} - 2y_1 - 2y_2 \tag{6.47}$$

$$\gamma_{12} = (8/3)(-y_1 + 2y_{1112} - 2y_{1222} + y_2) \tag{6.48}$$

$$\gamma_{13} = (8/3)(-y_1 + 2y_{1113} - 2y_{1333} + y_3) \tag{6.49}$$

$$\gamma_{23} = (8/3)(-y_2 + 2y_{2223} - 2y_{2333} + y_3) \tag{6.50}$$

$$\begin{aligned} \delta_{12} &= (8/3)(-y_1 + 4y_{1112} - 6y_{12} + 4y_{1222} - y_2) \\ \delta_{13} &= (8/3)(-y_1 + 4y_{1113} - 6y_{13} + 4y_{1333} - y_3) \\ \delta_{23} &= (8/3)(-y_2 + 4y_{2223} - 6y_{23} + 4y_{2333} - y_3) \end{aligned} \tag{6.51}$$

$$\begin{aligned} \beta_{1123} = {} & 32(3y_{1123} - y_{1223} - y_{1233}) + (8/3)(6y_1 - y_2 - y_3) - 16(y_{12} + y_{13}) \\ & - (16/3)(5y_{1112} + 5y_{1113} - 3y_{1222} - 3y_{1333} - y_{2223} - y_{2333}) \end{aligned} \tag{6.52}$$

$$\begin{aligned} \beta_{1223} = {} & 32(3y_{1223} - y_{1123} - y_{1233}) + (8/3)(6y_2 - y_1 - y_3) - 16(y_{12} + y_{13}) \\ & - (16/3)(5y_{1222} + 5y_{2223} - 3y_{1112} - 3y_{2333} - y_{1113} - y_{1333}) \end{aligned} \tag{6.53}$$

$$\begin{aligned} \beta_{1233} = {} & 32(3y_{1233} - y_{1123} - y_{1223}) + (8/3)(6y_3 - y_1 - y_2) - 16(y_{13} + y_{23}) \\ & - (16/3)(5y_{1333} + 5y_{2333} - 3y_{1113} - 3y_{2223} - y_{1112} - y_{1222}) \end{aligned} \tag{6.54}$$

For q-component system

$$\begin{aligned}\hat{y} = & \sum_{1\leq i\leq q} \beta_i x_i + \sum_{1\leq i<j\leq q} \beta_{ij} x_i x_j + \sum_{1\leq i<j\leq q} \gamma_{ij} x_i x_j (x_i - x_j) \\ & + \sum_{1\leq i<j\leq q} \delta_{ij} x_i x_j (x_i - x_j)^2 + \sum_{1\leq i<j<k\leq q} \beta_{iijk} x_i^2 x_j x_k \\ & + \sum_{1\leq i<j<k\leq q} \beta_{ijjk} x_i x_j^2 x_k + \sum_{1\leq i<j<k\leq q} \beta_{ijkk} x_i x_j x_k^2 \\ & + \sum_{1\leq i<j<k<l\leq q} \beta_{ijkl} x_i x_j x_k x_l \end{aligned} \tag{6.55}$$

$$\beta_i = y_i \tag{6.56}$$

$$\beta_{ij} = 4y_{ij} - 2y_i - 2y_j \tag{6.57}$$

$$\gamma_{ij} = (8/3)(-y_i + 2y_{iiij} - 2y_{ijjj} + y_j) \tag{6.58}$$

$$\delta_{ij} = (8/3)(-y_i + 4y_{iiij} - 6y_{ij} + 4y_{ijjj} - y_j) \tag{6.59}$$

$$\begin{aligned}\beta_{iijk} = & 32(3y_{iijk} - y_{ijjk} - y_{ijkk}) + (8/3)(6y_i - y_j - y_k) - 16(y_{ij} + y_{ik}) \\ & -(16/3)(5y_{iiij} + 5y_{iiik} - 3y_{ijjj} - 3y_{ikkk} - y_{jjjk} - y_{jkkk})\end{aligned} \tag{6.60}$$

$$\begin{aligned}\beta_{ijjk} = & 32(3y_{ijjk} - y_{iijk} - y_{ijkk}) + (8/3)(6y_j - y_i - y_k) - 16(y_{ij} + y_{jk}) \\ & -(16/3)(5y_{ijjj} + 5y_{jjjk} - 3y_{iiij} - 3y_{jkkk} - y_{iiik} - y_{ikkk})\end{aligned} \tag{6.61}$$

$$\begin{aligned}\beta_{ijkk} = & 32(3y_{ijkk} - y_{iijk} - y_{ijjk}) + (8/3)(6y_k - y_i - y_j) - 16(y_{ik} + y_{jk}) \\ & -(16/3)(5y_{ikkk} + 5y_{jkkk} - 3y_{iiik} - 3y_{jjjk} - y_{iiij} - y_{ijjj})\end{aligned} \tag{6.62}$$

$$\begin{aligned}\beta_{ijkl} = & 256y_{ijkl} - 32(y_{iijk} + y_{iijl} + y_{iikl} + y_{ijjk} + y_{ijjl} + y_{jjkl} \\ & + y_{ijkk} + y_{ikkl} + y_{jkkl} + y_{ijll} + y_{jkll} + y_{ikll}) \\ & + (32/3)(y_{iiij} + y_{iiik} + y_{iiil} + y_{ijjj} + y_{jjjk} + y_{jjjl} \\ & + y_{ikkk} + y_{jkkk} + y_{kkkl} + y_{illl} + y_{jlll} + y_{klll})\end{aligned} \tag{6.63}$$

After the coefficients of the regression equation have been derived, the statistical analysis is necessary, i.e. should be tested for goodness of fit the equation and response values predicted by the equation bound into the confidence intervals. In experimentation following simplex-lattice designs there are no degrees of freedom to test the equation for adequacy because the designs are saturated. Thus, to test the adequacy the experiments are run at additional, so-called *test points*. The number of the control points and their coordinates are conditioned by the problem formulation and experiment nature. Besides, the control points are sought so as to improve the model in case of inadequacy.

The accuracy of response prediction is dissimilar at different points of the simplex. The variance of predicted response $s_{\hat{y}}^2$ is obtainable from the error accumulation law. We illustrate this by an example of the second-degree polynomial for the ternary system. In our reasoning we assume that (1) x_i can be observed without errors, (2) the replication variance s_y^2 is similar at all the design points, and (3) response values are the averages of n_i and n_{ij} replicate observations at appropriate points of the simplex. Then the variances of $\bar{y}_i$ and $\bar{y}_{ij}$ will be

$$s_{\bar{y}_i}^2 = s_y^2/n_i \tag{6.64}$$

and

$$s_{\bar{y}_{ij}}^2 = s_y^2/n_{ij} \tag{6.65}$$

In the reduced polynomial

$$\hat{y} = \beta_1 x_1 + \beta_2 x_2 + \beta_3 x_3 + \beta_{12} x_1 x_2 + \beta_{13} x_1 x_3 + \beta_{23} x_2 x_3$$

we replace coefficients by their expressions in terms of responses

$$\beta_i = \bar{y}_i, \qquad \beta_{ij} = 4\bar{y}_{ij} - 2\bar{y}_i - 2\bar{y}_j$$

We obtain then

$$\begin{aligned} \hat{y} &= \bar{y}_1 x_1 + \bar{y}_2 x_2 + \bar{y}_3 x_3 + (4\bar{y}_{12} - 2\bar{y}_1 - 2\bar{y}_2) x_1 x_2 \\ &\quad + (4\bar{y}_{13} - 2\bar{y}_1 - 2\bar{y}_3) x_1 x_3 + (4\bar{y}_{23} - 2\bar{y}_2 - 2\bar{y}_3) x_2 x_3 \\ &= \bar{y}_1 (x_1 - 2x_1 x_2 - 2x_1 x_3) + \bar{y}_2 (x_2 - 2x_1 x_2 - 2x_2 x_3) \\ &\quad + \bar{y}_3 (x_3 - 2x_1 x_3 - 2x_2 x_3) + 4\bar{y}_{12} x_1 x_2 + 4\bar{y}_{13} x_1 x_3 + 4\bar{y}_{23} x_2 x_3 \end{aligned} \tag{6.66}$$

Using the condition $x_1 + x_2 + x_3 = 1$ we transform the coefficients at $\bar{y}_i$

$$\begin{aligned} x_1 - 2x_1 x_2 - 2x_1 x_3 &= x_1 - 2x_1(x_2 + x_3) \\ &= x_1 - 2x_1(1 - x_1) = x_1(2x_1 - 1) \text{ and so on.} \end{aligned} \tag{6.67}$$

Thus

$$\begin{aligned} \hat{y} &= x_1(2x_1 - 1)\bar{y}_1 + x_2(2x_2 - 1)\bar{y}_2 \\ &\quad + x_3(2x_3 - 1)\bar{y}_3 + 4x_1 x_2 \bar{y}_{12} + 4x_1 x_3 \bar{y}_{13} \\ &\quad + 4x_2 x_3 \bar{y}_{23} \end{aligned} \tag{6.68}$$

Introducing the designation

$$a_i = x_i(2x_i - 1), \quad a_{ij} = 4x_i x_j \tag{6.69}$$

and using Eqs. (6.64) and (6.65) gives the expression for the variance $s_{\hat{y}}^2$

$$s_{\hat{y}}^2 = s_y^2 \left(\sum_{1 \leq i \leq q} \frac{a_i^2}{n_i} \sum_{1 \leq i < j \leq q} \frac{a_{ij}^2}{n_{ij}} \right) \tag{6.70}$$

For incomplete-third, third- and fourth-degree polynomials the relationships are derived in much the same way. So, for the incomplete-third-degree polynomial

$$s_{\hat{y}}^2 = s_y^2 \left[\sum_{1 \leq i \leq q} \frac{b_i^2}{n_i} \sum_{1 \leq i < j \leq q} \frac{b_{ij}^2}{n_{ij}} + \sum_{1 \leq i < j < k \leq q} \frac{b_{ijk}^2}{n_{ijk}} \right] \tag{6.71}$$

where

$$b_i = \frac{1}{2}\, x_i(6x_i^2-2x_i+1)-3 \sum_{j=1}^{q} x_j^2 \tag{6.72}$$

$$b_{ij} = 4x_i x_j(3x_i+3x_j-2) \tag{6.73}$$

$$b_{ijk} = 27x_i x_j x_k \tag{6.74}$$

For the third-degree polynomial

$$s_{\hat{y}}^2 = s_y^2\Bigg[\sum_{1\le i\le q}\frac{c_i^2}{n_i}+\sum_{1\le i<j\le q}\frac{c_{iij}^2}{n_{iij}} + \sum_{1\le i<j\le q} + \frac{c_{ijj}^2}{n_{ijj}} + \sum_{1\le i<j<k\le q}\frac{c_{ijk}^2}{n_{ijk}}\Bigg] \tag{6.75}$$

where

$$c_i = \frac{1}{2}\, x_i(3x_i-1)\,(3x_i-2) \tag{6.76}$$

$$c_{iij} = 9/2x_i x_j(3x_i-1) \tag{6.77}$$

$$c_{ijj} = \frac{9}{2}\, x_i x_j(3x_j-1) \tag{6.78}$$

$$c_{ijk} = 27x_i x_j x_k \tag{6.79}$$

For the fourth-degree polynomial

$$s_{\hat{y}}^2 = s_y^2\Bigg[\sum_{1\le i\le q}\frac{d_i^2}{n_i}+\sum_{1\le i<j\le q}\frac{d_{ij}^2}{n_{ij}}+\sum_{1\le i<j\le q}\frac{d_{iiij}^2}{n_{iiij}} + \sum_{1\le i<j\le q}\frac{d_{ijjj}^2}{n_{ijjj}}+\sum_{1\le i<j<k\le q}\frac{d_{iijk}^2}{n_{iijk}}+\sum_{1\le i<j<k\le q}\frac{d_{ijjk}^2}{n_{ijjk}} + \sum_{1\le i<j<k\le q}\frac{d_{ijkk}^2}{n_{ijkk}}+\sum_{1\le i<j<k<l\le q}\frac{d_{ijkl}^2}{n_{ijkl}}\Bigg] \tag{6.80}$$

where

$$d_i = 1/6x_i(4x_i-1)\,(4x_i-2)\,(4x_i-3)$$

$$d_{ij} = 4x_i x_j(4x_i-1)\,(4x_j-1)$$

$$d_{ijjj} = 8/3x_i x_j(4x_j-1)\,(4x_j-2)$$

$$d_{iiij} = 8/3x_i x_j(4x_i-1)\,(4x_i-2)$$

$$d_{iijk} = 32x_i x_j x_k(4x_i-1)$$

$$d_{ijjk} = 32x_i x_j x_k(4x_j-1)$$

$$d_{ijkk} = 32x_i x_j x_k(4x_k-1)$$

If the number of replicate observations at all the points of the design is equal, i.e. $n_i = n_{ij} = n$, then all the relations for $s_{\hat{y}}^2$ will take the form

$$s_{\hat{y}}^2 = s_y^2 \frac{\xi}{n} \tag{6.88}$$

where for the second-degree polynomial

$$\xi = \sum_{1 \leq i \leq q} a_i^2 + \sum_{1 \leq i < j \leq q} a_{ij}^2 \tag{6.89}$$

for the incomplete-third-degree polynomial

$$\xi = \sum_{1 \leq i \leq q} b_i^2 + \sum_{1 \leq i < j \leq q} b_{ij}^2 + \sum_{1 \leq i < j < k \leq q} b_{ijk}^2 \tag{6.90}$$

for the third-degree polynomial

$$\xi = \sum_{1 \leq i \leq q} c_i^2 + \sum_{1 \leq i < j \leq q} c_{iij}^2 + \sum_{1 \leq i < j \leq q} c_{ijj}^2 + \sum_{1 \leq i < j < k \leq q} c_{ijk}^2 + \sum_{1 \leq i < j \leq q} c_{ij}^2 \tag{6.91}$$

and for the fourth-degree polynomial

$$\xi = \sum_{1 \leq i \leq q} d_i^2 + \sum_{1 \leq i < j \leq q} d_{ij}^2 + \sum_{1 \leq i < j \leq q} d_{iiij}^2 + \sum_{1 \leq i < j \leq q} d_{ijjj}^2 + \sum_{1 \leq i < j < k \leq q} d_{iijk}^2 + \sum_{1 \leq i < j < k \leq q} d_{ijjk}^2 + \sum_{1 \leq i < j < k \leq q} d_{ijkk}^2 + \sum_{1 \leq i < j < k < l \leq q} d_{ijkl}^2 \tag{6.92}$$

As in Eqs. (6.89) through (6.92), ξ is only dependent on the mixture composition, for ternary systems the lines of constant ξ may be plotted beforehand for polynomials of various degrees (Figs. 6.5 and 6.6). Given the replication variance and the number of parallel observations n, the error for predicted values of the response is readily calculated at any point of composition-property diagram using an appropriate value of ξ taken from the curve. Adequacy is tested at each control point, for which purpose the following statistic is build

$$t = \frac{\Delta y}{\sqrt{s_{\bar{y}}^2 + s_{\hat{y}}^2}} = \frac{\Delta y \sqrt{n}}{s_y^2 \sqrt{1+\xi}} \tag{6.93}$$

where $\Delta y = |\bar{y}_{exp} - y_{theor}|$, n is the number of parallel observations at every point. The t statistic has the Student distribution and it is compared with the tabulated value of $t_{\alpha/l}(v)$ at a level of significance α, where l is the number of control points and v is the number of degrees of freedom for the replication variance.

The null hypothesis that the equation is adequate is accepted if $t_{exp} < t_{table}$ for all the control points. Adequacy can also be tested at several points using the χ^2 test [39].

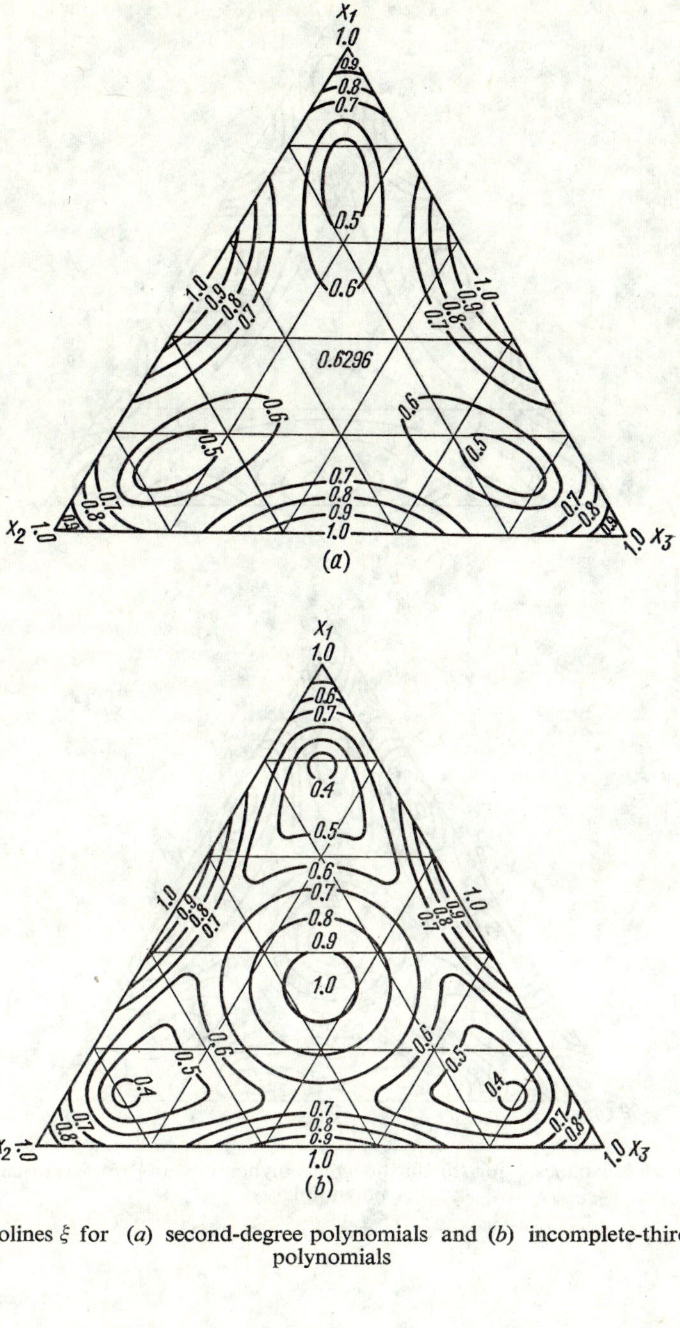

Fig. 6.5. Isolines ξ for (*a*) second-degree polynomials and (*b*) incomplete-third-degree polynomials

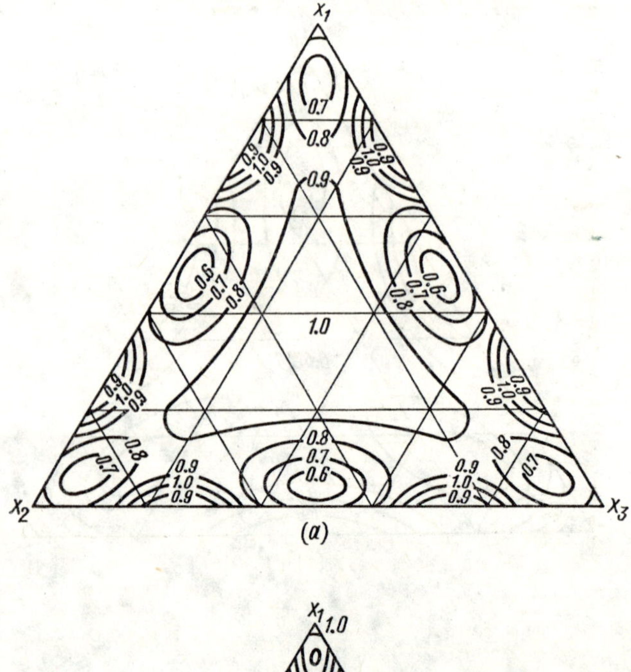

Fig. 6.6. Isolines ξ for (*a*) third-degree polynomials and (*b*) fourth-degree polynomials

The confidence interval for response values is

$$\hat{y}-\Delta \leqslant y \leqslant \hat{y}+\Delta \tag{6.94}$$

$$\Delta = t_{\alpha/k,\,\nu} s_{\hat{y}} \tag{6.95}$$

where k is the number of polynomial coefficients determined. By Eq. (6.88)

$$\Delta = t_{\alpha/k,\,\nu} \frac{s_y}{\sqrt{n}} \xi^{1/2} \tag{6.96}$$

With ternary systems, in building the confidence intervals, recourse may be made to contour charts (see Figs. 6.5 and 6.6) substituting for ξ in the isolines the value

$$t_{\alpha/k,\,\nu}\, s_y \xi^{1/2}/\sqrt{n}$$

Example 6.1. To study variation of reactivity and porosity of coke with charge composition, coals of four process groups were taken designated as x_1, x_2, x_3, and x_4. Experiments were conducted on a laboratory pilot plant. The characteristic of coke reactivity, y_{rct}, was the rate constant of the reaction $C + CO_2 = 2CO$, determined at 1 000 °C. The coke porosity, y_{por}, was given by the ratio of the true and apparent densities. Assuming that the response surfaces of physical and chemical characteristics of the mixtures at hand can be approximated by polynomials of not very high degree, we shall seek the regression equation in the form of the second-degree polynomial.

Solution. In solving the problem, the simplex-lattice design {4, 2} has been utilized. The second-order design matrix for the quaternary system and experimental results (each observation was repeated twice) are summerized in Table 6.5.

Table 6.5. Design Matrix and Measurements

N	x_1	x_2	x_3	x_4	Response	y_{rct}	y_{por}
1	1	0	0	0	y_1	1.48	54.0
2	0	1	0	0	y_2	0.32	55.2
3	0	0	1	0	y_3	0.50	43.3
4	0	0	0	1	y_4	0.53	45.3
5	0.5	0.5	0	0	y_{12}	0.63	53.1
6	0.5	0	0.5	0	y_{13}	0.92	48.0
7	0.5	0	0	0.5	y_{14}	1.08	49.0
8	0	0.5	0.5	0	y_{23}	0.39	46.3
9	0	0.5	0	0.5	y_{24}	0.38	47.1
10	0	0	0.5	0.5	y_{34}	0.54	44.0

Coefficients of the equations are derived using Eq. (6.26), so, for reactivity *vs.* composition variation,

$$\beta_1 = 1.48 \qquad \beta_2 = 0.32 \qquad \beta_3 = 0.50 \qquad \beta_4 = 0.53$$

$$\beta_{12} = 4y_{12} - 2y_1 - 2y_2 = 4\times0.63 - 2\times1.48 - 2\times0.32 = -1.08$$

$$\beta_{13} = 4y_{13} - 2y_1 - 2y_3 = 4\times0.92 - 2\times1.48 - 2\times0.50 = -0.22$$

$$\beta_{14} = 4y_{14} - 2y_1 - 2y_4 = 4\times1.08 - 2\times1.48 - 2\times0.53 = 0.30$$

$$\beta_{23} = 4y_{23}-2y_2-2y_3 = 4\times0.39-2\times0.32-2\times0.5 = -0.08$$
$$\beta_{24} = 4y_{24}-2y_2-2y_4 = 4\times0.38-2\times0.32-2\times0.53 = -0.18$$
$$\beta_{34} = 4y_{34}-2y_3-2y_4 = 4\times0.54-2\times0.50-2\times0.53 = 0.1$$

Thus the second-degree polynomial for the reactivity of the quaternary mixture has the form

$$\hat{y}_{\text{rct}} = 1.48x_1+0.32x_2+0.50x_3+0.53x_4-1.08x_1x_2 -0.22x_1x_3+0.3x_1x_4-0.08x_2x_3-0.18x_2x_4 +0.1x_3x_4. \quad (6.97)$$

For the porosity

$$\beta_1 = 54.0, \quad \beta_2 = 55.2, \quad \beta_3 = 43.3, \quad \beta_4 = 45.3$$
$$\beta_{12} = -6.4, \quad \beta_{13} = -2.6, \quad \beta_{14} = -2.6$$
$$\beta_{23} = -11.8, \quad \beta_{24} = -12.6, \quad \beta_{34} = -1.2$$

and the regression equation will be

$$\hat{y}_{\text{por}} = 54.0x_1+55.2x_2+43.3x_3+45.3x_4-6.4x_1x_2 -2.6x_1x_3-2.6x_1x_4-11.8x_2x_3-12.6x_2x_4-1.2x_3x_4 \quad (6.98)$$

N	Response	y_{rct}	$\hat{y}_{\text{rct}}$	Δy_{rct}	y_{por}	$\hat{y}_{\text{por}}$	Δy_{por}	ξ	t_{rct}	t_{por}
1	y_{1112}	0.77	0.99	0.22	53.5	53.1	0.4	0.72	3.16	0.3
2	y_{1113}	1.15	1.19	0.04	51.0	50.8	0.2	0.72	0.575	0.15
3	y_{1114}	1.25	1.20	0.05	50.3	51.3	1.0	0.72	0.72	0.72
4	y_{2223}	0.31	0.34	0.03	49.0	50.1	1.1	0.72	0.43	0.80
5	y_{2224}	0.39	0.34	0.05	52.3	50.4	1.9	0.72	0.72	1.37
6	y_{3334}	0.55	0.52	0.03	45.0	43.6	1.4	0.72	0.43	0.5
7	y_{1222}	0.35	0.41	0.06	57.2	53.8	2.6	0.72	0.86	0.7
8	y_{1333}	0.75	0.70	0.05	44.0	45.5	1.5	0.72	0.72	1.2
9	y_{1444}	0.94	0.82	0.12	48.9	47.0	1.9	0.72	1.72	0.4
10	y_{2333}	0.51	0.44	0.07	43.4	44.0	0.7	0.72	1.0	0.75
11	y_{2444}	0.36	0.45	0.09	46.3	45.4	0.9	0.72	0.74	0.8
12	y_{3444}	0.57	0.50	0.07	44.5	44.6	0.1	0.72	1.34	0.6
13	y_{1123}	0.82	0.77	0.05	52.4	49.8	2.6	0.59	0.74	1.95
14	y_{1124}	0.90	0.81	0.09	51.5	50.1	1.4	0.59	1.3	1.2
15	y_{1134}	1.17	1.00	0.17	47.0	48.4	1.4	0.59	2.54	0.3
16	y_{1223}	0.49	0.51	0.02	50.6	49.5	1.1	0.59	0.3	0.5
17	y_{1224}	0.52	0.53	0.01	48.0	49.8	1.8	0.59	0.15	0.4
18	y_{1334}	0.76	0.75	0.01	46.7	45.8	0.9	0.59	0.15	1.3
19	y_{2234}	0.44	0.40	0.04	48.4	46.5	1.9	0.59	0.6	0.7
20	y_{2334}	0.48	0.46	0.02	46.8	44.4	2.4	0.59	0.3	1.0
21	y_{1233}	0.58	0.63	0.05	45.7	46.7	1.0	0.59	0.74	1.4
22	y_{1244}	0.59	0.68	0.09	47.0	47.6	0.6	0.59	1.34	0.8
23	y_{1344}	0.78	0.80	0.02	49.0	46.4	2.6	0.59	0.3	1.2
24	y_{2344}	0.42	0.44	0.02	46.0	44.8	1.2	0.59	0.3	0.8
25	y_{1234}	0.60	0.65	0.05	48.0	47.0	1.0	0.44	0.78	0.79

To test the adequacy of equations derived, 25 test points are used (see the table below). The coordinates of these are selected so that a fourth-degree polynomial (see Fig. 6.4*d*) can be built, if the regression equations (6.97) and (6.98) fail to fit adequately.

For each test point the *t*-statistic is constructed

$$t = \frac{\Delta y \sqrt{n}}{s_y \sqrt{1+\xi}}$$

The replication errors are: in coke reactivity, $s_{y,\,\text{rct}} = 0.075$; in porosity, $s_{y,\,\text{por}} = 1.5$. The number of degrees of freedom was $\nu_y = 35$. The number of replicate observations at each point is $n = 2$. For each test point the quantity

$$\xi = \sum_{1 \leq i \leq q} a_i^2 + \sum_{1 \leq i < j \leq q} a_{ij}^2$$

is determined, where $a_i = x_i(2x_i - 1)$; $a_{ij} = 4x_i x_j$.

At a significance level $\alpha = 0.05$, and $\nu_y = 35$, $t_{0.05}(35) = 3.6$. Thus, both equations are found to adequately fit the experiment.

The equations derived enable the reactivity and porosity of coke to be determined for any charge composition of coals belonging to the above groups. Using the estimated regression equations, Eqs. (6.97) and (6.98), the lines of constant value are projected on the sections $x_1 = 0.0$ and $x_1 = 0.3$ (Fig. 6.7).

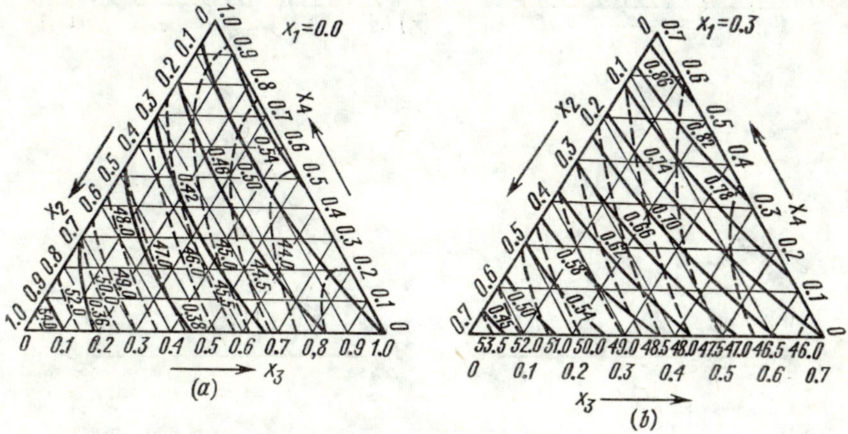

Fig. 6.7. Isolines for the properties (*a*) $x_1 = 0$, (*b*) $x_1 = 0.3$

Scheffe's simplex-lattice designs find their most advantageous use in the description of uniphase systems, for uniphase regions of complex system or where the property studied is conditioned by one phase only. Attempts at adopting the simplex-lattice method to plot property-composition variations over the whole of a multiphase system more often than not are a failure. Points of a simplex-lattice design may not coincide with critical points of the diagram, and the analytical description may overlook the areas of stepwise variation of properties. So, all the attempts at constructing the temperature dependence for the onset of crystallization to involve the whole eutectic type system Pb-Cd-Bi have not been a success, though the polynomials have been derived of second to fourth degree inclusive (Fig. 6.8*a* and *b*) [38].

When studying the dependence of properties on composition for the multiphase system one should take into account prior information on the structure of the system in hand. The liquidus surface in the eutectic-type system comprises three intersecting surfaces of primary crystallization of each phase. It has been suggested [38] to describe each of these surfaces analytically using simplex-lattice designs, and find then the lines of their

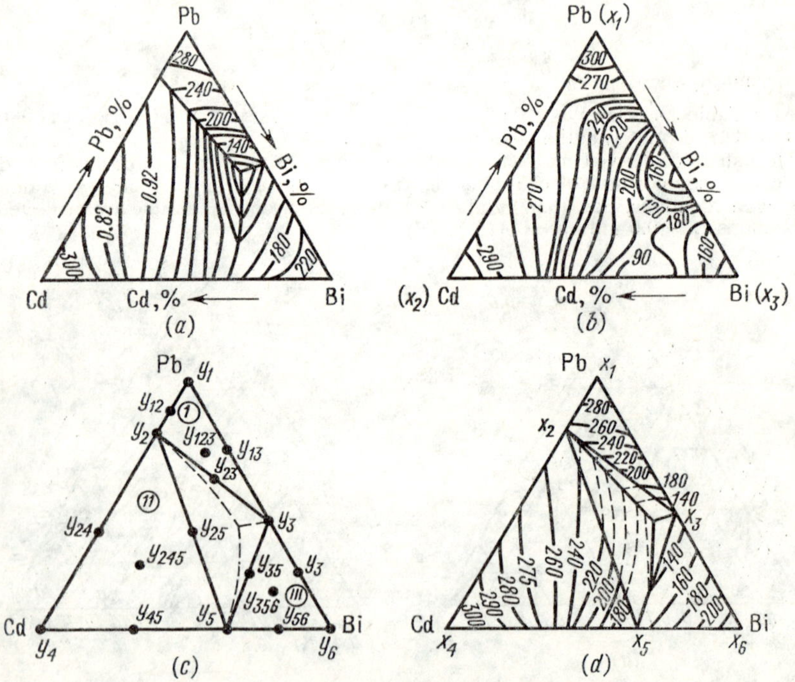

Fig. 6.8. Eutectic-type systems

intersection and the point where these lines intersect. Primary crystallization surfaces can be singled out using an auxiliary triangle whose verticles are the points of binary eutectics in binary constitution diagrams (Fig. 6.8*c*). The newly formed triangles, I, II, and III are viewed as starting.

Accordingly, for the Pb-Cd-Bi system discussed, inside every triangle an incomplete-cubic simplex-lattice design was realized (Table 6.6). From experimental results (Table 6.6) incomplete cubic polynomials were found:

triangle I

$$\hat{y} = 327x_1+248x_2+127x_3-46x_1x_2+4x_1x_3-30x_2x_3 + 108x_1x_2x_3 \tag{6.99}$$

Table 6.6. Design Matrix for the Incomplete-cubic Polynomials in Triangles I, II, and III [38]

N	Coded factors						Natural scale factors			Liquidus temperature, y, °C
	x_1	x_2	x_3	x_4	x_5	x_6	Pb	Cd	Bi	
1	1	0	0	0	0	0	100	0	0	$y_1 = 327$
2	0	1	0	0	0	0	82	18	0	$y_2 = 248$
3	0	0	1	0	0	0	45	0	55	$y_3 = 127$
4	1/2	1/2	0	0	0	0	91	9	0	$y_{12} = 276$
5	1/2	0	1/2	0	0	0	72.5	0	27.5	$y_{13} = 228$
6	0	1/2	1/2	0	0	0	63.5	9	27.5	$y_{23} = 180$
7	1/3	1/3	1/3	0	0	0	75.7	6	18.3	$y_{123} = 230$
8	0	0	0	1	0	0	0	100	0	$y_4 = 321$
9	0	0	0	0	1	0	0	40	60	$y_5 = 149$
10	0	1/2	0	1/2	0	0	41	59	0	$y_{24} = 278$
11	0	1/2	0	0	1/2	0	41	29	30	$y_{25} = 220$
12	0	0	0	1/2	1/2	0	0	70	30	$y_{45} = 254$
13	0	1/3	0	1/3	1/3	0	27	53	20	$y_{245} = 257$
14	0	0	0	0	0	1	0	0	100	$y_6 = 271$
15	0	0	1/2	0	1/2	0	22.5	20	57.5	$y_{35} = 127$
16	0	0	1/2	0	0	1/2	22.5	0	77.5	$y_{36} = 204$
17	0	0	0	0	1/2	1/2	0	20	80	$y_{56} = 185$
18	0	0	1/3	0	1/3	1/3	15	13.3	71.7	$y_{356} = 160$

triangle II

$$\hat{y} = 248x_2+321x_4+149x_5-26x_2x_4+86x_2x_5+76x_4x_5 +69x_2x_4x_5 \tag{6.100}$$

triangle III

$$\hat{y} = 127x_3+149x_5+271x_6-44x_3x_5+20x_3x_6+100x_5x_6 +39x_3x_5x_6 \tag{6.101}$$

where y is the liquidus temperature, °C; x_1 is Pb; x_2 is the alloy of Pb with 18% Cd; x_3 is the alloy of Bi with 45% Pb; x_4 is Cd; x_5 is the alloy of Bi with 40% Cd; x_6 is Bi.

All the polynomials were found adequate. Then a graphic extrapolation (Fig. 6.8*d*) was carried out that enabled crystallization lines for binary eutectics in the ternary alloys and the coordinates of the ternary eutectic point to be determined rather accurately.

In simplex-lattice designs, to obtain low degree polynomials coefficients are derived from experiment which for the most part represents not all the components. Surely, the results of experiments with straight components are not rich in information on the properties of the system under study.

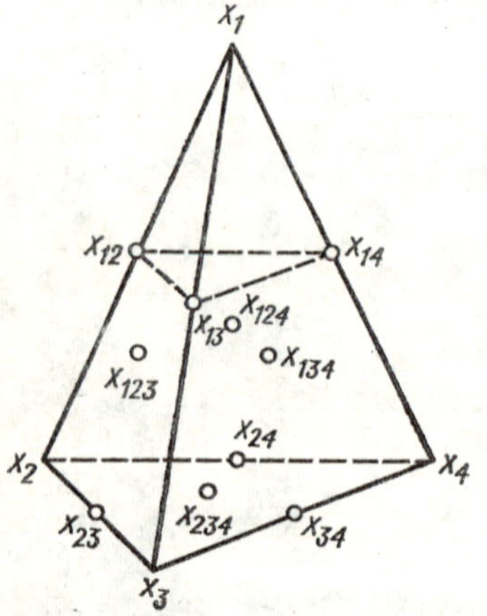

Fig. 6.9. Lambrakis's design

For systems with number of components $q \geqslant 4$, Lambrakis's designs can be used [40], which are essentially conventional Scheffe's simplex lattices where instead of straight components the experiments are run at q points with the coordinates

$$x_1 = x_2 = \ldots = x_q = \frac{1}{q-1}$$

For example, in deriving the second-degree polynomial for a quaternary system, the four points with the coordinates $x_1 = x_2 = x_3 = x_4 = 1/3$ (Fig. 6.9) should be substituted for the four points with the coordinates $x_1 = x_2 = x_3 = x_4 = 1$ (Fig. 6.4*a*). Accordingly, Lambrakis's design includes instead of four observations at vertices of a tetrahedron four observations at the centers of triangles forming that tetrahedron (x_{123}, x_{124}, x_{134}, and x_{234}), and six observations at the centers of the tetrahedron faces (x_{12}, x_{13}, x_{14}, x_{23}, and x_{34}).

6.3. Simplex-Centroid Design

Scheffe's simplex-centroid designs [41] contain $2^q - 1$ points, q of which fall on straight components, C_q^2 on binary mixtures, C_q^3 on ternary mixtures, and so forth, and one observation on a q-component mixture. Simplex-centroid designs, consist of the points whose coordinates are $(1, 0, \ldots, 0)$, $(1/2, 1/2, 0, \ldots, 0)$, $\ldots$, $(1/q, 1/q, \ldots, 1/q)$, and all points which can be obtained from these by permutations of coordinates. Thus, the design contains a point at the center (centroid) of the simplex and the centroids of all the component simplexes of lesser dimension, its proper faces.

Polynomials obtained from simplex-centroid designs contain as many coefficients as there are points in the design, and for the q-component mixture they have the form

$$\hat{y} = \sum_{1 \leqslant i \leqslant q} \beta_i x_i + \sum_{1 \leqslant i < j \leqslant q} \beta_{ij} x_i x_j + \sum_{1 \leqslant i < j < k \leqslant q} + \beta_{ijk} x_i x_j x_k + \beta_{12 \ldots q} x_1 x_2 \ldots x_q \qquad (6.102)$$

For a given number of components q, there exists only one simplex-centroid design. The simplex-lattice design intended to derive the polynomial of incomplete third degree is the simplex-centroid design for ternary systems (see Fig. 6.4*b*).

By way of example, we build the simplex-centroid design for a quaternary system ($q = 4$). The number of observations in the design is $N = 2^q - 1$

$= 2^4 - 1 = 15$. The arrangement of points over the concentration tetrahedron is shown in Fig. 6.4*c*, and the respective simplex-centroid design is shown in Table 6.7.

Table 6.7. Matrix of Simplex-Centroid Design for a Quaternary System

N	x_1	x_2	x_3	x_4	y	N	x_1	x_2	x_3	x_4	y
1	1	0	0	0	y_1	9	0	1/2	0	1/2	y_{24}
2	0	1	0	0	y_2	10	0	0	1/2	1/2	y_{34}
3	0	0	1	0	y_3	11	1/3	1/3	1/3	0	y_{123}
4	0	0	0	1	y_4	12	1/3	1/3	0	1/3	y_{124}
5	1/2	1/2	0	0	y_{12}	13	1/3	0	1/3	1/3	y_{134}
6	1/2	0	1/2	0	y_{13}	14	0	1/3	1/3	1/3	y_{234}
7	1/2	0	0	1/2	y_{14}	15	1/4	1/4	1/4	1/4	y_{1234}
8	0	1/2	1/2	0	y_{23}						

The polynomial of Eq. (6.102) for $q = 4$ includes 15 terms and has the form

$$\begin{aligned}\hat{y} = {} & \beta_1 x_1 + \beta_2 x_2 + \beta_3 x_3 + \beta_4 x_4 + \beta_{12} x_1 x_2 + \beta_{13} x_1 x_3 \\ & + \beta_{14} x_1 x_4 + \beta_{23} x_2 x_3 + \beta_{24} x_2 x_4 + \beta_{34} x_3 x_4 \\ & + \beta_{123} x_1 x_2 x_3 + \beta_{124} x_1 x_2 x_4 + \beta_{134} x_1 x_3 x_4 \\ & + \beta_{234} x_2 x_3 x_4 + \beta_{1234} x_1 x_2 x_3 x_4 \end{aligned} \qquad (6.103)$$

Making recourse to the saturation property of the design and substituting in succession the coordinates of experimental points 1 through 15 into the polynomial of Eq. (6.103), we determine the polynomial coefficients

$$\beta_1 = y_1 \quad \beta_2 = y_2 \quad \beta_3 = y_3 \quad \beta_4 = y_4 \qquad (6.104)$$

$$\begin{aligned} \beta_{12} &= 4y_{12} - 2y_1 - 2y_2 \\ \beta_{13} &= 4y_{13} - 2y_1 - 2y_3 \\ \beta_{14} &= 4y_{14} - 2y_1 - 2y_4 \\ \beta_{23} &= 4y_{23} - 2y_2 - 2y_3 \\ \beta_{24} &= 4y_{24} - 2y_2 - 2y_4 \\ \beta_{34} &= 4y_{34} - 2y_3 - 2y_4 \end{aligned} \qquad (6.105)$$

$$\begin{aligned} \beta_{123} &= 27y_{123} - 12(y_{12} + y_{13} + y_{23}) + 3(y_1 + y_2 + y_3) \\ \beta_{124} &= 27y_{124} - 12(y_{12} + y_{14} + y_{24}) + 3(y_1 + y_2 + y_4) \end{aligned} \qquad (6.106)$$

$$\beta_{134} = 27y_{134}-12(y_{13}+y_{14}+y_{34})+3(y_1+y_3+y_4)$$
$$\beta_{234} = 27y_{234}-12(y_{23}+y_{24}+y_{34})+3(y_2+y_3+y_4)$$
$$\beta_{1234} = 256y_{1234}-108(y_{123}+y_{124}+y_{134}+y_{234}) +32(y_{12}+y_{13}+y_{14}+y_{23}+y_{24}+y_{34}) -4(y_1+y_2+y_3+y_4) \quad (6.107)$$

In a similar manner, we have for the polynomial of Eq. (6.102) for the q-component mixture

$$\beta_i = y_i \quad (6.108)$$

$$\beta_{ij} = 4y_{ij}-2y_i-2y_j = 2[2y_{ij}-(y_i+y_j)] \quad (6.109)$$

$$\beta_{ijk} = 27y_{ijk}-12(y_{ij}+y_{ik}+y_{jk})+3(y_i+y_j+y_k) = 3[9y_{ijk}-4(y_{ij}+y_{ik}+y_{jk})+(y_i+y_j+y_k)] \quad (6.110)$$

$$\beta_{ijkm} = 256y_{ijkm}-108(y_{ijk}+y_{ijm}+y_{ikm}+y_{jkm}) +32(y_{ij}+y_{ik}+y_{im}+y_{jk}+y_{jm}+y_{km}) -4(y_i+y_j+y_k+y_m) = 4[64y_{ijkm} -27(y_{ijk}+y_{ijm}+y_{ikm})+8(y_{ij}+y_{ik}+y_{im}+y_{jk} +y_{jm}+y_{km})-(y_i+y_j+y_k+y_m)] \quad (6.111)$$

In the general case, the formula for coefficients of regression equation obtained following the simplex-centroid design, takes the form [41]

$$\beta_{ij\,\ldots} = r\sum_{t=1}^{r}(-1)^{r-t}\,t^{r-1}\,S_t \quad (6.112)$$

where r is the number of indices at coefficients β_{ij}; S_t is the sum of experimental results for all the t-component mixtures taken in equal proportions $(1/t)$. For example, for the β_{ijk} coefficients we have $r = 3$ (i, j, k) and three sums

$$y_i+y_j+y_k = S_1 \qquad 1/t = 1 \quad (6.113)$$
$$y_{ij}+y_{jk}+y_{ik} = S_2 \qquad 1/t = 1/2 \quad (6.114)$$
$$y_{ijk} = S_3 \qquad 1/t = 1/3 \quad (6.115)$$

Thus

$$\beta_{ijk} = 3[(-1)^{3-1}1^{3-1}S_1+(-1)^{3-2}2^{3-1}S_2+(-1)^{3-3}3^{3-1}S_3] = 3[(y_i+y_j+y_k)-4(y_{ij}+y_{ik}+y_{jk})+9y_{ijk}] \quad (6.116)$$

Adequacy of a regression equation derived by the simplex-centroid design is tested and the confidence intervals of property values predicted by the equation are assigned in much the same way as in the case of simplex-lattice method.

Example 6.2 [42]. We seek to determine how the activity (y_1) and durability (y_2) of platinum catalyst supported by a non-porous metal carrier depend on the catalyst composition at 350 °C. The total mass of components was maintained constant from experiment to experiment. Taking it to be unity, we can write $\sum_{i=1}^{3} x_i = 1$, where x_1, the component of Pt/Al_2O_3, was a ground, dead reforming catalyst; x_2 and x_3, also components, were inorganic oxides of metals belonging to Groups II and III of the periodic table.

Solution. The simplex-centroid design for $q = 3$ is applied. The design matrix and experimental results are arrayed in Table 6.8.

Table 6.8. Simplex-centroid Design for $q = 3$ and Experimental Results

N	x_1	x_2	x_3	y_1, %	y_2, %
1	1	0	0	97.4	62
2	0	1	0	3.0	73
3	0	0	1	4.7	47
4	0.5	0.5	0	70.0	64
5	0.5	0	0.5	66.0	55
6	0	0.5	0.5	6.8	72
7	0.333	0.333	0.333	95.4	67

Using Eqs. (6.39) and (6.40), the coefficients of regression equations are derived for both the catalyst activity

$$\hat{y}_1 = 97.4x_1+3.0x_2+4.7x_3+79.3x_1x_2+59.9x_1x_3 +11.8x_2x_3+1\,175.35x_1x_2x_3 \qquad (6.117)$$

and durability

$$\hat{y}_2 = 62x_1+73x_2+47x_3-14x_1x_2+2x_1x_3+48x_2x_3 +63x_1x_2x_3 \qquad (6.118)$$

The replication error in measuring the catalyst activity is $s_{y1} = 3.24$, and durability $s_{y2} = 2.37$. The adequacy of the regression equations (6.117) and (6.118) is tested using the Student t at test points 8, 9, and 10 (see the table below).

N	x_1	x_2	x_3	y_1	$\hat{y}_1$	y_2	$\hat{y}_2$
8	0.333	0.667	0	46	52	72	66
9	0.667	0.333	0	96	84	63	70
10	0.580	0.320	0.097	91	98	62	65

At all the test points, the values of t-test are less than the tabulated value at the significance level $\alpha = 0.05$. Figure 6.10 presents the lines of constant catalyst activity y_1 and durability y_2 plotted from Eq (6.117) and (6.118). The greatest activity of the catalyst corresponds to the area where the values of the component $x_1 > 0.4$. The durability of 65 per cent appears quite satisfactory. Of greatest interest are the points lying where the equal yield curves for $\hat{y}_2 = 65\%$ and $\hat{y}_1 = 100\%$ intersect. Trial 10 (see the table and Fig. 6.10) executed within the specified area gave a good agreement, within experimental error, of experimental results with theory.

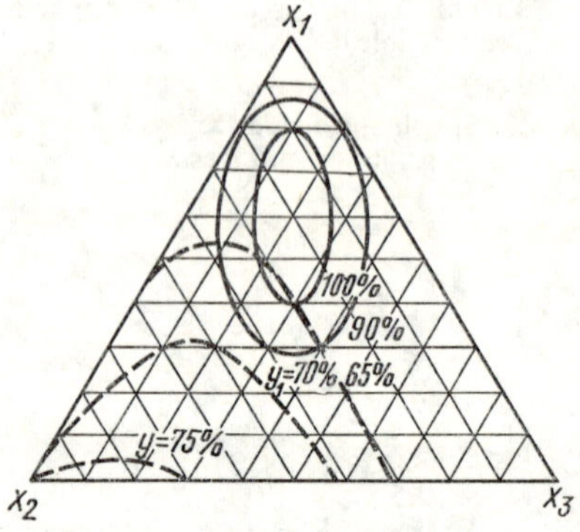

Fig. 6.10. Isolines for y_1 (—) and y_2 (– – –)

6.4. Design of Experiment in Local Regions of Diagrams

In studying the composition-property diagrams of q-component mixtures it is often necessary to explore the composition-dependence of a property not over the whole range of variation of component concentration $0 \leqslant x_i \leqslant 1$, but within a local interval of the diagram

$$0 \leqslant a_i \leqslant x_i \leqslant b_i \leqslant 1 \qquad i = 1, 2, \ldots, q$$

1. *The region studied is a simplex.* The local area of interest in the diagram may be an irregular simplex with unknown vertex coordinates $A_1(x_1^{(1)}, x_2^{(1)}, \ldots, x_q^{(1)})$, $A_2(x_1^{(2)}, x_2^{(2)}, \ldots, x_q^{(2)})$, $\ldots$, $A_q(x_1^{(q)}, x_2^{(q)}, \ldots, x_q^{(q)})$.

So that the designs used to study complete diagrams may be applied to this case, a renormalization is performed and compositions at vertices A_j ($j = 1, 2, \ldots, q$) are taken to be independent pseudocomponents so that for all the range of the local simplex the condition be met

$$\textstyle\sum_{i=1}^{q} z_i = 1 \tag{6.119}$$

The experimental design is in the pseudocomponent coordinates. All the designs descussed earlier can be build in the new variables $z_1, z_2, \ldots, z_q$ that satisfy the condition of Eq. (6.119). To conduct the experiments it is required to convert the pseudocomponents z_i into the initial components x_i. For an uth design point this convertion is by the formula

$$x_i^{(u)} = x_i^{(1)} + z_2^{(u)}(x_i^{(2)} - x_i^{(1)}) + z_3^{(u)}(x_i^{(3)} - x_i^{(1)}) + \ldots + z_q^{(u)}(x_i^{(q)} - x_i^{(1)}) \tag{6.120}$$

where x_i is the ith component content of the vertex $z_j(A_j)$.

After the design has been realized the coefficients of the regression equation are calculated in pseudocomponent coordinates

$$\hat{y} = f(z_1, z_2, \ldots, z_q) \qquad (6.121)$$

using earlier relationships for pertinent designs, and then adequacy of fit is tested. For Eq. (6.121) to be applied in practice, it is written in the initial coordinate system using an affine transformation of the form [37]

$$\begin{aligned} z_1 &= z_1^{(1)}+x_2(z_1^{(2)}-z_1^{(1)})+x_3(z_1^{(3)}-z_1^{(1)})+\ldots+x_q(z_1^{(q)}-z_1^{(1)}) \qquad (6.122)\\ z_2 &= z_2^{(1)}+x_2(z_2^{(2)}-z_2^{(1)})+x_3(z_2^{(3)}-z_2^{(1)})+\ldots+x_q(z_2^{(q)}-z_2^{(1)})\\ &\ldots\ldots\ldots\ldots\ldots\ldots\ldots\ldots\\ z_{q-1} &= z_{q-1}^{(1)}+x_2(z_{q-1}^{(2)}-z_{q-1}^{(1)})+x_3(z_{q-1}^{(3)}-z_{q-1}^{(1)})+\ldots\\ &\quad +x_q(z_{q-1}^{(q)}-z_{q-1}^{(1)}) \end{aligned}$$

Values of $z_i^{(j)}$ are calculated solving $(q-1)$ sets of equations below

$$\left.\begin{aligned} x_1^{(1)} z_1^{(1)}+x_2^{(1)} z_1^{(2)}+x_3^{(1)} z_1^{(3)}+\ldots+x_q^{(1)} z_1^{(q)} &= 1\\ x_1^{(2)} z_1^{(1)}+x_2^{(2)} z_1^{(2)}+x_3^{(2)} z_1^{(3)}+\ldots+x_q^{(2)} z_1^{(q)} &= 0\\ \ldots\ldots\ldots\ldots\ldots\ldots\ldots\ldots\\ x_1^{(q)} z_1^{(1)}+x_2^{(q)} z_1^{(2)}+x_3^{(q)} z_1^{(3)}+\ldots+x_q^{(q)} z_1^{(q)} &= 0 \end{aligned}\right\}$$

$$\left.\begin{aligned} x_1^{(1)} z_2^{(1)}+x_2^{(1)} z_2^{(2)}+x_3^{(1)} z_2^{(3)}+\ldots+x_q^{(1)} z_2^{(q)} &= 0\\ x_1^{(2)} z_2^{(1)}+x_2^{(2)} z_2^{(2)}+x_3^{(2)} z_2^{(3)}+\ldots+x_q^{(2)} z_2^{(q)} &= 1\\ \ldots\ldots\ldots\ldots\ldots\ldots\ldots\ldots\\ x_1^{(q)} z_2^{(1)}+x_2^{(q)} z_2^{(2)}+x_3^{(q)} z_2^{(3)}+\ldots+x_q^{(q)} z_2^{(q)} &= 0 \end{aligned}\right\}$$

$$\ldots\ldots\ldots\ldots\ldots\ldots\ldots\ldots \qquad (6.123)$$

$$\left.\begin{aligned} x_1^{(1)} z_{q-1}^{(1)}+x_2^{(1)} z_{q-1}^{(2)}+x_3^{(1)} z_{q-1}^{(3)}+\ldots+z_q^{(1)} z_{q-1}^{(q)} &= 0\\ x_1^{(2)} z_{q-1}^{(1)}+x_2^{(2)} z_{q-1}^{(2)}+x_3^{(2)} z_{q-1}^{(3)}+\ldots+x_q^{(2)} z_{q-1}^{(q)} &= 0\\ \ldots\ldots\ldots\ldots\ldots\ldots\ldots\ldots\\ x_1^{(q)} z_{q-1}^{(1)}+x_2^{(q)} z_{q-1}^{(2)}+x_3^{(q)} z_{q-1}^{(3)}+\ldots+x_q^{(q)} z_{q-1}^{(q)} &= 1 \end{aligned}\right\}$$

where $z_i^{(j)}$ is the pseudocomponent z_i content of vertices of the initial simplex; $x_i^{(j)}$ is the ith component content of vertices $z_j(A_j)$ $(j = 1, 2, \ldots, q)$. As such a coordinate conversion is only possible for equations in independent variables, the initial regression equation shall be transformed eliminating one variable, e. g. the last one, qth as follows

$$z_q = 1-\sum_{i=1}^{q-1} z_i \qquad (6.124)$$

Example 6.3 [43]. The boiling of the ternary mixture $H_2O—K_2HPO_4—K_2CO_3$ is studied. It is required to define the regression equation for the boiling point (y, °C) on the mixture composition (per cent). Not all the concentration triangle is covered, but only a

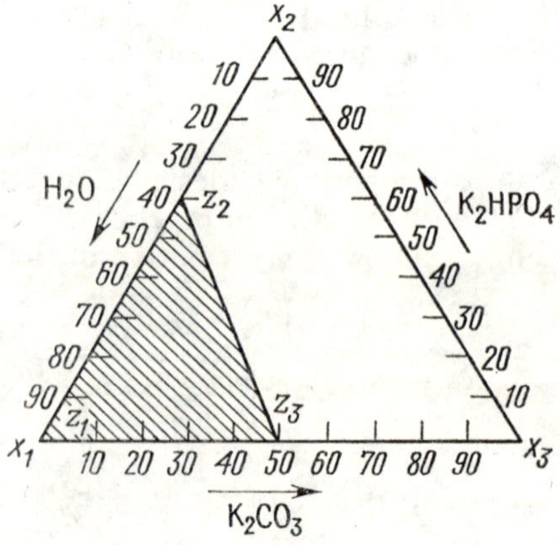

Fig. 6.11. Region of boiling point study in the system $K_2HPO_4 - K_2CO_3 - H_2O$

Table 6.9. Design Matrix and Experimental Results

N	z_1	z_2	z_3	x_1	x_2	x_3	$\bar{y}$
1	1	0	0	100	0	0	99.9
2	0	1	0	40	60	0	113.5
3	0	0	1	50	0	50	115.7
4	0.5	0.5	0	70	30	0	103.1
5	0.5	0	0.5	75	0	25	104.8
6	0	0.5	0.5	45	30	25	114.8
7	0.333	0.333	0.333	63.33	20	16.67	105.6
8	0.75	0.25	0	85	15	0	101.5
9	0.25	0.75	0	55	45	0	107.2
10	0.75	0	0.25	87.5	0	12.5	101.6
11	0.25	0	0.75	62.5	0	37.5	107.7
12	0	0.75	0.25	42.5	45	12.5	112.5
13	0	0.25	0.75	47.5	15	37.5	116.4
14	0.5	0.25	0.25	72.5	15	12.5	103.4
15	0.25	0.5	0.25	57.5	30	12.5	104.4
16	0.25	0.25	0.5	60	15	25	109.0
17	0.2	0.2	0.6	58	12	30	108.3*
18	0.5	0.125	0.375	73.75	7.5	18.75	103.3*
19	0.4	0.15	0.45	68.5	9	22.5	104.2*
20	0.3	0.175	0.525	63.25	10.5	26.25	106.2*

* Test points.

subarea of unsaturated solutions at 20 °C, i. e. a local section of the diagram in the form of a triangle with the vertices z_1 (100, 0, 0), z_2 (40, 60, 0), and z_3 (50, 0, 0, 50) (see Fig. 6.11).

Solution. To deduce the regression equation a simplex-lattice design is performed with pseudocomponents z_1, z_2, and z_3; and the content of initial components is then determined from Eq. (6.120). The regression equations of the second and incomplete third order are found to be inadequate. Using the property of composition of simplex-lattice designs, the design matrix is augmented further to yield fourth-order regression equations (Table 6.9). The experimental conditions are expressed in terms of pseudocomponents z_j and in the natural scale x (per cent). The mean values $\bar{y}$ of temperature measurements are determined from two replicate observations. The replication error is $s_y = 0.86$. The number of degrees of freedom for the error is $\nu_y = 20$.

The coefficients of the fourth-order regression equation are calculated by Eq. (6.46) using the property of saturated design matrix. The regression equation in pseudocomponent variables has the form

$$\begin{aligned}\hat{y} = {} & 99.88z_1 + 113.51z_2 + 115.69z_3 - 14.22z_1z_2 - 12.13z_1z_3 \\ & + 0.91z_2z_3 + 6.18z_1z_2(z_1 - z_2) + 10.12z_1z_3(z_1 - z_3) \\ & - 15.34z_2z_3(z_2 - z_3) + 6.90z_1z_2(z_1 - z_2)^2 - 17.61z_1z_3(z_1 - z_3)^2 \\ & + 6.32z_2z_3(z_2 - z_3)^2 + 1.07z_1^2z_2z_3 - 274.61z_1z_2^2z_3 \\ & + 142.21z_1z_2z_3^2 \qquad (6.125)\end{aligned}$$

The table below summarizes the results of the regression equation testing for adequacy of fit.

N	$\bar{y}$	$\hat{y}$	Δy	ξ	t
17	108.3	110.7	2.4	1.3	2.16
18	103.3	100.9	2.4	1.0	2.27
19	104.2	107.0	2.8	1.0	2.66
20	106.3	108.7	2.4	1.1	2.16

The tabulated value of Student's t is $t_{0.012,20} = 2.8$. Equation (6.125) is an adequate fit to the experiment. In Eq. (6.125), we convert from pseudocomponents z_j to initial variables x_j. For the problem in hand, the sets of simultaneous equations (6.123) take the form

$$\begin{aligned} 1z_1^{(1)} + 0z_1^{(2)} + 0z_1^{(3)} &= 1 \\ 0.4z_1^{(1)} + 0.6z_1^{(2)} + 0z_1^{(3)} &= 0 \\ 0.5z_1^{(1)} + 0z_1^{(2)} + 0.5z_1^{(3)} &= 0 \\ 1z_2^{(1)} + 0z_2^{(2)} + 0z_2^{(3)} &= 0 \\ 0.4z_2^{(1)} + 0.6z_2^{(2)} + 0z_2^{(3)} &= 1 \\ 0.5z_2^{(1)} + 0z_2^{(2)} + 0.5z_2^{(3)} &= 0 \end{aligned} \qquad (6.126)$$

Solving these we obtain

$$\begin{aligned} z_1^{(1)} &= 1 & z_2^{(1)} &= 0 \\ z_1^{(2)} &= 0.7 & z_2^{(2)} &= 1.7 \\ z_1^{(3)} &= -1 & z_2^{(3)} &= 0 \end{aligned} \qquad (6.127)$$

Substituting the above solutions into Eqs. (6.122) we arrive at formulas relating natural coordinates x_j to coordinates z_j

$$\begin{aligned} z_1 &= 1-1.7x_2-2x_3 \\ z_2 &= 1.7x_2 \\ z_3 &= 1-z_1-z_2 = 2x_3 \end{aligned} \qquad (6.128)$$

And, substituting Eq. (6.128) into Eq. (6.125), we get the regression equation in initial coordinates

$$\begin{aligned} \hat{y} = {} & 99.88+20.82x_2-7.63x_3+92.88x_2x_3-107.83x_2^2 \\ & +279.28x_3^2-1373.69x_2^2x_3-243.59x_2x_3^2 \\ & +2230.35x_2^2x_3^2+312.78x_2^3-965.12x_3^3 \\ & +2146.05x_2^3x_3-179.60x_2x_3^3-212.96x_2^4+1127.1x_3^4 \end{aligned} \qquad (6.129)$$

To use the regression equation more conveniently in Fig. 6.12, the isotherms are plotted.

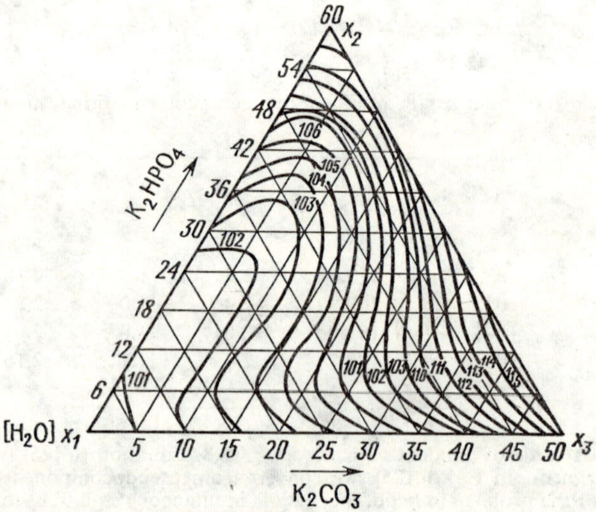

Fig. 6.12. Isotherms

2. *The space studied is a polyhedron.* With constraints on the component concentration variation, in the general case, the space studied forms a polyhedron. In building the experimental design we should somehow distribute the points over the polyhedron subject to the condition

$$0 \leq a_i \leq x_i \leq b_i \leq 1 \qquad (6.130)$$

This excludes the degenerate cases

$$\textstyle\sum_{i=1}^{q} a_i \geq 1 \quad \text{and} \quad \sum_{i=1}^{q} b_i \leq 1 \qquad (6.131)$$

It has been suggested in reference [44] to select the points as follows:

(1) All the possible combinations of the two levels a_i and b_i are put down for each and every component, but in each combination the content of one component is omitted. The number of these combinations for a q-component mixture is $q2^{q-1}$.

(2) Among all the combinations those are selected whose sum of components is less than unity and which meet the limitations of Eq. (6.130). Into the combinations selected the omitted components are added in amounts defined by the relationship $\sum_{i=1}^{q} x_i = 1$. Design points thus obtained and satisfying Eq. (6.130) lie at vertices of the bounding polyhedron.

(3) To the design points obtained are added center points (centroids) of two-, three-, ..., and $(q-1)$-dimensional faces of the polyhedron and its center point.

As the q increases the number of combinations of experimental conditions grows rapidly and outnumbers the coefficients of the second-degree polynomial normally utilized with these designs. Coefficients of the second-order regression equation are obtained using the least-squares method. As the design of experiment is unsaturated, the regression equation can be tested for adequacy of fit with F-ratio.

Now we consider the building of McLean and Anderson's design for the investigation and optimization of luminance of luminous mixtures, whose components are magnesium (x_1), soda (x_2), strontium nitrate (x_3), and binder (x_4) [44]. The mixture composition is subject to the following constraints

$$0.40 \leqslant x_1 \leqslant 0.60$$

$$0.10 \leqslant x_2 \leqslant 0.50$$

$$0.10 \leqslant x_3 \leqslant 0.50$$

$$0.03 \leqslant x_4 \leqslant 0.08$$

Table 6.10 summarizes all the possible combinations of mixture composition with one of the components missing.

Accordingly, 8 design points, i. e. polyhedron vertices (Fig. 6.13), are obtained. These points are to be supplemented by the coordinates of center

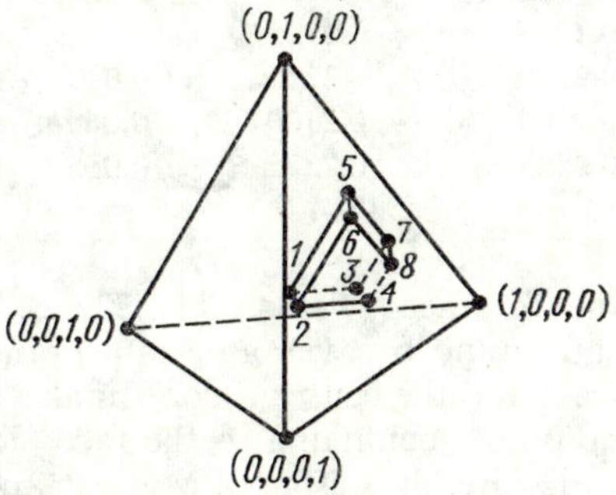

Fig. 6.13. The Design of McLean and Anderson

Table 6.1. Selection of Polyhedron Vertices in McLean and Anderson's Design

N	Component content x_1	x_2	x_3	x_4	N	Component content x_1	x_2	x_3	x_4
1	0.40	0.10	0.10	—	17 (1)	0.40	0.10	0.47*	0.03
2	0.40	0.10	0.50	—	18 (2)	0.40	0.10	0.42*	0.08
3	0.40	0.50	0.10	—	19	0.40	0.50	—	0.03
4	0.40	0.50	0.50	—	20	0.40	0.50	—	0.08
5	0.60	0.10	0.10	—	21 (3)	0.60	0.10	0.27*	0.03
6	0.60	0.10	0.50	—	22 (4)	0.60	0.10	0.22*	0.08
7	0.60	0.50	0.10	—	23	0.60	0.50	—	0.03
8	0.60	0.50	0.50	—	24	0.60	0.50	—	0.08
9 (5)	0.40	0.47*	0.10	0.03	25	—	0.10	0.10	0.03
10 (6)	0.40	0.42*	0.10	0.08	26	—	0.10	0.10	0.08
11	0.40	—	0.50	0.03	27	—	0.10	0.50	0.03
12	0.40	—	0.50	0.08	28	—	0.10	0.50	0.08
13 (7)	0.60	0.27*	0.10	0.03	29	—	0.50	0.10	0.03
14 (8)	0.60	0.22*	0.10	0.08	30	—	0.50	0.10	0.08
15	0.60	—	0.50	0.03	31	—	0.50	0.50	0.03
16	0.60	—	0.50	0.08	32	—	0.50	0.50	0.08

* Amount of component added.

Table 6.11. Selection of Face Center Points in McLean and Anderson's Design

N	Component content x_1	x_2	x_3	x_4	Points of face
(9)	0.50	0.10	0.345	0.055	(1), (2), (3), (4)
(10)	0.50	0.345	0.10	0.055	(5), (6), (7), (8)
(11)	0.40	0.2725	0.2725	0.055	(1), (2), (5), (6)
(12)	0.60	0.1725	0.1725	0.055	(3), (4), (7), (8)
(13)	0.50	0.2350	0.2350	0.030	(1), (3), (5), (7)
(14)	0.50	0.2100	0.2100	0.080	(2), (4), (6), (8)
(15)	0.50	0.2225	0.2225	0.055	Polyhedron center point

points of all the polyhedron faces (Table 6.11). The coordinates of the polyhedron center point are found by averaging appropriate coordinates of all the eight design vertices; and the centroid coordinates of faces by averaging the coordinates of the points belonging to the face (Table 6.11, Fig. 6.13).

For the quaternary mixture, the design of McLean and Anderson, together with the experimental results, is provided in Table 6.12.

Table 6.12. McLean-Anderson's Design for Quaternary Mixture

N	x_1	x_2	x_3	x_4	y	N	x_1	x_2	x_3	x_4	y
1	0.40	0.10	0.47	0.03	75	9	0.50	0.10	0.345	0.055	220
2	0.40	0.10	0.42	0.08	180	10	0.50	0.345	0.10	0.055	200
3	0.60	0.10	0.27	0.03	195	11	0.40	0.2725	0.2725	0.055	190
4	0.60	0.10	0.22	0.08	300	12	0.60	0.1725	0.1725	0.055	310
5	0.40	0.47	0.10	0.03	145	13	0.50	0.235	0.235	0.030	200
6	0.40	0.42	0.10	0.08	230	14	0.50	0.210	0.210	0.080	410
7	0.60	0.27	0.10	0.03	220	15	0.50	0.2225	0.2225	0.055	425
8	0.60	0.22	0.10	0.08	350						

The coefficients of the reduced second-degree polynomial are found by the method of least squares.

Here the regression equation will be

$$\hat{y} = -1.558x_1 - 2.851x_2 - 2.426x_3 + 14.372x_4 + 8.300x_1x_2 + 8.076x_1x_3 - 6.625x_1x_4 + 3.213x_2x_3 - 16.998x_2x_4 - 17.127x_3x_4 \qquad (6.132)$$

As the dependence of the property on components is described adequately by the second-order regression equation, the possibility presented itself to find optimal conditions through the use of non-linear programming. Subject to restrictions of Eq . (6.131), the conditions providing the maximum luminance are found to be:

$$\hat{y}_{max} = 397.48 \quad \text{at} \quad x_1 = 0.5233, \quad x_2 = 0.2299, \quad x_3 = 0.1608, \quad x_4 = 0.080$$

As the number of mixture components increases, the number of points in the design of McLean and Anderson grows rapidly. A reduction in the number of observations may be achieved by eliminating some of the face center points or such points the elimination of which results in the rest being distributed over the space under study more or less uniformly.

6.5. D-Optimal Designs

The most important among the known criteria of design optimality is the requirement of *D* and *G*-optimality. A design is said to be *D*-optimal when it minimizes the volume of the scatter ellipsoid for estimates of regression equation coefficients. The property of *G*-optimality provides the least maximum variance of predicted response values in a region under investigation. Simplex-lattice designs exhibit the properties of *D*- and *G*-optimality in the building of second- and incomplete third-degree polynomials only. Scheffe's designs of higher degree are not *D*-optimal [45]. A *D*-optimal simplex lattice for the third-degree polynomial was deduced later by Kiefer [48]. If we consider a set of designs with the coordinates of points

$$\begin{gathered} x_i = 1 \quad x_j = x_k = 0 \\ x_i = 1 - x_j = b \quad x_k = 0 \quad b < 1/2 \\ x_i = x_j = x_k = 1/3 \end{gathered} \tag{6.133}$$

then to produce a third-degree polynomial a design will be *D*-optimal at $b = (1-\sqrt{5})/2$, i. e. the points in the faces of simplex are taken with coordinates $x_i = 0.2764$ and $x_j = 0.7236$.

Table 6.13. D-optimal Design for a Ternary System Third-Degree Polynomial

N	x_1	x_2	x_3	y	N	x_1	x_2	x_3	y
1	1	0	0	y_1	6	0.7236	0	0.2764	y_{113}
2	0	1	0	y_2	7	0.2764	0	0.7236	y_{133}
3	0	0	1	y_3	8	0	0.7236	0.2764	y_{223}
4	0.7236	0.2764	0	y_{112}	9	0	0.2764	0.7236	y_{233}
5	0.2764	0.7236	0	y_{122}	10	0.3333	0.3333	0.3333	y_{123}

Table 6.13 tabulates the *D*-optimal design for the derivation of a ternary system third-degree polynomial. Following this design the coefficients are obtained for a third-degree polynomial having the same form as that from a conventional simplex lattice

$$\hat{y} = \sum_{1 \le i \le q} \beta_i x_i + \sum_{1 \le i < j \le q} \beta_{ij} x_i x_j + \sum_{1 \le i < j \le q} \gamma_{ij} x_i x_j (x_i - x_j) + \sum_{1 \le i < j < k \le q} \beta_{ijk} x_i x_j x_k \tag{6.134}$$

Formulas for the polynomial coefficients are derived by substituting the coordinates of points into the regression equation

$$\begin{aligned} \beta_i &= y_i \\ \beta_{ij} &= (5/2)\,(y_{iij} + y_{ijj} + y_i - y_j) \\ \gamma_{ij} &= (5/2)\,[5(y_{iij} - y_{ijj}) - y_i + y_j] \\ \beta_{ijk} &= 27 y_{ijk} - (15/2)\,(y_{iij} + y_{ijj} + y_{iik} + y_{ikk} + y_{jkk} \\ &\quad + y_{jjk}) + 6(y_i + y_j + y_k) \end{aligned} \tag{6.135}$$

The adequacy test and the assignment of confidence intervals using a *D*-optimal design (Table 6.13) are accomplished along the same lines, as in the simplex-lattice method. The variation of ξ with composition is given in Fig. 6.14.

Table 6.14. D-optimal Design for a Ternary System Fourth-Degree Polynomial

N	x_1	x_2	x_3	y	N	x_1	x_2	x_3	y
1	1	0	0	y_1	9	0.8273	0	0.1727	y_{1113}
2	0	1	0	y_2	10	0.1727	0	0.8273	y_{1333}
3	0	0	1	y_3	11	0	0.8273	0.1727	y_{2223}
4	0.5	0.5	0	y_{12}	12	0	0.1727	0.8273	y_{2333}
5	0.5	0.	0.5	y_{13}	13	0.567	0.2165	0.2165	y_{1123}
6	0	0.5	0.5	y_{23}	14	0.2165	0.567	0.2165	y_{1223}
7	0.8273	0.1727	0	y_{1112}	15	0.2165	0.2165	0.567	y_{1233}
8	0.1727	0.8273	0	y_{1222}					

In constructing the fourth-order polynomial for the ternary system the design will be *D*-optimal at

$$x_i = (7-\sqrt{21})/14 \qquad x_j = 1-x_i \qquad x_k = 0 \tag{6.136}$$

or

$$x_i = 0.1727 \qquad x_j = 0.8273 \qquad x_k = 0$$

Moreover, in the fourth-order *D*-optimal design there are points with coordinates

$$x_i = x_j = (7-\sqrt{5})/22 \qquad x_k = 1-(x_i+x_j) \tag{6.137}$$

or

$$x_i = x_j = 0.2165 \qquad x_k = 0.567$$

In Table 6.14 a fourth-order *D*-optimal design for a ternary system is presented.

According to this design the coefficients are obtained for a regression equation of the form

$$\begin{aligned}\hat{y} = {} & \beta_1 x_1+\beta_2 x_2+\beta_3 x_3+\beta_{12} x_1 x_2+\beta_{13} x_1 x_3+\beta_{23} x_2 x_3 \\ & +\gamma_{12} x_1 x_2 (x_1-x_2)+\gamma_{13} x_1 x_3 (x_1-x_3) \\ & +\gamma_{23} x_2 x_3(x_2-x_3)+\delta_{12} x_1 x_2 (x_1-x_2)^2 \\ & +\delta_{13} x_1 x_3(x_1-x_3)^2+\delta_{23} x_2 x_3 (x_2-x_3)^2 \\ & +\beta_{1123} x_1^2 x_2 x_3+\beta_{1223} x_1 x_2^2 x_3+\beta_{1233} x_1 x_2 x_3^2\end{aligned} \tag{6.138}$$

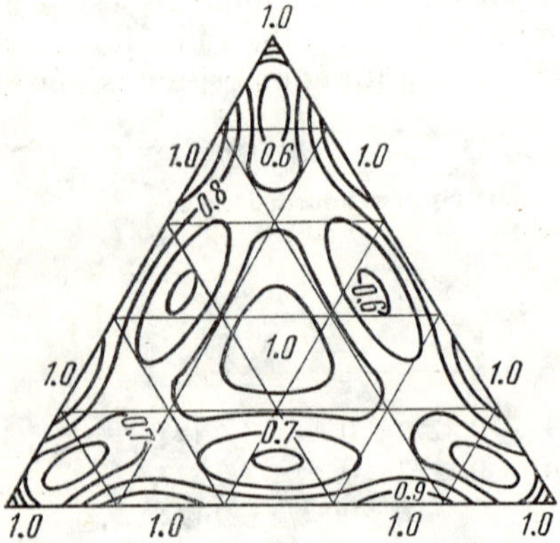

Fig. 6.14. Isolines ξ for *D*-optimal third-order design

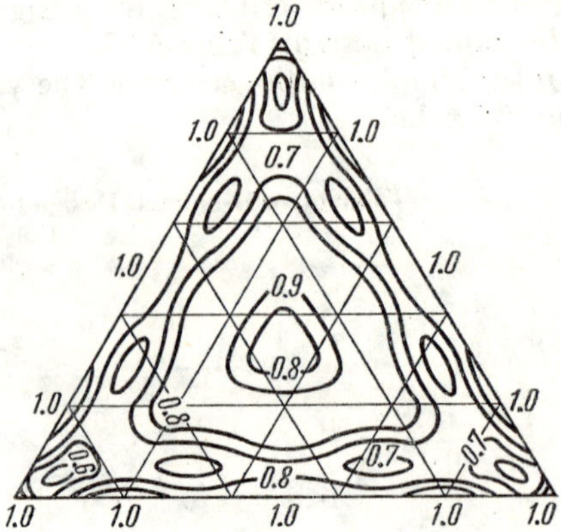

Fig. 6.15. Isolines ξ for *D*-optimal fourth-order design

Substituting the coordinates of points into the regression equation gives the relationships for calculation of the coefficients of the fourth-degree polynomial

$$\beta_i = y_i \tag{6.139}$$

$$\beta_{ij} = 4y_{ij} - 2y_i - 2y_j \tag{6.140}$$

$$\gamma_{ij} = (7/6)[3(-y_i + y_j) + \sqrt{21}\,(y_{iiij} - y_{ijjj})] \tag{6.141}$$

$$\delta_{ij} = (7/6)[-3(y_i + y_j) - 8y_{ij} + 7(y_{iiij} + y_{ijjj})] \tag{6.142}$$

$$\begin{aligned} \beta_{iijk} = {} & 26.657y_i - 6.167(y_j + y_k) - 16.96(y_{ij} + y_{ik}) \\ & + 0.511y_{ik} - 32.18(y_{iiij} + y_{iiik}) \\ & + 17.196(y_{ijjj} + y_{ikkk}) + 5.72(y_{jjjk} + y_{jkkk}) \\ & + 84.11y_{iijk} - 23.237(y_{ijjk} + y_{ijkk}) \end{aligned} \tag{6.143}$$

$$i \neq j \neq k \quad i, j, k = 1, 2, 3$$

To test for adequacy in defining the composition dependence of ξ use can be made of a contour chart (Fig. 6.15). It is owing to the *D*-optimality of the design that in the contour chart lines of constant value for $\xi > 1.0$ are absent (cf. Fig. 6.6*b* for the simple simplex lattice). But the preparation of mixtures with components contents of 0.8273 and 0.1727 is more difficult than of those with component ratio 3/4 : 1/4.

A material limitation of the above *D*-optimal third- and fourth-order designs is the absence of composition nature. Figure 6.16 shows the arrangement of points in *D*-optimal designs for ternary systems. A transition can be made, if need be, from the second-order design to a fourth-order design with the *D*-optimality retained. In doing so, the point at the center (Fig. 6.16) can be used as a test one. In changing over from the third-order design to the fourth-order design (Fig. 6.16) it is required either to remake the experiment, or follow a design that does not show the property of *D*-optimality.

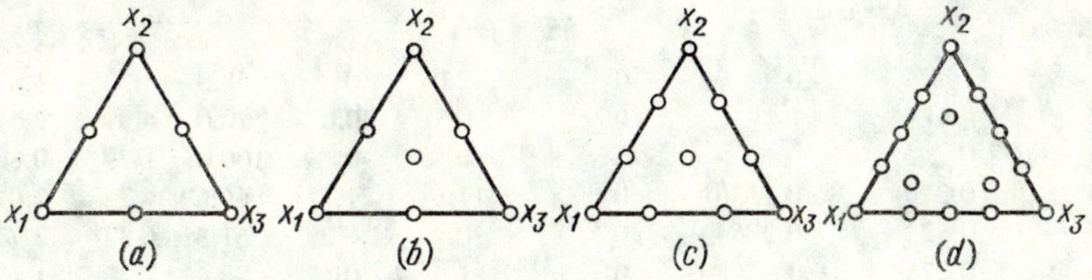

Fig. 6.16. Arrangement of points in the *D*-optimal designs of: (*a*) second-order, (*b*) incomplete second order, (*c*) third order, (*d*) fourth-order

Example 6.4 [49]. The variation of viscosity (y) of solutions in the system $(NH_4)_2HPO_4$—K_2CO_3—H_2O with composition and temperature is studied. The experimental design is conducted within a local region of the concentration triangle bounded by a saturation line at 0 °C (Fig. 6.17). The local region of the diagram was a triangle with the vertices z_1 (42, 0, 58), z_2 (0, 30, 70), and z_3 (0, 0, 100).

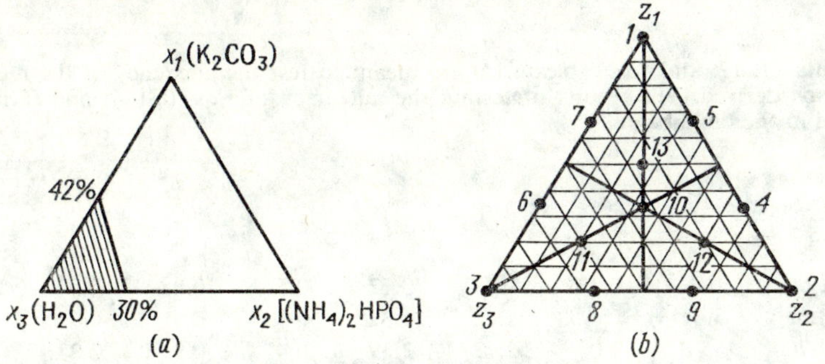

Fig. 6.17. Region of viscosity study in the system ($(NH_4)_2HPO_4-K_2CO_3-H_2O$) (*a*), and experimental design (*b*)

Solution. The third-order *D*-optimal design is prepared relative to pseudocomponents z_1, z_2, and z_3; and the content of initial components at the design points is determined by Eq. (6.120). Table 6.15 presents the experimental conditions both in terms of pseudocomponents z_j and on the natural scale (per cent).

The sample variance here is $s_y = 0.53$, and the number of degrees of freedom is $\nu_y = 13$. From Eq. (6.135) for viscosity at 0 °C the coefficients have been calculated for the third-order regression equation

$$\hat{y}_1 = 8.33z_1 + 4.99z_2 + 1.79z_3 - 6.95z_1z_2 - 9.05z_1z_3 - 1.37z_2z_3 + 17.90z_1z_2(z_1 - z_2) + 9.90z_1z_3(z_1 - z_3) + 12.37z_2z_3(z_2 - z_3) + 18.06z_1z_2z_3 \quad (6.144)$$

and at 30 °C

$$\hat{y}_2 = 3.83z_1 + 2.54z_2 + 0.80z_3 - 8.77z_1z_2 - 3.10z_1z_3 - 0.87z_2z_3 + 5.27z_1z_2(z_1 - z_2) + 6.55z_1z_3(z_1 - z_3) + 5.77z_2z_3(z_2 - z_3) + 3.00z_1z_2z_3 \quad (6.145)$$

Table 6.15. Design Matrix and Viscosity Measurements

N	z_1	z_2	z_3	x_1	x_2	x_3	y_1 (0 °C)	y_2 (30 °C)
1	1	0	0	42.0	0	58.0	8.33	3.83
2	0	1	0	0	30.0	70.0	4.99	2.54
3	0	0	1	0	0	100.0	1.79	0.80
4	0.2764	0.7236	0	11.6	21.71	66.69	4.22	2.09
5	0.7236	0.2764	0	30.4	8.29	61.31	6.32	2.77
6	0.2764	0	0.7236	11.6	0	88.4	2.20	1.13
7	0.7236	0	0.2764	30.4	0	69.6	4.30	2.26
8	0	0.2764	0.7236	0	8.29	91.71	2.30	1.09
9	0	0.7236	0.2764	0	21.71	78.29	3.93	1.90
10	0.333	0.333	0.333	14.0	10.0	76.0	3.59	1.64
11	0.22	0.22	0.56	9.1	6.5	84.4	2.00	1.23
12	0.22	0.56	0.22	9.1	17.0	73.9	3.68	1.82
13	0.56	0.22	0.22	23.9	6.5	69.6	4.70	2.12

Points 11, 12, and 13 (Tables 6.15) are meant to test the adequacy of the regression equations derived. The results of testing the adequacy of Eqs. (6.144) and (6.145) are arrayed in the table below.

N	y_1	$\hat{y}_1$	Δy_1	y_2	$\hat{y}_2$	Δy_2	ξ	t_1	t_2
11	2.00	1.68	0.32	1.23	0.72	0.51	0.8	0.77	1.22
12	3.68	3.68	0	1.82	1.82	0	0.8	0	0
13	4.70	5.70	1.00	2.12	2.59	0.47	0.9	2.33	1.09

The table value of Student's t is $t_{0.016,13} = 2.85$. For all the test points the values of t-ratio were found to be less than table one, hence regression equations (6.144) and (6.145) adequately fit the experiment. Consider the reducing of regression equations (6.144) and (6.145) into one, assuming a linear temperature dependence of the coefficients β (see the table below).

T	T_1	β_1	β_2	β_3	β_{12}	β_{13}	β_{23}	γ_{12}	γ_{13}	γ_{33}	γ_{123}
0 °C	−1	8.33	4.99	1.79	−6.95	−9.05	−1.37	17.90	9.90	12.37	13.06
30 °C	+1	3.83	2.54	0.80	−8.77	−3.10	−0.87	5.27	6.55	5.77	3.00

To facilitate the computation of the temperature dependence of β we make resort to the linear transformation similar to Eq. (5.3). In this case

$$T_1 = (T - 15)/15 \tag{6.146}$$

Now let us consider the linear dependence of β_j on the dimensionless parameter T_1

$$\beta_j = \beta_{1j} + \beta_{2j} T_1 \tag{6.147}$$

Then

$$\beta_{1j} = \frac{\beta_j^{(0)} + \beta_j^{(30)}}{2} \tag{6.148}$$

$$\beta_{2j} = \frac{\beta_j^{(30)} - \beta_j^{(0)}}{2} \tag{6.149}$$

where $\beta_j^{(0)}$ is a suitable coefficient in Eq. (6.144); $\beta_j^{(30)}$ is a corresponding coefficient in Eq. (6.145). Substituting Eqs. (6.146), (6.148), and (6.149) into Eq. (6.147), we obtain

$$\beta_j = \beta_j^{(0)} + \frac{\beta_j^{(30)} - \beta_j^{(0)}}{2\times 15} T \tag{6.150}$$

The final regression equation for the variation of solution viscosity with temperature and composition has the form

$$\begin{aligned} \hat{y} = {} & (8.33 - 0.150T)z_1 + (4.99 - 0.082T)z_2 \\ & + (1.79 - 0.001T)z_3 + (-6.95 + 0.106T)z_1 z_2 \\ & + (-9.05 + 0.347T)z_1 z_3 + (1.37 + 0.166T)z_2 z_3 \\ & + (17.90 - 0.069T)z_1 z_2 (z_1 - z_2) + (9.90 - 0.008T)z_1 z_3 (z_1 - z_3) \\ & + (12.37 - 0.017T)z_2 z_3 (z_2 - z_3) + (13.06 - 0.720T)z_1 z_2 z_3 \end{aligned} \tag{6.151}$$

Next we derive equations relating pseudocomponents z_j to the natural variable x_j. For the problem at hand the set of equations (6.123) will be

$$\begin{aligned} 1 &= 0.42z_1^{(1)} + 0z_1^{(2)} + 0.58z_1^{(3)} \\ 0 &= 0z_1^{(1)} + 0.3z_1^{(2)} + 0.7z_1^{(3)} \\ 0 &= 0z_1^{(1)} + 0z_1^{(2)} + 1z_1^{(3)} \end{aligned} \tag{6.152}$$

$$\begin{aligned} 0 &= 0.42z_2^{(1)} + 0z_2^{(2)} + 0.58z_2^{(3)} \\ 1 &= 0z_2^{(1)} + 0.3z_2^{(2)} + 0.7z_2^{(3)} \\ 0 &= 0z_2^{(1)} + 0z_2^{(2)} + 1z_2^{(3)} \end{aligned}$$

The solutions to the sets of Eq. (6.152) are

$$\begin{aligned} z_1^{(1)} &= 2.38 \qquad & z_2^{(1)} &= 0 \\ z_1^{(2)} &= 0 \qquad & z_2^{(2)} &= 3.33 \\ z_1^{(3)} &= 0 \qquad & z_2^{(3)} &= 0 \end{aligned} \tag{6.153}$$

Substituting Eq. (6.153) into the set of equations (6.122), we arrive at

$$\begin{aligned} z_1 &= 2.38(1 - x_2 - x_3) \\ z_2 &= 3.33x_2 \\ z_3 &= 1 - z_1 - z_2 = 2.38x_3 - 0.95x_2 - 1.38 \end{aligned} \tag{6.154}$$

Example 6.5. We seek to determine an optimal composition of a multicomponent solvent utilized to remove hydrocarbons from yeast. The major index of purification here is the hydrocarbon content in biomass upon extraction (y). For technological and economic reasons, the experimental design is accomplished in a local section of the concentration triangle (Fig. 6.18). In the region covered the mixture contains, in per cent: acetone, $x_1 \leqslant 74$; hexane, $x_2 \leqslant 90$; and water, $x_3 \leqslant 10$. The local portion of the diagram is a triangle with the vertices: $z_1(9.5, 89.5, 1)$, $z_2(58.5, 40, 1.5)$, and z_3 (74, 16, 10).

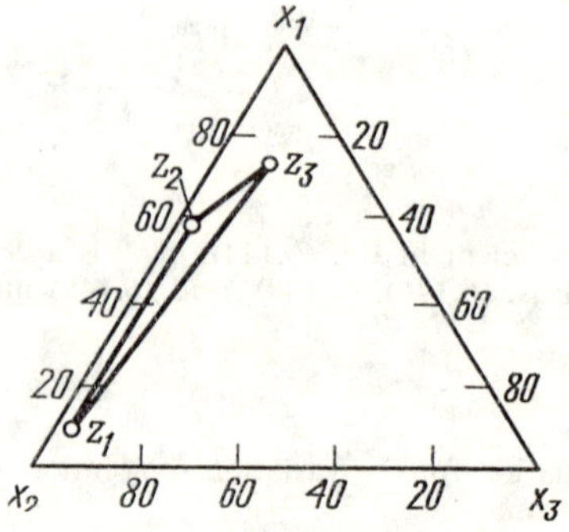

Fig. 6.18. Composition of multicomponent solvent; the region of study and experimental design

Solution. A fourth-order *D*-optimal design is produced with reference to pseudocomponents z_1, z_2, and z_3 (Table 6.16). The pseudocomponents satisfy the principal condition of Eq. (6.1) for Scheffe's designs.

Table 6.16. Fourth-order D-optimal Design

N	z_1	z_2	z_3	x_1	x_2	x_3	y
1	1	0	0	9.5	89.5	1	.1
2	0	1	0	58.5	40	1.5	.3
3	0	0	1	74	16	10.0	.04
4	0.5	0.5	0	34	64.7	1.3	.08
5	0.5	0	0.5	41.7	52.8	5.5	.06
6	0	0.5	0.5	66.2	28.2	5.8	.06
7	0.176	0.824	0	49.9	48.7	1.4	.05
8	0.824	0.176	0	18.12	80.79	1.09	.09
9	0.176	0	0.824	62.6	29.0	8.4	.12
10	0.824	0	0.176	20.85	76.55	2.6	.1
11	0	0.176	0.824	71.22	20.30	8.49	.2
12	0	0.824	0.176	61.25	35.75	3.0	.11
13	0.216	0.216	0.568	56.7	37.12	6.18	.11
14	0.216	0.568	0.216	51.2	45.65	3.15	.091
15	0.568	0.216	0.216	34.0	62.97	3.03	.11
16	0.333	0.333	0.333	47.3	48.5	4.2	.108

The conversion to initial components at any point within the local simplex studied is carried out from Eq. (6.120).

According to this design, an experiment is run with mixtures, each observation being repeated twice. Using Eqs. (6.139) –(6.143) the coefficients of fourth-order regression equation are calculated in pseudocomponents

$$\begin{aligned}\hat{y} = {} & 0.1z_1 + 0.3z_2 + 0.04z_3 - 0.48z_1z_2 - 0.04z_1z_3 - 0.44z_2z_3 \\ & + 0.914z_1z_2(z_1 - z_2) - 0.312z_1z_3(z_1 - z_3) - 1.39z_2z_3(z_2 - z_3) \\ & - 1.003z_1z_2(z_1 - z_2)^2 + 0.747z_1z_3(z_1 - z_3)^2 + 0.782z_2z_3(z_2 - z_3)^2 \\ & + 1.398z_1^2z_2z_3 + 8.416z_1z_2^2z_3 - 4.703z_1z_2z_3^2\end{aligned} \tag{6.155}$$

The sets of Eq. (6.123) under the constraints on the component content in the solvent have the form

$$\left.\begin{aligned} 9.5z_1^{(1)} + 89.5z_1^{(2)} + 1z_1^{(3)} &= 1 \\ 58.5z_1^{(1)} + 40z_1^{(2)} + 1.5z_1^{(3)} &= 0 \\ 74z_1^{(1)} + 16z_1^{(2)} + 10z_1^{(3)} &= 0 \end{aligned}\right\}$$

$$\left.\begin{aligned} 9.5z_2^{(1)} + 89.5z_2^{(2)} + 1z_2^{(2)} &= 0 \\ 58.5z_2^{(1)} + 40z_2^{(2)} + 1.5z_2^{(3)} &= 1 \\ 74z_2^{(1)} + 16z_2^{(2)} + 10z_2^{(3)} &= 1 \end{aligned}\right\} \tag{6.156}$$

The solutions to Eq. (6.156) are

$$z_1^{(1)} = 0.0092 \quad z_1^{(2)} = 0.0116 \quad z_1^{(3)} = 0.0495$$
$$z_2^{(1)} = 0.0215 \quad z_2^{(2)} = 0.00051 \quad z_2^{(3)} = -0.1583$$

Using the solutions found we obtain the formulas relating natural coordinates x to z

$$\begin{aligned} z_1 &= -0.92 + 0.0208x_2 + 0.058x_3 \\ z_3 &= 2.15 - 0.022x_2 - 0.18x_3 \\ z_3 &= 1 - z_1 - z_2 = -0.23 + 0.001x_2 + 0.121x_3 \end{aligned} \tag{6.157}$$

Adequacy of the regression equation obtained was tested by the Student test at five test points, the results being given in the table below.

x_1	x_2	x_3	y	$\hat{y}$	t	$t_{0.05}$ (21)
47.3	48.5	4.2	.1022	.108	0.365	2.83
53	44	3.0	.079	.072	0.392	2.83
19	79	2.0	.1	.07	1.778	2.83
37.9	58.7	3.4	.13	.12	0.547	2.83
44.5	54.0	1.5	.04	.05	0.57	2.83

It is seen that Eq. (6.155) adequately fits the experiment at the significance level $\alpha = 0.05$. The quality of the resultant product is considered satisfactory, if the content of residual hydrocarbons in biomass is under 0.05 per cent. With the aim to elucidate the solvent compositions meeting this requirement the lines of constant $\hat{y}$ are plotted to Eq. (6.155); the curves are shown in Fig. 6.19. The solvent compositions meeting the requirement that $\hat{y} < 0.05$ per cent lie within the shaded area of the simplex.

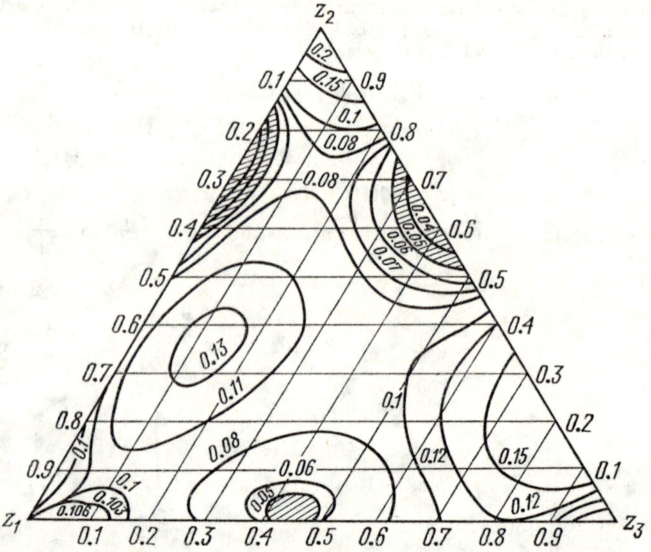

Fig. 6.19. Isolines of residual hydrocarbons content of biomass

6.6. Designs with Minimization of Systematic Bias

Draper and Lawrence [50] have proposed designs in which, unlike the simplex lattices, all the points are located within the investigation area, i. e. experiments are conducted with the q-component mixtures only. These designs make allowance for the absence of prior information on the response surface and that it is desirable to approximate an unknown response surface by low-degree polynomials. Data points are chosen so that to provide the best representation of a complex surface by simple polynomials.

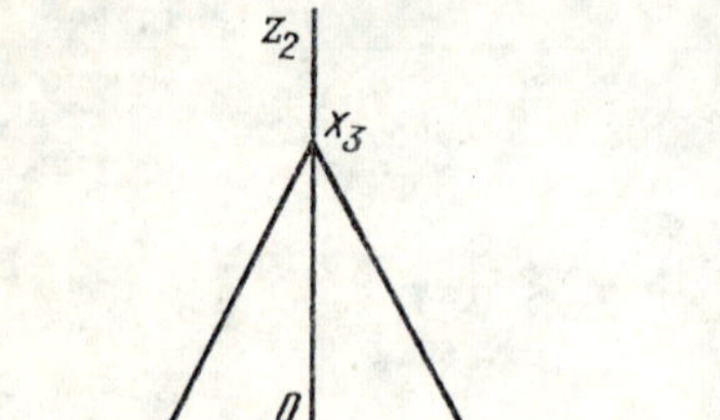

Fig. 6.20. Coordinate system for the designs of Draper and Lawrence

In building a polynomial of degree n_1, design points are to be selected so that a minimal systematic error results, which occurs due to the higher degree of response function polynomial, n_2 compared with the degree of estimating polynomial, n_1. Principles underlying the selection of suitable designs have been put forward earlier in Box and Draper [51].

Draper and Lawrence built designs for ternary and quaternary systems and polynomials of degrees $n_1 = 1$, $n_2 = 2$, $n_1 = 2$, and $n_2 = 3$. To make the generation of design more convenient these authors introduce a new reference system.

With ternary mixtures, the new coordinate system is selected in the plane of concentration triangle (x_1, x_2, x_3) so that the origin coincides with the centroid of the triangle, one of the triangle axes lies on the axis z_2, two others being symmetrical about this axis (Fig. 6.20). The triangular coordinate system (x_1, x_2, x_3) is related to the rectangular one (z_1, z_2) as follows:

$$z_1 = \frac{1}{2}(-x_1+x_2)$$

$$z_2 = \frac{\sqrt{3}}{6}(-x_1-x_2+2x_3) \qquad (6.158)$$

$$x_1 = \frac{1}{3}(-3z_1-z_2\sqrt{3}+m)$$

$$x_2 = \frac{1}{3}(3z_1-z_2\sqrt{3}+m) \qquad (6.159)$$

$$x_3 = \frac{1}{3}(2z_2\sqrt{3}+m)$$

where m is the length of a side of the concentration triangle. When the whole of the diagram is explored, $m = 1$; and when local areas of the diagram, $m < 1$.

Design points for ternary systems are selected (in the coordinates z_1, z_2) from the following sets:

(1) vertices of a triangle similar to a given concentration triangle and centered on the origin, with side of length p:

$$\left(0,\ +\frac{1}{\sqrt{3}}p\right) \quad \left(\pm\frac{1}{2}p,\ -\frac{\sqrt{3}}{6}p\right)$$

(2) vertices of a triangle similar to a given concentration triangle and centered on the origin, with side of length q:

$$\left(0,\ -\frac{1}{\sqrt{3}}q\right) \quad \left(\pm\frac{1}{2}q,\ \frac{\sqrt{3}}{6}q\right)$$

(3) vertices of a square centered on the origin, with sides $2a$ parallel to axes ($\pm a$, $\pm a$);

(4) points on coordinate axes ($\pm b$, 0), (0, $\pm b$);

(5) vertices of a rectangle (c, d), ($-c$, $-d$), (c, $-d$), and ($-c$, d).

After one or another design of Draper-Lawrence has been constructed for a ternary system, first-degree polynomials are derived in two independent variables z_1 and z_2 ($n_1 = 1$ for $n_2 = 2$)

$$\hat{y} = b_0+b_1z_1+b_2z_2 \qquad (6.160)$$

or second-order polynomials ($n_1 = 2$ for $n_2 = 3$)

$$\hat{y} = b_0+b_1z_1+b_2z_2+b_{12}z_1z_2+b_{11}z_1^2+b_{22}z_2^2 \qquad (6.161)$$

Draper and Lawrence suggested the designs containing from 6 to 9 points to derive the first-order polynomials for ternary systems ($q = 3$). Parameters for some of Draper-Lawrence's designs (in fractions of m) at $q = 3$, $n_1 = 1$, and $n_2 = 2$ are given in Table 6.17. If the number of design points is more than that of a selected set, then an appropriate number of points is added at the center of triangle (with coordinates $z_1 = 0$, $z_2 = 0$).

Table 6.17. Parameters of Draper-Lawrence's Designs for $q = 3$, $n_1 = 1$, $n_2 = 2$

Set	Center points	Design points total, N	Parameters	
(1, 2)	0	6	$p = 0.621$	$q = 0.339$
(1, 2)	1	7	$p = 0.662$	$q = 0.381$
(1, 2)	2	8	$p = 0.699$	$q = 0.421$
(1, 2)	3	9	$p = 0.733$	$q = 0.457$
(1, 3)	0	7	$p = 0.616$	$q = 0.160$
(1, 4)	0	7	$p = 0.616$	$b = 0.226$
(1, 5)	0	7	$p = 0.616$	$c = \sqrt{0.051m^2 - d^2}$*
(1, 1, 2)	0	9	$p_1 = 0.606$ $q = 0.364$	$p_2 = 0.500$
(1, 2, 2)	0	9	$p = 0.727$ $q_2 = 0.200$	$q_1 = 0.425$

* Value of d^2 is selected at random.

By way of example, we consider a Draper-Lawrence design (1, 2) containing six points (Table 6.16). Points of set 1 at $m = 1$ have the coordinates (z_1, z_2):

$$\left(0.0, \frac{0.621}{\sqrt{3}}\right) \quad \left(\frac{0.621}{2}, -\frac{0.621\sqrt{3}}{6}\right) \quad \left(-\frac{0.621}{2}, -\frac{0.621\sqrt{3}}{6}\right)$$

or

$$(0.0, 0.366); \quad (0.3105, -0.18); \quad (-0.3105, -0.18)$$

Points of set 2 have the coordinates

$$\left(0.0, -\frac{0.339}{\sqrt{3}}\right) \quad \left(+1/2\times 0.339, \frac{\sqrt{3}\,0.339}{6}\right)$$

$$\left(-1/2\times 0.339, \frac{\sqrt{3}\times 0.339}{6}\right)$$

or

$$(0.0, -0.196) \quad (0.170, 0.098), \quad (-0.170, 0.098)$$

We make a transition from point coordinates in the system (z_1, z_2) to those in the triangle $x_1-x_2-x_3$ using Eqs. (6.159). Let us map, for example, the first point with coordinates $z_1 = 0.0$, $z_2 = 0.366$ $(m = 1)$.

For this point

$$x_1 = \frac{1}{3}(-0.366\sqrt{3}+1) = 0.12$$

$$x_2 = \frac{1}{3}(-0.366\sqrt{3}+1) = 0.12$$

$$x_3 = \frac{1}{3}(2\times 0.366\sqrt{3}+1) = 0.76$$

To test the calculation we sum up

$$x_1+x_2+x_3 = 0.12+0.12+0.76 = 1.0$$

The arrangement of the points in the concentration triangle is shown in Fig. 6.21, the experimental design being arrayed in Table 6.18.

Table 6.18. Design Matrix (1, 2) for $q = 3$, $n_1 = 1$, $n_2 = 2$

N	z_1	z_2	x_1	x_2	x_3
1	0	0.366	0.12	0.12	0.76
2	0.311	−0.18	0.127	0.748	0.125
3	−0.311	−0.18	0.748	0.127	0.125
4	0.0	−0.196	0.447	0.447	0.106
5	0.170	0.098	0.106	0.447	0.447
6	−0.170	0.098	0.447	0.106	0.447

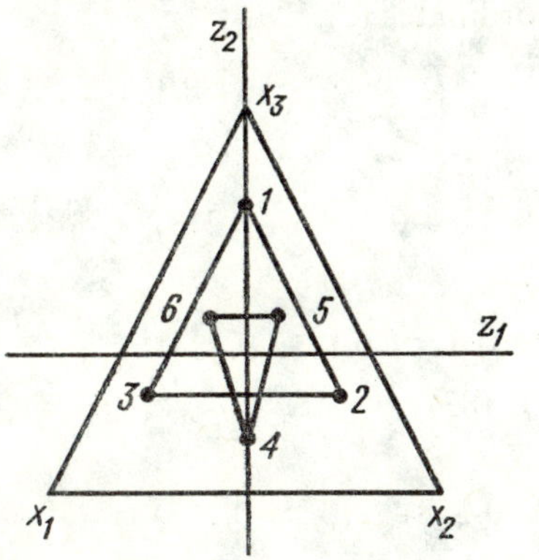

Fig. 6.21. Draper-Lawrence's design (1, 2)

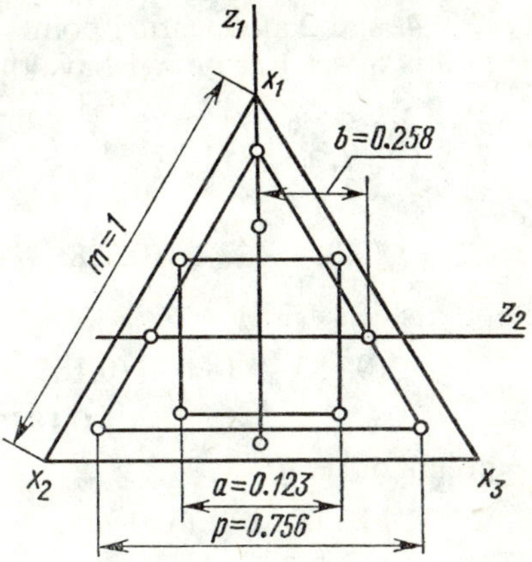

Fig. 6.22. Draper-Lawrence's design (1, 3, 4)

Then, Draper and Lawrence suggested for second-order polynomials of Eq. (6.161) as applied to ternary systems, the designs containing from 8 to 15 design points. Parameters for the Draper-Lawrence designs (in fractions of m) at $q = 3$, $n_1 = 2$, and $n_2 = 3$ are summerized in Table 6.19.

Table 6.19. Parameters of Draper-Lawrence's Designs at $q = 3$, $n_1 = 2$, $n_2 = 3$

Set	Center points n_0	N	Parameters			
(1, 2)	1	7	$p = 0.670$	$q = 0.385$		
(1, 2)	2	8	$p = 0{,}698$	$q = 0.421$		
(1, 2)	3	9	$p = 0.723$	$q = 0.450$		
(1, 1, 2)	0	9	$p_1 = 0.715$	$p_2 = 0.233$	$q = 0.430$	
(1, 1, 2)	1	10	$p_1 = 0.729$	$p_2 = 0.323$	$q = 0.445$	
(1, 1, 2)	2	11	$p_1 = 0.738$	$p_2 = 0.398$	$q = 0.462$	
(1, 1, 2)	3	12	$p_1 = 0.743$	$p_2 = 0.465$	$q = 0.450$	
(1, 1, 2)	4	13	$p_1 = 0.742$	$p_2 = 0.532$	$q = 0.485$	
(1, 2, 2)	0	9	$p = 0.716$	$q_1 = 0.342$	$q = 0.342$	
(1, 2, 2)	1	10	$p = 0.739$	$q_1 = 0.367$	$q_2 = 0.367$	
(1, 1, 1, 2)	0	12	$p_1 = 0.751$	$p_2 = 0.422$	$p_3 = 0.189$	$q = 0.47$
(1, 1, 2, 2)	0	12	$p_1 = 0.748$	$p_2 = 0.445$	$q_1 = 0.468$	$q_2 = 0.156$
(1, 2, 2, 2)	0	12	$p = 0.782$	$q_1 = 0.348$	$q_2 = 0.348$	$q_3 = 0.348$
(1, 3, 4)	2	13	$p = 0.756$	$a = 0.183$	$b = 0.258$	
(1, 3, 5)	2	13	$p = 0.756$	$a = 0.300$	$c = 0.547$	$d = 0.130$
(1, 4, 5)	2	13	$p = 0.756$	$b = 0.212$	$c = 0.130$	$d = 0.257$
(1, 5, 5)	2	13	$p = 0.756$	$c_1 = 0.094$	$d_1 = 0.272$	$c_2 = 0.172$ $d_2 = 0.125$
(1, 1, 2, 5)	0	13	$p_1 = 0.297$	$p_2 = 0.756$	$q = 0.295$	$c = 0.111$ $d = 0.268$
(1, 1, 2, 5)	0	13	$p_1 = 0.478$	$p_2 = 0.756$	$q = 0.477$	$c = 0.045$ $d = 0.109$
(1, 1, 2, 5)	1	14	$p = 0.369$	$p_2 = 0.766$	$q = 0.319$	$c = 0.112$ $d = 0.270$
(1, 1, 2, 5)	1	14	$p_1 = 0.514$	$p_2 = 0.762$	$q = 0.481$	$c = 0.058$ $d = 0.140$
(1, 1, 2, 5)	2	15	$p_1 = 0.545$	$p_2 = 0.766$	$q = 0.480$	$c = 0.071$ $d = 0.171$

Consider then a design (1, 3, 4) containing 13 points: 11 points of the sets 1, 3, 4, and 2 augmenting points at the center of the triangle (Fig. 6.22). The points of set 1 at $m = 1$ have the coordinates (z_1, z_2):

$$\left(0, \frac{0.756}{\sqrt{3}}\right) \quad \left(\frac{0.756}{2}, -\frac{0.756\sqrt{3}}{6}\right) \quad \left(-\frac{0.756}{2}, \frac{0.756\sqrt{3}}{6}\right)$$

or

$$(0.0,\ 0.437) \quad (0.378,\ -0.218) \quad (-0.378,\ -0.218)$$

The points of set 3:

$$(0.183,\ 0.183) \quad (0.183,\ -0.183) \quad (-0.183,\ 0.183) \quad (-0.183,\ -0.183)$$

The points of set 4:

$$(0.258,\ 0) \quad (-0.258,\ 0) \quad (0,\ 0.258) \quad (0,\ -0.258)$$

The experimental design is tabulated in Table 6.20.

Table 6.20. Design Matrix (1. 3, 4) for $q = 3, n_1 = 2, n_2 = 3, N = 13$

Design points	z_1	z_2	x_1	x_2	x_3
1	0	0.437	0.081	0.081	0.838
2	0.378	−0.218	0.081	0.837	0.082
3	−0.378	−0.218	0.837	0.081	0.082
4	0.183	0.183	0.044	0.410	0.546
5	0.183	−0.183	0.256	0.622	0.122
6	−0.183	0.183	0.410	0.045	0.545
7	−0.183	−0.183	0.622	0.256	0.122
8	0.258	0	0.076	0.591	0.333
9	−0.258	0	0.592	0.075	0.333
10	0	0.258	0.184	0.184	0.632
11	0	−0.258	0.482	0.482	0.036
12	0	0	0.333	0.333	0.333
13	0	0	0.333	0.333	0.333

The coordinates x_1, x_2, x_3 are related to z_1, z_2 by Eq. (6.159). The coefficients of the second-order regression equation $\hat{y} = f(z_1, z_2)$ are derived by the method of least squares. For the adequacy of fit, t-test is applied to experimental data at test points. The equation is adequate if the t-test values for all the test points are less than the table value. The t-test values are to be found from Eq. (6.93). Here values of ξ are taken from appropriate contour charts. Using the designs of Draper and Lawrence, the composition-dependence of ξ can only be calculated with a digital computer. Such a contour chart for the design (1, 3, 4) of Table 6.20 is shown in Fig. 6.23. As is seen in the figure, the regression equation yields the worst predicted points for the property near the vertices of the concentration triangle (ξ = 1.5-3).

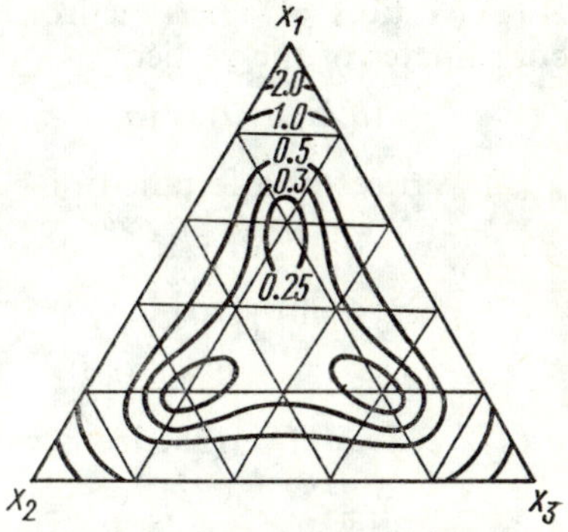

Fig. 6.23. Isolines ξ in Draper-Lawrence's design (1, 3, 4)

To develop designs for quaternary systems, Draper and Lawrence have also introduced a coordinate system (z_1, z_2, z_3). The origin of the new system coincides with the centroid of the concentration tetrahedron (x_1, x_2, x_3, x_4), and the coordinate axes are directed so that the four vertices of the tetrahedron in the new coordinate system form a half-replicate of the full factorial design 2^3 with the defining contrast $1 = z_1 z_2 z_3$. In the new system

(z_1, z_2, z_3) the coordinates of the tetrahedron vertices are in the general case:

$$(m, m, -m) \quad (m, -m, m) \quad (-m, m, m) \quad (-m, -m, -m)$$

and for the tetrahedron edge $m = 1$:

$$(1, 1, -1) \quad (1, -1, 1) \quad (-1, 1, 1) \quad (-1, -1, -1)$$

The coordinate systems (x_1, x_2, x_3, x_4) and (z_1, z_2, z_3) are related to each other as follows:

$$\begin{aligned} z_1 &= x_1+x_2-x_3-x_4 \\ z_2 &= x_1-x_2+x_3-x_4 \\ z_3 &= -x_1+x_2+x_3-x_4 \end{aligned} \tag{6.162}$$

and

$$\begin{aligned} x_1 &= \frac{1}{4}(z_1+z_2-z_3+m) \\ x_2 &= \frac{1}{4}(z_1-z_2+z_3+m) \\ x_3 &= \frac{1}{4}(-z_1+z_2+z_3+m) \\ x_4 &= \frac{1}{4}(-z_1-z_2-z_3+m) \end{aligned} \tag{6.163}$$

Design points for quaternary systems are chosen (in coordinates z_1, z_2, z_3) from the following sets:

(1) vertices of a tetrahedron similar to the concentration one with the coordinates of the vertices

$$(a, a, -a) \quad (a, -a, a) \quad (-a, a, a) \quad (-a, -a, -a)$$

(2) vertices of a tetrahedron

$$(b, b, b) \quad (b, -b, -b) \quad (b, b, -b) \quad (-b, -b, b)$$

(3) points on the axes

$$(\pm h, 0, 0) \quad (0, \pm h, 0) \quad (0, 0, \pm h)$$

(4) vertices of tetrahedrons with the coordinates

$$\begin{array}{llll} (-r, -s, -t) & (-r, s, t) & (r, -s, t) & (r, s, -t) \\ (-t, r, s) & (t, -r, s) & (t, r, -s) & (-t, -r, -s) \\ (-s, t, r) & (s, -t, r) & (s, t, -r) & (-s, -t, -r) \end{array}$$

After one or another Draper-Lawrence's design has been constructed for quaternary systems, polynomials are obtained in three independent variables z_1, z_2, and z_3 of first degree ($n_1 = 1$ for $n_2 = 2$)

$$\hat{y} = b_0+b_1z_1+b_2z_2+b_3z_3 \tag{6.164}$$

or second order ($n_1 = 2$ for $n_2 = 3$)

$$\hat{y} = b_0+b_1z_1+b_2z_2+b_3z_3+b_{12}z_1z_2+b_{13}z_1z_3 \\ +b_{23}z_2z_3+b_{11}z_1^2+b_{22}z_2^2+b_{33}z_3^2 \quad (6.165)$$

Parameters (in fractions of m) for some designs of Draper and Lawrence containing no more than 12 points, at $q = 4$, $n_1 = 1$, $n_2 = 2$, are provided in Table 6.21.

Table 6.21. Parameters of Draper-Lawrence's Design for $q = 4$, $n_1 = 1$, $n_2 = 2$

Set	Center points	Design points total	Parameters	
(1, 2)	0	8	$a = 0.548$	$b = 0.315$
(1, 2)	1	9	$a = 0.567$	$b = 0.344$
(1, 2)	2	10	$a = 0.602$	$b = 0.371$
(1, 2)	3	11	$a = 0.626$	$b = 0.397$
(1, 2)	4	12	$a = 0.650$	$b = 0.421$
(1, 3)	0	10	$a = 0.550$	$h = 0.628$
(1, 3)	1	11	$a = 0.568$	$h = 0.674$
(1, 3)	2	12	$a = 0.585$	$h = 0.718$
(4)	0	12	$r = 0.539$	$s = 0.248$
			$t = 0.500$	
(4)	6	12	$r = 0.616$	$s = 0.360$
			$t = 0.300$	

At $N \geqslant 12$ the values of parameters r, s, t, and N are to be found from the set of equations

$$r^2+s^2+t^2 = Nm^2/20 \\ rst = Nm^3/180 \quad (6.166)$$

where m is the edge of the concentration tetrahedron.

Design parameters (in fractions of m) for quaternary mixtures with $n_1 = 2$, $n_2 = 3$, and $18 \leqslant N \leqslant 24$ are presented in Table 6.22.

All the above designs are built assuming that there only exists a systematic bias. In actual practice, however, besides the systematic error the experimental data also contain random errors.

When minimizing the total error [50] the basic configuration of designs [48] may be retained with the coordinates of design points multiplied by a quantity $\theta > 1$, i. e. for a ternary system the points $(\theta z_1, \theta z_2)$ and for a quaternary one with $(\theta z_1, \theta z_2, \theta z_3)$ should be used. The parameter θ depends on the random error and the polynomial coefficients, being close to unity if the random error is predominant. As for each particular experiment the exact value of θ is rather difficult to find, it may, to a rather rough approxi-

Table 6.22. Parameters of Draper-Lawrence's Design for $q = 4$, $n_1 = 2$, and $n_2 = 3$

Set	Center points	Design points total	Parameters				
(1, 1, 4)			a_1	a_2	r	s	t
(1, 1, 4)	0	20	0.673	0.0945	0.684	0.260	0.0524
(1, 1, 4)	1	21	0.679	0.179	0.694	0.270	0.0564
(1, 1, 4)	2	22	0.685	0.248	0.702	0.274	0.0532
(1, 1, 4)	3	23	0.690	0.315	0.708	0.268	0.0406
(1, 1, 4)	4	24	0.694	0.393	0.710	0.242	0.00912
			a	b	r	s	t
(1, 2, 4)	1	21	0.676	0.165	0.696	0.274	0.0784
(1, 2, 4)	2	22	0.680	0.220	0.706	0.281	0.106
(1, 2, 4)	3	23	0.683	0.272	0.717	0.274	0.144
(1, 2, 4)	4	24	0.685	0.317	0.727	0.226	0.225
			a	h	r	s	t
(1, 3, 4)	0	22	0.682	0.319	0.0807	0.291	0.702
(1, 3, 4)	1	23	0.686	0.390	0.0925	0.306	0.708
(1, 3, 4)	2	24	0.690	0.459	0.104	0.321	0.710
			a_1	a_2	b	h	
(1, 1, 2, 3)	0	18	0.292	0.667	0.279	0.765	
(1, 1, 3, 3)	1	19	0.337	0.672	0.292	0.776	
(1, 1, 2, 3)	2	20	0.380	0.674	0.305	0.786	
(1, 1, 2, 3)	3	21	0.420	0.676	0.318	0.795	
(1, 1, 2, 3)	4	22	0.460	0.674	0.332	0.805	
(1, 1, 2, 3)	5	23	0.501	0.669	0.346	0.814	
(1, 1, 2, 3)	6	24	0.548	0.656	0.359	0.822	
			a_1	a_2	a_3	b	h
(1, 1, 1, 2, 3)	0	22	0.679	0.442	0.132	0.326	0.805
(1, 1, 1, 2, 3)	1	23	0.683	0.455	0.191	0.332	0.814
(1, 1, 1, 2, 3)	2	24	0.691	0.441	0.288	0.340	0.822
			a_1	a_2	b_1	b_2	h
(1, 1, 2, 2, 3)	0	22	0.677	0.451	0.126	0.321	0.805
(1, 1, 2, 2, 3)	1	23	0.677	0.479	0.181	0.315	0.814
(1, 1, 2, 2, 3)	2	24	0.672	0.517	0.275	0.275	0.822
(1, 1, 2, 3, 3)	0	24	0.680	0.494	0.329	0.317	0.818

mation, be taken equal to 1.1 for ternary and 1.2 for quaternary systems. By way of example, to minimize the total error we transform the design (1, 3, 4) for $q = 3$, $n_1 = 2$, and $n_2 = 3$, given in Table 6.20. The coordinates of the first point for $\theta = 1.1$ are to be found as follows:

$$x_1 = \frac{1}{3}(-3\theta z_1 - \theta z_2\sqrt{3} + m) = \frac{1}{3}(-1.1 \times 0.437\sqrt{3}+1) = 0.056$$

$$x_2 = \frac{1}{3}(+3\theta z_1 - \theta z_2\sqrt{3} + m)$$

$$= \frac{1}{3}(+3 \times 1.1 \times 0.0 - 1.1 \times 0.437\sqrt{3}+1) = 0.056$$

$$x_3 = \frac{1}{3}(2\theta z_2\sqrt{3} + m) = \frac{1}{3}(2 \times 1.1 \times 0.437\sqrt{3}+1) = 0.888$$

In full this minimizing the total error design (1, 3, 4) for $q = 3$, $n_1 = 2$, $n_2 = 3$ is provided in Table 6.23.

Table 6.23. Design Matrix (1, 3, 4) for $q = 3$, $n_1 = 2$, $n_2 = 3$, $\theta = 1.1$

N	x_1	x_2	x_3	N	x_1	x_2	x_3
1	0.056	0.056	0.888	8	0.05	0.617	0.333
2	0.056	0.888	0.056	9	0.617	0.05	0.333
3	0.888	0.056	0.056	10	0.17	0.17	0.66
4	0.016	0.418	0.566	11	0.5	0.5	0
5	0.248	0.651	0.101	12	0.333	0.333	0.333
6	0.418	0.016	0.566	13	0.333	0.333	0.333
7	0.651	0.248	0.101				

Example 6.6 [52]. The dependence of viscosity at 30 °C on composition is studied for the liquid complex fertilizer consisting of diammonium phosphate, potash, and water. For the investigation, the region of unsaturated solutions for both salts at 30 °C is selected (Fig. 6.24), the side of the concentration triangle being 0.5.

Solution. A Draper-Lawrence design containing 13 points is performed (Table 6.20). It is convenient to treat the subregion studied as a concentration triangle in the new coordinate system (x'_1, x'_2, x'_3),

$$x'_1 + x'_2 + x'_3 = 0.5$$

The coordinates x'_j and x_j are related by

$$\begin{aligned} x'_1 &= 2x_1 \\ x'_2 &= 2x_2 \\ x'_3 &= 1 - x'_1 - x'_2 = 1 - 2x_1 - 2x_2 \end{aligned} \qquad (6.167)$$

We also have, by Eq. (6.159),

$$z_1 = (1/2)(-x'_1 + x'_2) = (-x_1 + x_2)$$

$$z_2 = \frac{\sqrt{3}}{6}(-x'_1 + x'_2 + 2x'_3) = \frac{\sqrt{3}}{3}(1 - 3x_1 - 3x_2) = \frac{\sqrt{3}}{3}(x_3 - 2x_1 - 2x_3) \qquad (6.168)$$

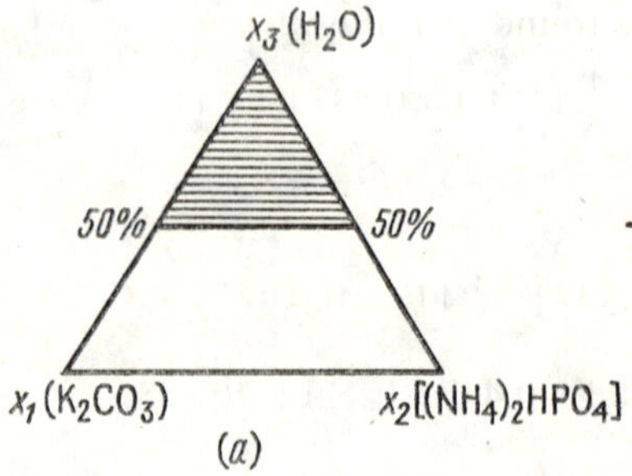

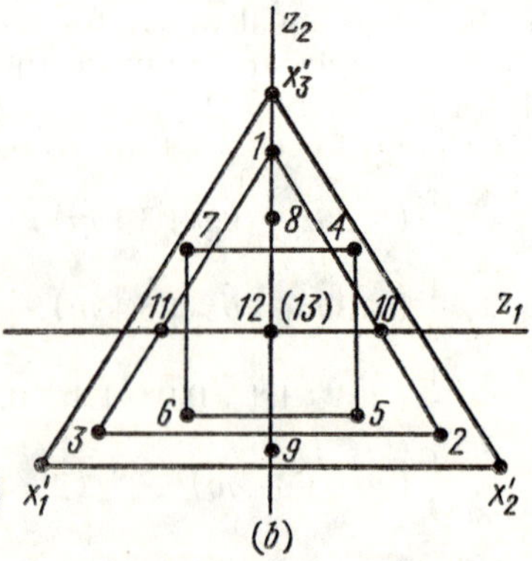

Fig. 6.24. Region of viscosity study in the system $(NH_4)_2HPO_4-K_2CO_3-H_2O$ at 30 °C (*a*), and experimental design (*b*)

The design of the experiment and measurements of viscosity (y) for two parallel experiments are given in Table 6.24.

Table 6.24. Design Matrix and Experimental Data

N	z_1	z_2	x_1'	x_2'	x_3'	x_1	x_2	x_3	$\bar{y}$
1	0	0.437	.081	.081	.838	.040	.040	.920	1.033
2	0.378	−0.218	.081	.837	.082	.040	.418	.542	4.873
3	−0.378	−0.218	.837	.081	.082	.418	.040	.542	4.722
4	0.183	0.183	.044	.410	.546	.022	.205	.772	1.481
5	0.183	−0.183	.256	.622	.122	.128	.311	.561	3.294
6	−0.183	0.183	.410	.045	.545	.311	.128	.561	2.996
7	−0.183	−0.183	.622	.256	.122	.205	.023	.772	2.160
8	0.258	0	.075	.592	.333	.092	.092	.816	1.430
9	−0.258	0	.592	.075	.333	.241	.241	.518	3.624
10	0	0.258	.184	.184	.632	.038	.296	.666	2.423
11	0	−0.258	.482	.482	.036	.296	.038	.666	2.165
12	0	0	.333	.333	.333	.167	.167	.666	2.191
13	0	0	.333	.333	.333	.167	.167	.666	2.207

From this table the coefficients of the regression equation, analogous to Eq. (6.161), are obtained by the computer-aided method of least squares,

$$\hat{y} = 1.54 - 0.94z_1 - 1.01z_2 - 8.93z_1z_2 + 10.48z_1^2 + 0.76z_2^2$$

The equation derived is found to be an adequate fit to the experiment. By Eqs. (6.168) the estimated regression equation in the natural scale takes the form

$$\hat{y} = 1.54 + 2.1x_1 + 0.22x_2 - 0.58x_3 + 1.18x_1^2 + 21.81x_2^2 - 18.93x_1x_2 + 4.14x_1x_3 - 6.17x_2x_3 + 0.25x_3^2$$

6.7. Factorial Experiments with Mixture Using Ratios

In certain practical applications it is advantageous to consider variation of properties not with absolute amounts of components, but with their ratios. If the percentage of each component is not zero, then given upper and lower constraints for the components, ratios of components may be utilized to build conventional factorial designs [53]. The number of ratios in a q-component system subject to

$$x_1+x_2+\ldots+x_q = 1$$

will be $q-1$;

$$z_1 = \frac{x_1}{x_2} \quad z_2 = \frac{x_3}{x_2} \quad \ldots \quad z_i = \frac{x_i}{x_j} \quad \ldots \quad z_{q-1} = \frac{x_{q-1}}{x_q} \qquad (6.169)$$

Thus, using the component ratios as independent factors the dimensionality of the problem is reduced by one, and hence the number of experiments is also decreased.

Figure 6.25 shows Kenworthy's designs [53] 2^2 and 2^3 for handling the variation of property with component ratios $z_1 = x_1/x_3$ and $z_2 = x_3/x_2$.

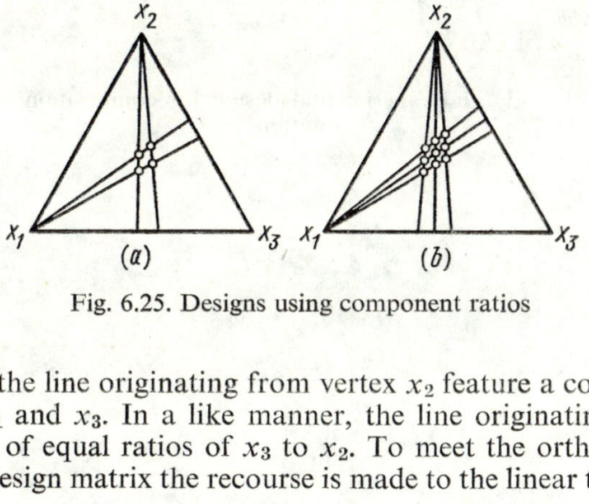

Fig. 6.25. Designs using component ratios

The points on the line originating from vertex x_2 feature a constant ratio of components x_1 and x_3. In a like manner, the line originating from vertex x_1 is the locus of equal ratios of x_3 to x_2. To meet the orthogonality condition for the design matrix the recourse is made to the linear transformation of Eq. (5.3)

Example 6.7. We seek the functional relationship between the yield of sodium and potassium bicarbonates and the composition of initial sylvinite solution. The factors controlling the potassium utilization, y, in the carbonization process are chosen to be the percent ratios of two of three components making up the system.

$$z_1 = [NaCl]/[KCl] \quad \text{and} \quad z_2 = [H_2O]/[NaCl]$$

Solution. To derive the regression equation, we shall use a second order orthogonal design for $k = 2$, $N = 9$, and the star arm $\alpha = 1$ (Fig. 6.26). The region where independent factors will be studied is given in the table below, and the design matrix in Table 6.25.

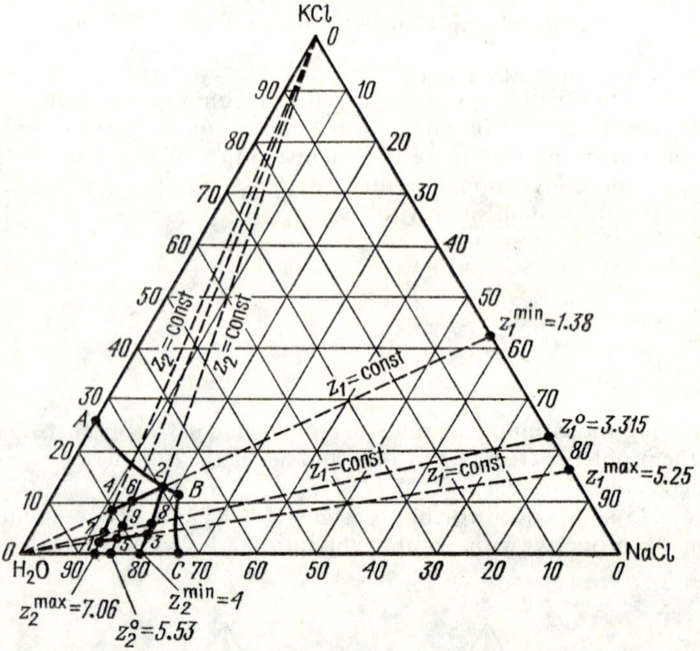

Fig. 6.26. Region of study and experimental design for composition of initial sylvinite solution

	z_1	z_2
z_j^0	3.315	5.53
Δz_j	1.935	1.53
+1	5.25	7.06
−1	1.38	4.0

Sample variances are homogeneous. The error mean square is

$$s_e^2 = \frac{\sum^{9} s_i^2}{9} = \frac{30.28}{9} = 3.37$$

$$v_e = N(m-1) = 9$$

where m is the number of replicate observations.

Table 6.25. Design Matrix

N	x_0	x_1	x_2	x_1x_2	x_1'	x_2'	y'	y''	$\bar{y}$	s_y^2	$\hat{y}$	$\bar{y}-\hat{y}$	$(\bar{y}-\hat{y})^2$
1	+1	+1	+1	+1	+1/3	+1/3	10	14	12	8	10.74	1.26	1.59
2	+1	−1	−1	+1	+1/3	+1/3	77.5	79.5	78.5	2	77.74	0.76	0.58
3	+1	+1	−1	−1	+1/3	+1/3	27	28.6	27.8	1.28	27.84	0.04	0.001
4	+1	−1	+1	−1	+1/3	+1/3	59.5	62.5	61	4.5	60.64	0.36	0.13
5	+1	+1	0	0	+1/3	−2/3	18.5	17.5	18	0.5	19.3	1.3	1.69
6	+1	−1	0	0	+1/3	−2/3	68.2	67.8	68	0.08	69.19	1.19	1.42
7	+1	0	+1	0	−2/3	+1/3	30	34	32	8	32.59	0.59	0.35
8	+1	0	−1	0	−2/3	+1/3	51	49	50	2	49.69	0.31	0.09
9	+1	0	0	0	−2/3	−2/3	43	40.2	41.6	3.92	41.14	0.46	0.21

The coefficients of the regression equation and their errors are computed from Eqs. (5.56) and (5.57)

$$b_0 = 43.21 \qquad b_{12} = +0.425$$
$$b_1 = -24.95 \qquad b_{11} = +3.1$$
$$b_2 = -8.55 \qquad b_{22} = +1.1$$
$$s_{b_j} = 0.75 \qquad s_{b_{uj}} = 0.92 \qquad s_{b_{jj}} = 1.3$$

For Student's t the table value is $t_{0.05}(9) = 2.26$. The coefficients b_{12} and b_{22} are found to be not significant. Deleting them, the regression equation takes the form

$$\hat{y} = 41.14 - 24.95x_1 - 8.55x_2 + 3.1x_1^2$$

The adequacy variance is obtained from the formula

$$s_a^2 = \frac{\sum_{i=1}^{N} (\bar{y}_i - \hat{y}_i)^2}{N-l} = 2.46$$

where l is the number of significant coefficients.

The equation is an adequate fit to the experiment, as $s_a^2 < s_e^2$. The analysis of the regression equation indicates that the potassium utilization in the process of production of $KHCO_3$ from sylvinite is the more the higher is the content of KCl in the initial sylvinite and the less water is used to dissolve the initial mixture of dry NaCl and KCl.

REFERENCES

1. KAFAROV, V., Cybernetic Methods in Chemistry and Chemical Engineering, Mir Publishers, Moscow, 1976.
2*. NALIMOV, V., CHERNOVA, N., Statistical Methods of Extremal Experimental Design, Nauka Publishers, Moscow, 1965.
3*. AKHNAZAROVA, V., KAFAROV, V., Statistical Methods of Design and Processing of Experiment, "MKhTI" Press, Moscow, 1972.
4. HIMMELBLAU, D. M., Process Analysis by Statistical Methods, John Wiley and Sons, Inc., New York, 1970.
5. DRAPER, N. R. and H. SMITH, Applied Regression Analysis, John Wiley and Sons, New York, 1966.
6*. NALIMOV, V., The Use of Mathematical Statistics in Material Analysis, Physmatgiz Publishers, 1960.
7*. STEPANOV, N., ERLYKINA, M., FILIPPOV, G., Methods of Linear Algebra in Physical Chemistry, Moscow University Press, 1976.
8. BOX, G., WILSON, K., *J. Roy. Statist. Soc.*, Ser. B, 1951.
9. HICKS, C. R., Fundamental Concepts in the Design of Experiments, Holt, Rinehart and Winston, New York, 1964.
10*. RUZINOV, L., Statistical Method of Optimization of Chemical Processes, Khimiya Publishers, Moscow, 1972.
11*. MARKOVA, E., LISENKOV, A., Design of Experiment under Nonhomogeneity, Nauka Publishers, Moscow, 1973.
12. CRAMÉR, H., Mathematical Methods of Statistics, Princeton University Press, Princeton. N. J. 1946.
13*. SMIRNOV, N., DUNIN-BARKOVSKY, I., Probability Theory and Mathematical Statistics for Engineering Applications, Nauka Publishers, Moscow, 1969.
14. HUITSON, A., The Analysis of Variance, Charles Griffin Co., Ltd., London, 1966.
15*. MARKOVA, E., KARTASHOVA, T. M., et al., "*Zav. lab.*", No. 7, 1967.
16*. MARKOVA, E., "*Zav. lab.*", No. 7, 1968.
17. BRANDON, D., *B. I. S. A. Journal*, 6, No. 7, 1959.
18*. VOLTER, B., MARKOVA, E., "*Khim. prom.*", No. 4, 1961.
19*. KONONOVA, G., et al., "*Khim. prom.*", No. 4, 1971.
20. BOX, G., HUNTER, J., *Ann. Math. Statistics*, 28, No. 1, 1957.
21*. SHOKIN, I., et al., "*Khim. prom.*", No. 5, 1972.
22*. POLIEVKTOVA, E., et al., "*Trudy MKhTI*", Ser. 65, 39, 1970.
23. HARRINGTON, E., *Industr. Quality Control*, 21, No. 10, 1965.
24*. ZUBAKOVA, L., et al., "*ZhPKh*", XVIII, No. 10, 1975.
25. SPENDLEY, W., Hext, G., Himsworth, F., *Technometrics*, 4, 1962.
26*. GORSKY, V., BRODSKY, V., "*Zav. lab.*", 31 , No. 7, 1965.
27. PLACKETT, R., BURMAN, J., *Biometrika*, 33, No. 4, 1946 .
28. SATTERTHWAITE, F., *Technometrics*, 1, No. 2, 1959.
29*. SLOBODCHIKOVA, R., et al., "*Zav. lab.*", No. 1, 1966.
30*. PISARENKO, V., POGORELOV, A., Design of Kinetic Studies, Nauka Publishers, Moscow, 1969.

---

*The asterisked entries are in Russian.

31. FEDOROV, V. V., Theory of Optimal Experiments, Academic Press, N.-Y., 1972.
32*. POLAK, L., ed., Use of Computational Mathematics in Chemical and Physical Kinetics, Nauka Publishers, Moscow, 1969.
33*. NASRETDINOVA, M., et al., "*VMS*", No. 12, 1974.
34. BOX, G., HUNTER, W., *Technometrics*, 7, 23, 1965.
35*. GAGARIN, S., et al., "*Zav. lab.*", 34, 65, 1968.
36. SCHEFFE, H., *J. Roy. Statist. Soc.*, Ser. B, 20, No. 2, 1958.
37*. ZEDGINIDZE, I., Mathematical Design of Experiment to Study and Optimize Mixture Properties, Metsniereba Publishers, Tbilisi, 1971.
38*. NOVIK, F., "*Zav. lab.*", 34, No. 10, 1968.
39*. CHEMLEVA, T., MIKESHINA, N., In collected papers: New Ideas in Experimental Design, Edited by V. Nalimov, Nauka, Moscow, 1969.
40. LAMBRAKIS, D., *J. Roy. Statist. Soc.*, Ser. B, 30, No. 1, 1968.
41. SCHEFFE, H., *J. Roy. Statist. Soc.*, Ser. B, 25, No. 2, 1963.
42*. FISHER, R., MIKHEEVA, T., "*Prom. i sanit. ochistka gazov*", No. 6, 1973.
43*. USHAROVA, L., et al., "*Trudy MKhTI*", Ser. 85, 1975.
44. McLEAN, R., ANDERSON, V., *Technometrics*, 8, No. 3, 1966.
45. KIEFER, J., *J. Roy. Statist. Soc.*, Ser. B, 30, No. 2, 1968.
46*. GOLIKOVA, T., MIKESHINA, N., In collected papers: New Ideas in Experimental Design, Edited by V. Nalimov, Nauka Publishers, Moscow, 1969.
47*. ANDRUKOVICH, P., Ibid.
48. KIEFER, J. ,*Ann. Math. Statistics*, 32, No. 1, 1961.
49*. SYROMYATNIKOVA, M., et al., "*Trudy MKhTI*", Ser. 85, p. 3-4, 1975.
50. DRAPER, N., LAWRENCE, W., *J. Roy. Statist. Soc.*, Ser. B, 27, No. 3, 1965.
51. BOX, G., DRAPER, N., *J. Amer. Statist. Ass.*, 54, No. 287, 1959.
52*. SYROMYATNIKOVA, M., et al., "*Trudy MKhTI*", Ser. 85, p. 5-7, 1975.
53. KENWORTHY, O., *Industr. Quality Control*, 19, No. 12, 1963.

APPENDIX

Table 1. Error function, erf (*x*)

$$\Phi(x) = \frac{1}{\sqrt{\pi 2}} \int_0^x e^{-\frac{x^2}{2}} \mathrm{d}x$$

x	$\Phi(x)$	x	$\Phi(x)$	x	$\Phi(x)$	x	$\Phi(x)$
.00	.0000	.34	.1331	.68	.2517	1.02	.3461
.01	.0040	.35	.1368	.69	.2549	1.03	.3485
.02	.0080	.36	.1406	.70	.2580	1.04	.3508
.03	.0120	.37	.1443	.71	.2611	1.05	.3531
.04	.0160	.38	.1480	.72	.2642	1.06	.3554
.05	.0199	.39	.1517	.73	.2673	1.07	.3577
.06	.0239	.40	.1554	.74	.2703	1.08	.3599
.07	.0279	.41	.1591	.75	.2734	1.09	.3621
.08	.0319	.42	.1628	.76	.2764	1.10	.3643
.09	.0359	.43	.1664	.77	.2794	1.11	.3665
.10	.0398	.44	.1700	.78	.2823	1.12	.3686
.11	.0438	.45	.1736	.79	.2852	1.13	.3708
.12	.0478	.46	.1772	.80	.2881	1.14	.3729
.13	.0517	.47	.1808	.81	.2910	1.15	.3749
.14	.0557	.48	.1844	.82	.2939	1.16	.3770
.15	.0596	.49	.1879	.83	.2967	1.17	.3790
.16	.0636	.50	.1915	.84	.2995	1.18	.3810
.17	.0675	.51	.1950	.85	.3023	1.19	.3830
.18	.0714	.52	.1985	.86	.3051	1.20	.3849
.19	.0753	.53	.2019	.87	.3078	1.21	.3869
.20	.0793	.54	.2054	.88	.3106	1.22	.3883
.21	.0832	.55	.2088	.89	.3133	1.23	.3907
.22	.0871	.56	.2123	.90	.3159	1.24	.3925
.23	.0910	.57	.2157	.91	.3186	1.25	.3944
.24	.0948	.58	.2190	.92	.3212	1.26	.3962
.25	.0987	.59	.2224	.93	.3238	1.27	.3980
.26	.1026	.60	.2257	.94	.3264	1.28	.3997
.27	.1064	.61	.2291	.95	.3289	1.29	.4015
.28	.1103	.62	.2324	.96	.3315	1.30	.4032
.29	.1141	.63	.2357	.97	.3340	1.31	.4049
.30	.1179	.64	.2389	.98	.3365	1.32	.4066
.31	.1217	.65	.2422	.99	.3389	1.33	.4082
.32	.1255	.66	.2454	1.00	.3413	1.34	.4099
.33	.1293	.67	.2486	1.01	.3438	1.35	.4115

Table 1 *(concluded)*

x	$\Phi(x)$	x	$\Phi(x)$	x	$\Phi(x)$	x	$\Phi(x)$
1.36	.4131	1.67	.4525	1.97	.4756	2.54	.4945
1.37	.4147	1.68	.4535	1.98	.4761	2.56	.4948
1.38	.4162	1.69	.4545	1.99	.4767	2.58	.4951
1.39	.4177	1.70	.4554	2.00	.4772	2.60	.4953
1.40	.4192	1.71	.4564	2.02	.4783	2.62	.4956
1.41	.4207	1.72	.4573	2.04	.4793	2.64	.4959
1.42	.4222	1.73	.4582	2.06	.4803	2.66	.4961
1.43	.4236	1.74	.4591	2.08	.4812	2.68	.4963
1.44	.4251	1.75	.4599	2.10	.4821	2.70	.4965
1.45	.4265	1.76	.4608	2.12	.4830	2.72	.4967
1.46	.4279	1.77	.4616	2.14	.4838	2.74	.4969
1.47	.4292	1.78	.4625	2.16	.4846	2.76	.4971
1.48	.4306	1.79	.4633	2.18	.4854	2.78	.4973
1.49	.4319	1.80	.4641	2.20	.4861	2.80	.4974
1.50	.4332	1.81	.4649	2.22	.4868	2.82	.4976
1.51	.4345	1.82	.4656	2.24	.4875	2.84	.4977
1.52	.4357	1.83	.4664	2.26	.4881	2.86	.4979
1.53	.4370	1.84	.4671	2.28	.4887	2.88	.4980
1.54	.4382	1.85	.4678	2.30	.4893	2.90	.4981
1.55	.4394	1.86	.4686	2.32	.4898	2.92	.4982
1.56	.4406	1.87	.4693	2.34	.4904	2.94	.4984
1.57	.4418	1.88	.4699	2.36	.4909	2.96	.49846
1.58	.4429	1.89	.4706	2.38	.4913	2.98	.49856
1.59	.4441	1.90	.4713	2.40	.4918	3.00	.49865
1.60	.4452	1.91	.4719	2.42	.4922	3.20	.49931
1.61	.4463	1.92	.4726	2.44	.4927	3.40	.49966
1.62	.4474	1.93	.4732	2.46	.4931	3.60	.49984
1.63	.4484	1.94	.4738	2.48	.4934	3.80	.499928
1.64	.4495	1.95	.4744	2.50	.4938	4.00	.499968
1.65	.4505	1.96	.4750	2.52	.4941	5.00	.499997
1.66	.4515						

Table 2. Quantiles for the Normal Distribution

α	$1-\frac{\alpha}{2}$	$u_{1-\frac{\alpha}{2}}$	α	$1-\frac{\alpha}{2}$	$u_{1-\frac{\alpha}{2}}$	α	$1-\frac{\alpha}{2}$	$u_{1-\frac{\alpha}{2}}$
.80	.60	.25	.15	.925	1.44	.01	.995	2.58
.50	.75	.67	.10	.95	1.64	.005	.9975	2.81
.40	.80	.84	.05	.975	1.96	.002	.999	3.09
.30	.85	1.04	.04	.980	2.05	.001	.9995	3.29
.25	.875	1.15	.02	.990	2.33	.0001	.99995	3.89
.20	.90	1.28						

Table 3. Quantiles for Student's *t* Distribution

d.f., ν	Level of significance, α						
	0.20	0.10	0.05	0.02	0.01	0.005	0.001
1	3.08	6.31	12.71	31.82	63.66	127.32	636.62
2	1.89	2.92	4.30	6.97	9.93	14.09	31.60
3	1.64	2.35	3.18	4.54	5.84	7.45	12.94
4	1.53	2.13	2.78	3.75	4.60	5.60	8.61
5	1.48	2.02	2.57	3.37	4.03	4.77	6.86
6	1.44	1.94	2.45	3.14	3.71	4.32	5.96
7	1.42	1.90	2.37	3.00	3.50	4.03	5.41
8	1.40	1.86	2.31	2.90	3.36	3.83	5.04
9	1.38	1.83	2.26	2.82	3.25	3.69	4.78
10	1.37	1.81	2.23	2.76	3.17	3.58	4.59
11	1.36	1.80	2.20	2.72	3.11	3.50	4.44
12	1.36	1.78	2.18	2.68	3.06	3.43	4.32
13	1.35	1.77	2.16	2.65	3.01	3.37	4.22
14	1.34	1.76	2.15	2.62	2.98	3.33	4.14
15	1.34	1.75	2.13	2.60	2.95	3.29	4.07
16	1.34	1.75	2.12	2.58	2.92	3.25	4.02
17	1.33	1.74	2.11	2.57	2.90	3.22	3.97
18	1.33	1.73	2.10	2.55	2.88	3.20	3.92
19	1.33	1.73	2.09	2.54	2.86	3.17	3.88
20	1.33	1.73	2.09	2.53	2.85	3.15	3.85
21	1.32	1.72	2.08	2.52	2.83	3.14	3.82
22	1.32	1.72	2.07	2.51	2.82	3.12	3.79
23	1.32	1.71	2.07	2.50	2.81	3.10	3.77
24	1.32	1.71	2.06	2.49	2.80	3.09	3.75
25	1.32	1.71	2.06	2.48	2.79	3.08	3.73
26	1.32	1.71	2.06	2.48	2.78	3.07	3.71
27	1.31	1.70	2.05	2.47	2.77	3.06	3.69
28	1.31	1.70	2.05	2.47	2.76	3.05	3.67
29	1.31	1.70	2.04	2.46	2.76	3.04	3.66
30	1.31	1.70	2.04	2.46	2.75	3.03	3.65
40	1.30	1.68	2.02	2.42	2.70	2.97	3.55
60	1.30	1.67	2.00	2.39	2.66	2.91	3.46
120	1.29	1.66	1.98	2.36	2.62	2.86	3.37
∞	1.28	1.64	1.96	2.33	2.58	2.81	3.29

Table 4. Quantiles for the Chi-Square Distribution

d.f., v	Level of significance, α							
	0.99	0.98	0.95	0.90	0.80	0.70	0.50	0.30
1	0.00016	0.0006	0.0039	0.016	0.064	0.148	0.455	1.07
2	0.020	0.040	0.103	0.211	0.446	0.713	1.386	2.41
3	0.115	0.185	0.352	0.584	1.005	1.424	2.366	3.66
4	0.30	0.43	0.71	1.06	1.65	2.19	3.36	4.9
5	0.55	0.75	1.14	1.61	2.34	3.00	4.35	6.1
6	0.87	1.13	1.63	2.20	3.07	3.83	5.35	7.2
7	1.24	1.56	2.17	2.83	3.82	4.67	6.35	8.4
8	1.65	2.03	2.73	3.49	4.59	5.53	7.34	9.5
9	2.09	2.53	3.32	4.17	5.38	6.39	8.34	10.7
10	2.56	3.06	3.94	4.86	6.18	7.27	9.34	11.8
11	3.1	3.6	4.6	5.6	7.0	8.1	10.3	12.9
12	3.6	4.2	5.2	6.3	7.8	9.0	11.3	14.0
13	4.1	4.8	5.9	7.0	8.6	9.9	12.3	15.1
14	4.7	5.4	6.6	7.8	9.5	10.8	13.3	16.2
15	5.2	6.0	7.3	8.5	10.3	11.7	14.3	17.3
16	5.8	6.6	8.0	9.3	11.2	12.6	15.3	18.4
17	6.4	7.3	8.7	10.1	12.0	13.5	16.3	19.5
18	7.0	7.9	9.4	10.9	12.9	14.4	17.3	20.6
19	7.6	8.6	10.1	11.7	13.7	15.4	18.3	21.7
20	8.3	9.2	10.9	12.4	14.6	16.3	19.3	22.8
21	8.9	9.9	11.6	13.2	15.4	17.2	20.3	23.9
22	9.5	10.6	12.3	14.0	16.3	18.1	21.3	24.9
23	10.2	11.3	13.1	14.8	17.2	19.0	22.3	26.0
24	10.9	12.0	13.8	15.7	18.1	19.9	23.3	27.1
25	11.5	12.7	14.6	16.5	18.9	20.9	24.3	28.2
26	12.2	13.4	15.4	17.3	19.8	21.8	25.3	29.3
27	12.9	14.1	16.2	18.1	20.7	22.7	26.3	30.3
28	13.6	14.8	16.9	18.9	21.6	23.6	27.3	31.4
29	14.3	15.6	17.7	19.8	22.4	24.6	28.3	32.5
30	15.0	16.3	18.5	20.6	23.4	25.5	29.3	33.5

v \ α	0.20	0.10	0.05	0.02	0.01	0.005	0.002	0.001
1	1.64	2.7	3.8	5.4	6.6	7.9	9.5	10.8
2	3.22	4.6	6.0	7.8	9.2	10.6	12.4	13.8
3	4.64	6.3	7.8	9.8	11.3	12.8	14.8	16.3
4	6.0	7.8	9.5	11.7	13.3	14.9	16.9	18.5
5	7.3	9.2	11.1	13.4	15.1	16.3	18.9	20.5
6	8.6	10.6	12.6	15.0	16.8	18.6	20.7	22.5
7	9.8	12.0	14.1	16.6	18.5	20.3	22.6	24.3
8	11.0	13.4	15.5	18.2	20.1	21.9	24.3	26.1
9	12.2	14.7	16.9	19.7	21.7	23.6	26.1	27.9
10	13.4	16.0	18.3	21.2	23.2	25.2	27.7	29.6
11	14.6	17.3	19.7	22.6	24.7	26.8	29.4	31.3
12	15.8	18.5	21.0	24.1	26.2	28.3	31	32.9
13	17.0	19.8	22.4	25.5	27.7	29.8	32.5	34.5
14	18.2	21.1	23.7	26.9	29.1	31.3	34	36.1
15	19.3	22.3	25.0	28.3	30.6	32.8	35.5	37.7

Table 4 *(concluded)*

ν \ α	0.20	0.10	0.05	0.02	0.01	0.005	0.002	0.001
16	20.5	23.5	26.3	29.6	32.0	34.3	37	39.2
17	21.6	24.8	27.6	31.0	33.4	35.7	38.5	40.8
18	22.8	26.0	28.9	32.3	34.8	37.2	40	42.3
19	23.9	27.2	30.1	33.7	36.2	38.6	41.5	43.8
20	25.0	28.4	31.4	35.0	37.6	40.0	43	45.3
21	26.2	29.6	32.7	36.3	38.9	41.4	44.5	46.8
22	27.3	30.8	33.9	37.7	40.3	42.8	46	48.3
23	28.4	32.0	35.2	39.0	41.6	44.2	47.5	49.7
24	29.6	33.2	36.4	40.3	43.0	45.6	48.5	51.2
25	30.7	34.4	37.7	41.6	44.3	46.9	50	52.6
26	31.8	35.6	38.9	42.9	45.6	48.3	51.5	54.1
27	32.9	36.7	40.1	44.1	47.0	49.6	53	55.5
28	34.0	37.9	41.3	45.4	48.3	51.0	54.5	56.9
29	35.1	39.1	42.6	46.7	49.6	52.3	56	58.3
30	36.3	40.3	43.8	48.0	50.9	53.7	57.5	59.7

Table 5. Quantiles for the *F* Distribution, $F_{.80}$

ν_2	Degrees of freedom for numerator, ν_1								
	1	2	3	4	5	6	12	24	∞
1	9.5	12.0	13.1	13.7	14.0	14.3	14.9	15.2	15.6
2	3.6	4.0	4.2	4.2	4.3	4.3	4.4	4.4	4.5
3	2.7	2.9	2.9	3.0	3.0	3.0	3.0	3.0	3.0
4	2.4	2.5	2.5	2.5	2.5	2.5	2.5	2.4	2.4
5	2.2	2.3	2.3	2.2	2.2	2.2	2.2	2.2	2.1
6	2.1	2.1	2.1	2.1	2.1	2.1	2.0	2.0	2.0
7	2.0	2.0	2.0	2.0	2.0	2.0	1.9	1.9	1.8
8	2.0	2.0	2.0	1.9	1.9	1.9	1.8	1.8	1.7
9	1.9	1.9	1.9	1.9	1.9	1.8	1.8	1.7	1.7
10	1.9	1.9	1.9	1.8	1.8	1.8	1.7	1.7	1.6
11	1.9	1.9	1.8	1.8	1.8	1.8	1.7	1.6	1.6
12	1.8	1.8	1.8	1.8	1.7	1.7	1.7	1.6	1.5
13	1.8	1.8	1.8	1.8	1.7	1.7	1.6	1.6	1.5
14	1.8	1.8	1.8	1.7	1.7	1.7	1.6	1.6	1.5
15	1.8	1.8	1.8	1.7	1.7	1.7	1.6	1.6	1.5
16	1.8	1.8	1.7	1.7	1.7	1.6	1.6	1.5	1.4
17	1.8	1.8	1.7	1.7	1.7	1.6	1.6	1.5	1.4
18	1.8	1.8	1.7	1.7	1.6	1.6	1.5	1.5	1.4
19	1.8	1.8	1.7	1.7	1.6	1.6	1.5	1.5	1.4
20	1.8	1.8	1.7	1.7	1.6	1.6	1.5	1.5	1.4
22	1.8	1.7	1.7	1.6	1.6	1.6	1.5	1.4	1.4
24	1.7	1.7	1.7	1.6	1.6	1.6	1.5	1.4	1.3
26	1.7	1.7	1.7	1.6	1.6	1.6	1.5	1.4	1.3
28	1.7	1.7	1.7	1.6	1.6	1.6	1.5	1.4	1.3
30	1.7	1.7	1.6	1.6	1.6	1.5	1.5	1.4	1.3
40	1.7	1.7	1.6	1.6	1.5	1.5	1.4	1.4	1.2
60	1.7	1.7	1.6	1.6	1.5	1.5	1.4	1.3	1.2
120	1.7	1.6	1.6	1.5	1.5	1.5	1.4	1.3	1.1
∞	1.6	1.6	1.6	1.5	1.5	1.4	1.3	1.2	1.0

Table 6. Quantiles of the F Distribution, $F_{.95}$

ν_2	Degrees of freedom for numerator, ν_1 1	2	3	4	5	6	12	24	∞
1	161.4	199.5	215.7	224.6	230.2	234.0	243.9	249.0	254.3
2	18.5	19.0	19.2	19.3	19.3	19.3	19.4	19.5	19.5
3	10.1	9.6	9.3	9.1	9.0	8.9	8.7	8.6	8.5
4	7.7	6.9	6.6	6.4	6.3	6.2	5.9	5.8	5.6
5	6.6	5.8	5.4	5.2	5.1	5.0	4.7	4.5	4.4
6	6.0	5.1	4.8	4.5	4.4	4.3	4.0	3.8	3.7
7	5.6	4.7	4.4	4.1	4.0	3.9	3.6	3.4	3.2
8	5.3	4.5	4.1	3.8	3.7	3.6	3.3	3.1	2.9
9	5.1	4.3	3.9	3.6	3.5	3.4	3.1	2.9	2.7
10	5.0	4.1	3.7	3.5	3.3	3.2	2.9	2.7	2.5
11	4.8	4.0	3.6	3.4	3.2	3.1	2.8	2.6	2.4
12	4.8	3.9	3.5	3.3	3.1	3.0	2.7	2.5	2.3
13	4.7	3.8	3.4	3.2	3.0	2.9	2.6	2.4	2.2
14	4.6	3.7	3.3	3.1	3.0	2.9	2.5	2.3	2.1
15	4.5	3.7	3.3	3.1	2.9	2.8	2.5	2.3	2.1
16	4.5	3.6	3.2	3.0	2.9	2.7	2.4	2.2	2.0
17	4.5	3.6	3.2	3.0	2.8	2.7	2.4	2.2	2.0
18	4.4	3.6	3.2	2.9	2.8	2.7	2.3	2.1	1.9
19	4.4	3.5	3.1	2.9	2.7	2.6	2.3	2.1	1.8
20	4.4	3.5	3.1	2.9	2.7	2.6	2.3	2.1	1.8
22	4.3	3.4	3.1	2.8	2.7	2.6	2.2	2.0	1.8
24	4.3	3.4	3.0	2.8	2.6	2.5	2.2	2.0	1.7
26	4.2	3.4	3.0	2.7	2.6	2.4	2.1	1.9	1.7
28	4.2	3.3	2.9	2.7	2.6	2.4	2.1	1.9	1.6
30	4.2	3.3	2.9	2.7	2.5	2.4	2.1	1.9	1.6
40	4.1	3.2	2.9	2.6	2.5	2.3	2.0	1.8	1.5
60	4.0	3.2	2.8	2.5	2.4	2.3	1.9	1.7	1.4
120	3.9	3.1	2.7	2.5	2.3	2.2	1.8	1.6	1.3
∞	3.8	3.0	2.6	2.4	2.2	2.1	1.8	1.5	1.0

Table 7. Quantiles of the *F* Distribution, $F_{.99}$

ν_2	Degrees of freedom for numerator, ν_1									
	1	2	3	4	5	6	8	12	24	∞
1	4 052	5 000	5 403	5 625	5 764	5 859	5 982	6 106	6 235	6 366
2	98.5	99.0	99.2	99.3	99.3	99.3	99.4	99.4	99.5	99.5
3	34.1	30.8	29.5	28.7	28.2	27.9	27.5	27.1	26.6	26.1
4	21.2	18.0	16.7	16.0	15.5	15.2	14.8	14.4	13.9	13.5
5	16.3	13.3	12.1	11.4	11.0	10.7	10.3	9.9	9.5	9.0
6	13.7	10.9	9.8	9.2	8.8	8.5	8.1	7.7	7.3	6.9
7	12.3	9.6	8.5	7.9	7.5	7.2	6.8	6.5	6.1	5.7
8	11.3	8.7	7.6	7.0	6.6	6.4	6.0	5.7	5.3	4.9
9	10.6	8.0	7.0	6.4	6.1	5.8	5.5	5.1	4.7	4.3
10	10.0	7.6	6.6	6.0	5.6	5.4	5.1	4.7	4.3	3.9
11	9.7	7.2	6.2	5.7	5.3	5.1	4.7	4.4	4.0	3.6
12	9.3	6.9	6.0	5.4	5.1	4.8	4.5	4.2	3.8	3.4
13	9.1	6.7	5.7	5.2	4.9	4.6	4.3	4.0	3.6	3.2
14	8.9	6.5	5.6	5.0	4.7	4.5	4.1	3.8	3.4	3.0
15	8.7	6.4	5.4	4.9	4.6	4.3	4.0	3.7	3.3	2.9
16	8.5	6.2	5.3	4.8	4.4	4.2	3.9	3.6	3.2	2.8
17	8.4	6.1	5.2	4.7	4.3	4.1	3.8	3.5	3.1	2.7
18	8.3	6.0	5.1	4.6	4.3	4.0	3.7	3.4	3.0	2.6
19	8.2	5.9	5.0	4.5	4.2	3.9	3.6	3.3	2.9	2.4
20	8.1	5.9	4.9	4.4	4.1	3.9	3.6	3.2	2.9	2.4
22	7.9	5.7	4.8	4.3	4.0	3.8	3.5	3.1	2.8	2.3
24	7.8	5.6	4.7	4.2	3.9	3.7	3.3	3.0	2.7	2.2
26	7.7	5.5	4.6	4.1	3.8	3.6	3.3	3.0	2.6	2.1
28	7.6	5.5	4.6	4.1	3.8	3.5	3.2	2.9	2.5	2.1
30	7.6	5.4	4.5	4.0	3.7	3.5	3.2	2.8	2.5	2.0
40	7.3	5.2	4.3	3.8	3.5	3.3	3.0	2.7	2.3	1.8
60	7.1	5.0	4.1	3.7	3.3	3.1	2.8	2.5	2.1	1.6
120	6.9	4.8	4.0	3.5	3.2	3.0	2.7	2.3	2.0	1.4
∞	6.6	4.6	3.8	3.3	3.0	2.8	2.5	2.2	1.8	1.0

Table 8. Quantiles of the Cochran Distribution, $G_{.95}$
(Each figure in the body of the table is preceeded by a decimal point.)

n \ ν	1	2	3	4	5	6	7	8	9	10	16	36	144	∞
2	9985	9750	9392	9057	8772	8534	8332	8159	8010	7880	7341	6602	5813	5000
3	9669	8709	7977	7457	7071	6771	6530	6333	6167	6025	5466	4748	4031	3333
4	9065	7679	6841	6287	5895	5598	5365	5175	5017	4884	4366	3720	3093	2500
5	8412	6838	5981	5441	5065	4783	4564	4387	4241	4118	3645	3066	2513	2000
6	7808	6161	5321	4803	4447	4184	3980	3817	3682	3568	3135	2612	2119	1667
7	7271	5612	4800	4307	3974	3726	3535	3384	3259	3154	2756	2278	1833	1429
8	6798	5157	4377	3910	3595	3362	3185	3043	2926	2829	2462	2022	1616	1250
9	6385	4775	4027	3584	3286	3067	2901	2768	2659	2568	2226	1820	1446	1111
10	6020	4450	3733	3311	3029	2823	2666	2541	2439	2353	2032	1655	1308	1000
12	5410	3924	3264	2880	2624	2439	2299	2187	2098	2020	1737	1403	1100	0833
15	4709	3346	2758	2419	2195	2034	1911	1815	1736	1671	1429	1144	0889	0667
20	3894	2705	2205	1921	1735	1602	1501	1422	1357	1303	1108	0879	0675	0500
24	3434	2354	1907	1656	1493	1374	1286	1216	1160	1113	0942	0743	0567	0417
30	2929	1980	1593	1377	1237	1137	1061	1002	0958	0921	0771	0604	0457	0333
40	2370	1576	1259	1082	0968	0887	0827	0780	0745	0713	0595	0462	0347	0250
60	1737	1131	0895	0765	0682	0623	0583	0552	0520	0497	0411	0316	0234	0167
120	0998	0632	0495	0419	0371	0337	0312	0292	0279	0266	0218	0165	0120	0083
∞	0000	0000	0000	0000	0000	0000	0000	0000	0000	0000	0000	0000	0000	0000

20 – 24013

Table 9. Quantiles of the Cochran Distribution $G_{.99}$
(Each figure in the body of the table is preceded by a decimal point.)

n \ ν	1	2	3	4	5	6	7	8	9	10	16	36	144	∞
2	9999	9950	9794	9586	9373	9172	8998	8823	8674	8539	7949	7067	6062	5000
3	9933	9423	8831	8335	7933	7606	7335	7107	6912	6743	6059	5153	4230	3333
4	9676	8643	7814	7212	6761	6410	6129	5897	5702	5536	4884	4057	3251	2500
5	9279	7885	6957	6329	5875	5531	5259	5037	4854	4697	4094	3351	2644	2000
6	8828	7218	6258	5635	5195	4866	4608	4401	4229	4084	3529	2858	2229	1667
7	8376	6644	5685	5080	4659	4347	4105	3911	3751	3616	3105	2494	1929	1429
8	7945	6152	5209	4627	4226	3932	3704	3522	3373	3248	2779	2214	1700	1250
9	7544	5727	4810	4251	3870	3592	3378	3207	3067	2950	2514	1992	1521	1111
10	7175	5358	4469	3934	3572	3308	3106	2945	2813	2704	2297	1811	1376	1000
12	6528	4751	3919	3428	3099	2861	2680	2535	2419	2320	1961	1535	1157	0833
15	5747	4069	3317	2882	2593	2386	2228	2104	2002	1918	1612	1251	0943	0667
20	4799	3297	2654	2288	2048	1877	1748	1646	1567	1501	1248	0960	0709	0500
24	4247	2871	2295	1970	1759	1608	1495	1406	1338	1283	1060	0810	0595	0417
30	3632	2412	1913	1635	1454	1327	1232	1157	1100	1054	0867	0658	0480	0333
40	2940	1915	1508	1281	1135	1033	0957	0898	0853	0816	0668	0503	0363	0250
60	1251	1371	1069	0902	0796	0722	0668	0625	0594	0567	0461	0344	0245	0167
120	1225	0759	0585	0489	0429	0387	0357	0334	0316	0302	0242	0178	0125	0083
∞	0000	0000	0000	0000	0000	0000	0000	0000	0000	0000	0000	0000	0000	0000

Table 10. Duncan's New Multiple Ranges (5% Level)

d.f. n_D	p = number of means within range being tasted: 2	3	4	5	6	7	8	9	10	12	14	16	18	20	50	100
1	18.00	18.00	18.00	18.00	18.00	18.00	18.00	18.00	18.00	18.00	18.00	18.00	18.00	18.00	18.00	18.00
2	6.09	6.09	6.09	6.09	6.09	6.09	6.09	6.09	6.09	6.09	6.09	6.09	6.09	6.09	6.09	6.09
3	4.50	4.50	4.50	4.52	4.50	4.50	4.50	4.50	4.50	4.50	4.50	4.50	4.50	4.50	4.50	4.50
4	3.98	4.01	4.02	4.02	4.00	4.02	4.02	4.02	4.02	4.02	4.02	4.02	4.02	4.02	4.02	4.02
5	3.64	3.74	3.79	3.83	3.83	3.83	3.83	3.83	3.83	3.83	3.83	3.83	3.83	3.83	3.83	3.83
6	3.46	3.58	3.64	3.68	3.68	3.68	3.68	3.68	3.68	3.68	3.68	3.68	3.68	3.68	3.68	3.68
7	3.35	3.47	3.54	3.58	3.60	3.61	3.61	3.61	3.61	3.61	3.61	3.61	3.61	3.61	3.61	3.61
8	3.26	3.39	3.47	3.52	3.55	3.56	3.56	3.56	3.56	3.56	3.56	3.56	3.56	3.56	3.56	3.56
9	3.20	3.34	3.41	3.47	3.50	3.52	3.52	3.52	3.52	3.52	3.52	3.52	3.52	3.52	3.52	3.52
10	3.15	3.30	3.37	3.43	3.46	3.47	3.47	3.47	3.47	3.47	3.47	3.47	3.47	3.47	3.47	3.47
11	3.11	3.27	3.35	3.39	3.43	3.44	3.45	3.46	3.46	3.46	3.46	3.46	3.46	3.46	3.46	3.46
12	3.08	3.23	3.33	3.36	3.40	3.42	3.44	3.44	3.46	3.46	3.46	3.46	3.47	3.48	3.48	3.48
13	3.06	3.21	3.30	3.35	3.38	3.41	3.42	3.44	3.45	3.45	3.46	3.46	3.47	3.47	3.47	3.47
14	3.03	3.18	3.27	3.33	3.37	3.39	3.41	3.42	3.44	3.45	3.46	3.46	3.47	3.47	3.47	3.47
15	3.01	3.16	3.25	3.31	3.36	3.38	3.40	3.42	3.43	3.44	3.45	3.46	3.47	3.47	3.47	3.47
16	3.00	3.15	3.23	3.30	3.34	3.37	3.39	3.41	3.43	3.44	3.45	3.46	3.47	3.47	3.47	3.47
17	2.98	3.13	3.22	3.28	3.33	3.36	3.38	3.40	3.42	3.44	3.45	3.46	3.47	3.47	3.47	3.47
18	2.97	3.12	3.21	3.27	3.32	3.35	3.37	3.39	3.41	3.43	3.45	3.46	3.47	3.47	3.47	3.47
19	2.96	3.11	3.19	3.26	3.31	3.35	3.37	3.39	3.41	3.43	3.44	3.46	3.47	3.47	3.47	3.47
20	2.95	3.10	3.18	3.25	3.30	3.34	3.36	3.38	3.40	3.43	3.44	3.46	3.47	3.47	3.47	3.47
22	2.93	3.08	3.17	3.24	3.29	3.32	3.35	3.37	3.39	3.42	3.44	3.45	3.46	3.47	3.47	3.47
24	2.92	3.07	3.15	3.22	3.28	3.31	3.34	3.37	3.38	3.41	3.44	3.45	3.46	3.47	3.47	3.47
26	2.91	3.06	3.14	3.21	3.27	3.30	3.34	3.36	3.38	3.41	3.43	3.45	3.46	3.47	3.47	3.47
28	2.90	3.04	3.13	3.20	3.26	3.30	3.33	3.35	3.37	3.40	3.43	3.45	3.46	3.47	3.47	3.47
30	2.89	3.04	3.12	3.20	3.25	3.29	3.32	3.35	3.37	3.40	3.43	3.44	3.46	3.47	3.47	3.47
40	2.86	3.01	3.10	3.17	3.22	3.27	3.30	3.33	3.35	3.39	3.42	3.44	3.46	3.47	3.47	3.47
60	2.83	2.98	3.08	3.14	3.20	3.24	3.28	3.31	3.33	3.37	3.40	3.43	3.45	3.47	3.48	3.48
100	2.80	2.95	3.05	3.12	3.18	3.22	3.26	3.29	3.32	3.36	3.40	3.42	3.45	3.47	3.53	3.53
∞	2.77	2.92	3.02	3.09	3.15	3.19	3.23	3.26	3.29	3.34	3.38	3.41	3.44	3.47	3.61	3.67

Table 11. Random Balance Matrix

Observation No.	x_1	x_2	x_3	x_4	x_5	x_6	x_7	x_8	x_9	x_{10}	x_{11}	x_{12}
1	−1	+1	−1	+1	+1	+1	+1	−1	+1	+1	+1	+1
2	−1	−1	+1	−1	−1	−1	−1	−1	+1	+1	+1	−1
3	−1	+1	+1	−1	+1	−1	−1	−1	−1	−1	+1	+1
4	+1	−1	−1	−1	+1	+1	−1	−1	−1	+1	−1	−1
5	+1	−1	−1	+1	−1	−1	−1	+1	−1	+1	+1	+1
6	−1	+1	−1	−1	+1	−1	−1	−1	+1	−1	−1	−1
7	+1	+1	−1	−1	−1	+1	+1	−1	+1	−1	−1	+1
8	−1	−1	+1	+1	+1	−1	−1	+1	+1	+1	−1	−1
9	−1	−1	+1	+1	−1	−1	+1	+1	+1	−1	+1	+1
10	−1	+1	+1	−1	−1	−1	+1	+1	+1	+1	+1	−1
11	+1	+1	−1	+1	+1	+1	−1	−1	−1	+1	−1	+1
12	−1	−1	−1	−1	−1	+1	+1	+1	−1	+1	+1	−1
13	+1	+1	+1	+1	+1	+1	−1	−1	−1	+1	−1	+1
14	−1	−1	−1	+1	−1	+1	+1	−1	−1	−1	−1	−1
15	−1	−1	−1	+1	−1	+1	−1	+1	+1	−1	−1	+1
16	+1	−1	−1	−1	+1	−1	−1	−1	+1	+1	+1	−1
17	+1	+1	+1	+1	−1	+1	−1	+1	+1	+1	−1	+1
18	+1	−1	+1	−1	+1	−1	+1	+1	−1	−1	−1	−1
19	+1	−1	+1	−1	−1	−1	+1	+1	−1	−1	+1	+1
20	−1	+1	−1	+1	−1	+1	+1	+1	+1	+1	−1	−1
21	−1	+1	+1	−1	+1	+1	+1	−1	−1	−1	+1	+1
22	+1	−1	+1	+1	+1	−1	−1	−1	+1	−1	+1	−1
23	+1	+1	+1	+1	−1	+1	+1	+1	−1	−1	−1	+1
24	+1	+1	−1	−1	+1	−1	+1	+1	−1	−1	+1	−1

Table 12. Random Balance Matrix

Observation No.	x_1	x_2	x_3	x_4	x_5	x_6	x_7	x_8	x_9	x_{10}	x_{11}	x_{12}	x_{13}	x_{14}	x_{15}	x_{16}	x_{17}	x_{18}	x_{19}
1	−1	−1	+1	+1	+1	−1	+1	+1	+1	−1	+1	−1	−1	−1	−1	−1	−1	−1	−1
2	+1	−1	−1	+1	+1	−1	+1	+1	−1	−1	+1	+1	−1	+1	+1	−1	+1	+1	−1
3	−1	+1	−1	+1	+1	+1	−1	+1	+1	+1	+1	−1	+1	−1	−1	−1	+1	+1	+1
4	+1	+1	+1	+1	+1	−1	+1	−1	+1	−1	+1	+1	+1	−1	+1	+1	−1	+1	−1
5	−1	−1	−1	−1	+1	+1	−1	+1	+1	+1	−1	−1	−1	+1	+1	+1	+1	−1	+1
6	+1	−1	+1	−1	+1	−1	+1	−1	−1	−1	+1	−1	−1	+1	+1	−1	−1	+1	+1
7	−1	+1	+1	−1	+1	−1	+1	+1	−1	−1	−1	+1	+1	−1	−1	−1	+1	+1	−1
8	+1	+1	−1	−1	+1	+1	−1	−1	−1	−1	+1	−1	+1	−1	+1	+1	+1	−1	−1
9	−1	−1	+1	−1	−1	−1	−1	+1	+1	+1	−1	−1	+1	−1	+1	+1	+1	−1	+1
10	+1	−1	−1	−1	−1	−1	−1	−1	−1	−1	−1	−1	−1	−1	−1	−1	−1	−1	+1
11	−1	+1	−1	−1	−1	+1	+1	+1	+1	+1	+1	+1	−1	+1	+1	−1	−1	+1	−1
12	+1	+1	+1	−1	−1	−1	−1	+1	−1	+1	+1	−1	+1	+1	−1	+1	−1	−1	−1
13	−1	−1	−1	+1	−1	+1	+1	−1	+1	−1	−1	+1	+1	+1	−1	+1	−1	+1	+1
14	+1	−1	+1	+1	−1	−1	−1	−1	+1	+1	+1	−1	−1	−1	−1	+1	+1	−1	−1
15	−1	+1	+1	+1	−4	+1	+1	+1	−1	+1	+1	+1	−1	−1	+1	+1	−1	+1	+1
16	+1	+1	−1	+1	−1	+1	+1	+1	+1	+1	−1	+1	−1	−1	+1	−1	+1	−1	+1
17	−1	−1	+1	+1	−1	+1	−1	+1	−1	−1	+1	−1	+1	+1	−1	−1	+1	−1	−1
18	+1	−1	−1	+1	−1	+1	−1	−1	−1	+1	+1	+1	+1	+1	+1	−1	−1	−1	−1
19	−1	+1	−1	+1	−1	+1	−1	−1	−1	−1	+1	+1	−1	−1	+1	−1	+1	+1	+1
20	+1	+1	+1	+1	−1	−1	+1	−1	+1	−1	−1	+1	−1	−1	+1	+1	−1	−1	−1
21	−1	−1	−1	−1	−1	−1	+1	−1	−1	−1	−1	−1	+1	−1	−1	+1	+1	+1	−1
22	+1	−1	+1	−1	−1	+1	−1	+1	−1	−1	−1	−1	−1	+1	−1	+1	−1	+1	−1
23	−1	+1	+1	−1	−1	+1	−1	−1	−1	+1	−1	−1	−1	−1	+1	−1	+1	+1	+1
24	+1	+1	−1	−1	−1	+1	+1	+1	−1	+1	−1	+1	+1	+1	+1	−1	−1	+1	−1
25	−1	−1	+1	−1	+1	−1	+1	+1	−1	+1	+1	−1	−1	+1	−1	+1	+1	−1	−1
26	+1	−1	−1	−1	+1	−1	+1	+1	+1	−1	+1	+1	+1	+1	−1	−1	+1	−1	+1
27	−1	+1	−1	−1	+1	+1	−1	−1	+1	+1	−1	+1	−1	+1	−1	+1	+1	+1	+1
28	+1	+1	+1	−1	+1	−1	+1	−1	−1	−1	−1	−1	+1	+1	−1	−1	−1	+1	+1
29	−1	−1	−1	+1	+1	+1	−1	−1	+1	+1	+1	+1	+1	−1	−1	−1	−1	−1	+1
30	+1	−1	+1	+1	+1	−1	−1	+1	+1	+1	−1	−1	+1	+1	−1	+1	−1	−1	+1
31	−1	+1	+1	+1	+1	−1	−1	−1	+1	−1	−1	+1	−1	+1	+1	+1	−1	+1	−1
32	+1	+1	−1	+1	+1	+1	+1	−1	+1	+1	−1	+1	+1	−1	+1	+1	+1	−1	+1

INDEX